Georg Mandl

Rock Joints

The Mechanical Genesis

Georg Mandl

Rock Joints

The Mechanical Genesis

With 153 Figures

Springer

Authors

Prof. Dr. Georg Mandl

Wolf-Huber-Str. 12/4
6800 Feldkirch
Austria
Phone: 0043-5522-77580
Fax: 0043-5522-79075
E-mail: georg.mandl.sen@aon.at

Library of Congress Control Number: 2005921337

ISBN-10 3-540-24553-7 **Springer Berlin Heidelberg New York**
ISBN-13 978-3-540-24553-7 **Springer Berlin Heidelberg New York**

Springer is a part of Springer Science+Business Media
springeronline.com

Printed in The Netherlands

Cover design: Erich Kirchner, Heidelberg
Typesetting: Camera-Ready by Author
Production: Luisa Tonarelli
Printing: Krips bv, Meppel
Binding: Stürtz AG, Würzburg

Printed on acid-free paper 32/2132/LT – 5 4 3 2 1 0

Preface

This book developed from my annual course on the genesis of joints in rocks, in the department of rock engineering at the University of Technology, Graz, Austria. As joints are *fractures* which, barren or filled with fluid or minerals, interrupt the continuity of rock bodies, the development of these features is a mechanical process. Hence, the course and, in more detail this book, deal with the *mechanical* genesis of joints.

By considering jointing as a mechanical process of fracturing one hopes to obtain a simpler, more intelligible and coherent picture of the bewildering multiformity and complexity of joints and joint systems in the field, and to facilitate the engineering assessment of jointed rocks. However, one has to admit that the mechanics of jointing is still at the stage of being a loose patchwork of theoretical models, each of which cope with a special aspect of jointing; but it still leaves many gaps and unsolved problems.

Limited by time and didactic requirements, when lecturing I restricted myself to the mechanical aspects of jointing that I considered to be well understood. But then, in writing this book, doubts arose about theories which I had treated somewhat summarily in the course, and which, I felt, needed to be re-examined, and possibly amended or improved. Also, when aiming for a more coherent exposition, I occasionally introduced tentative suggestions and "guesstimations". All of this had the effect that some chapters became lengthier than originally intended – a shortcoming, which I have tried to remedy by inserting summaries at the ends of each chapter. I therefore advise any reader who is discouraged by the length of a chapter, to turn first to the summary to find out whether the chapter, or part of it, is of interest to him.

Throughout the book, I have used the extremely useful graphical method of Mohr's stress circle, by which much of the mathematics is avoided. For the reader who is less conversant with this method, an Appendix on Mohr's stress circle is reprinted from my former book "*Faulting in Brittle Rocks*" ("FBR", 2000, Springer). Although the present book sometimes overlaps with "FBR", it is intended to be completely self-contained.

I should further note, that the analyses in the book ignore the morphological details of the fracture surface; instead, the fracture surface is assumed as smooth, the way it appears when viewed on a scale much larger than the "micro"-scale of the surface features. Readers interested in the surface morphology of fractures and the associated mechanisms, the field of "Fractography", are referred to the expositions in D. Bahat, *Tectono-fractography* (1991, Springer) and to Section 2.2 in Bahat, A. Rabinowitch, B.V. Frid, *Tensile Fracturing in Rocks* (2004, Springer).

Another point to be mentioned concerns the list of references. Instead of compiling many pages of references, I inserted key references in the text, at the places where the related issues were discussed. So, the reader may at least be sure that this author has actually studied the papers he referred to.

In concluding this preface I wish to thank my friends Prof. Florian Lehner, Prof. Horst Neugebauer and Norbert Tschierske for their encouragement and support, and Mrs. Emma Moseley for correcting the manuscript.

But above all, I would like to thank Berta, who unselfishly endured the times of my seclusion in writing the book, or my bad mood when I was hit by software calamities.

Georg Mandl

Contents

Chapter 1

Introduction

The general theme of this book is the genesis of rock joints (germ. *Kluftgenese*); that is to say the physical or mechanical processes by which joints of various kinds are produced. What are rock joints, and why should we be concerned with their genesis? First, by definition, rock joints are fractures, hence discontinuities caused by the rupturing of the rock material. The ruptures may be restricted to individual grains or may cut continuously through rock bodies over distances varying from millimetres to kilometres. We shall be mainly concerned with the latter case, the "macroscopic" fractures whose dimensions are much larger than the characteristic grain size of the rock.

From a phenomenological point of view, the macroscopic rock fractures are basically divided in two major classes: "joints" and "faults". In the terminology of the International Society of Rock Mechanics, "a joint is a break of geological origin in the continuity of a body of rock occurring either single, or more frequently in a set or system, but not attended by a *visible* (italics by the author) movement parallel to the surface of discontinuity." In contrast, a fault is defined "as a fracture or fracture zone along which there has been displacement of the two sides relative to one another parallel to the fracture. (This displacement may be a few centimeters or many kilometers.)" This descriptive classification is schematized in Fig. 1.1. Joints may be barren fractures, or infilled by various materials, such as quartz, calcite, or other minerals. In this case the fractures are called veins (germ. *Gangspalten*) or dykes if filled by solidified magma.

Note that the distinction between joints and faults hinges on the term *visible* which, unfortunately, depends on the scale of observation. Thus, a joint may have formed by a parting of the rock strictly perpendicular to the fracture plane (Fig. 1.1A) or even by involving some shear displacement of the fracture walls that remains "invisible" at the scale of observation. In this book we shall focus on joints, as faulting, the most prominent deformation mechanism in the Earth's crust, has been dealt with in a variety of text books (e.g., G. Mandl (2000) *Faulting in Brittle Rocks*, Springer; henceforth referred to as FBR).

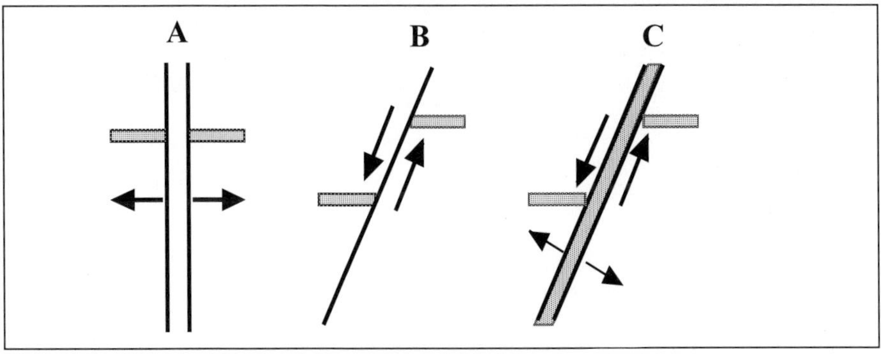

Fig. 1.1. Kinematic fracture types: joints (A), faults (B), dilational faults (C)

Joints are probably the most ubiquitous and, at the same time, most confusing features of crustal rocks. They vary greatly in appearance, dimensions, and arrangement, and occur in quite different tectonic environments. Figures 1.2–1.6 give a first impression of the wide range of the appearance of rock joints.

Fig. 1.2. Cooling joints in andesite, Mt. Rainier National Park, Wash. (A. Lachenbruch (1962) Special GSA paper, New York)

Fig. 1.3. Joint sets in limestone layers separated by marl beds (J. Ramsay and M. Huber (1987) *Modern Structural Geology*, Vol. 2, Academic Press)

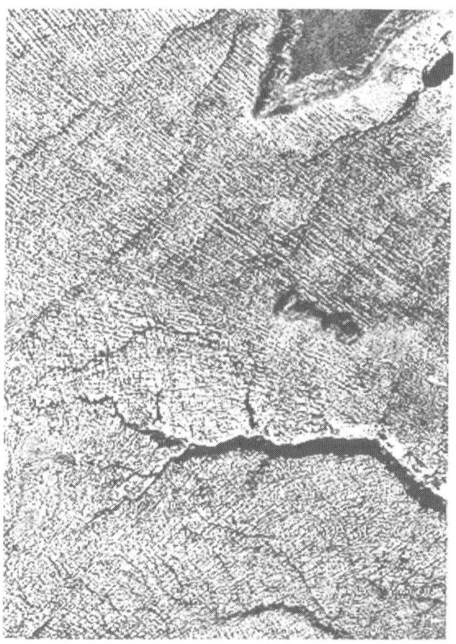

Fig. 1.4. Joint system in flat lying Entrada sandstone, Utah; aerial photograph, scale ca. 1:40000 (from D. Meier and P. Kronberg (1989) *Klüftung in Sedimentgesteinen*, Enke Verlag, Stuttgart)

2 cm

Fig. 1.5. Dyke-sill system in oil source rock

Figure 1.2 shows a polygonal joint pattern caused by shrinkage during the cooling of a volcanic rock. (Similar patterns of shrinkage fractures are readily observed in drying mud.) Figure 1.3 shows typical joint sets in an alternating sequence of limestone and marl beds. The joints cut roughly orthogonally across the limestone layers (germ. *bankrechte Klüfte*) while, interestingly, the marl beds have not been fractured at all. Quite a different view is presented in Fig. 1.4, as an example of regional systems of parallel joints that extend straight over a wide region. It is, in particular, the surprising straightness of the regional joints that is still somewhat of a mystery. Figure 1.5 brings us down to the other extreme of scale; it shows a core sample of a Tertiary oil source rock (diatomite) that was fractured by high-pressure hydrocarbons to provide escape routes along sill- and dyke-type fractures. Figure 1.6 gives a foretaste of the fascinating association of joints with anticlinal

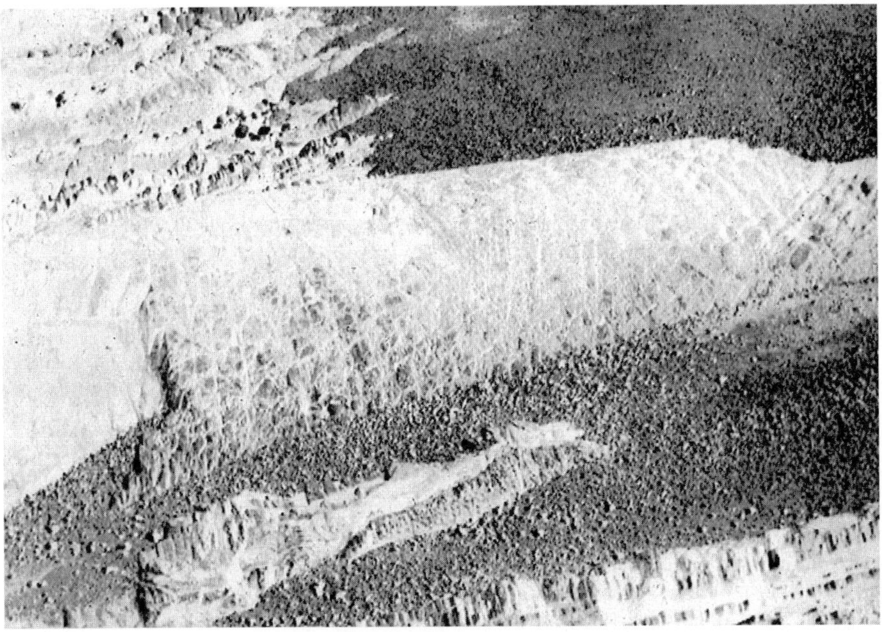

Fig. 1.6. Traces of anticlinal joint sets (Bude, Devon, SW-England)

bending. The traces of typical anticlinal joint systems can be seen on the nicely exposed anticlinal "whale back" (periclinal fold) in Carboniferous turbidite sediments.

These examples should give the reader some feeling for the abundance of rock joints. But it is not our aim in this course to indulge in a broad exposition and detailed descriptions of the various joint geometries. Rather, we shall inquire into the mechanical side, that is into the mechanical aspects that are common to various, or even all types of rock joints. In this way geometrically disparate or dissimilar joint phenomena may be revealed and understood as closely related by the underlying mechanical process, and complex arrangements of joints may be better understood. In the first place, this approach brings out the very crucial role of rock stresses and fluid pressures in the development of rock joints. Most rock joints owe their origin to stresses that were induced or imposed from outside (e.g. by the stretching of layers, or the rise of pore fluid pressure by external compression or fluid injection). We shall therefore fix our attention on the processes of jointing that are caused by exogenous loading of rock bodies, and leave aside jointing by the endogenous tensile rock stresses induced by the shrinkage of a cooling (Fig. 1.2) or desiccating rock whose outside boundaries remained fixed.

Why should structural geologists, engineering geologists and rock engineers be interested in a discourse on the mechanics of rock joints? Naturally, there is intellectual satisfaction to be gained from understanding how and why the various types of joints are formed in rocks. But besides this academic aspect, the study of the jointing mechanisms has a very practical side: Observations of joint systems can convey information on the tectonic stress field that was active at the time the joints were formed. And since joint systems commonly consist of several joint sets, often different in age, the associated stress fields may also differ and represent different geologic periods or episodes. Conversely, if the tectonic history of a region or locality is known, the character of ancient or recent stress fields may be inferred, and eventually a reasonable idea conceived of the joint structures to be expected in the interior of a rock body. On the other hand, if interior joints are identified as "recent" features, they can provide information on the present state of the local rock stresses. If the joints are open, the rock pressure on these joints is clearly zero. If the joint faces show traces of striation (germ. *Striemung*), this is evidence of the action of shear stresses. Also note that the stress field is changed near a free surface, which commonly gives rise to some extra jointing that should be "filtered out" when predictions are made on the joint structures further inside the rock.

Obviously, the better one knows a local joint system, in particular the orientation, type, density and interconnectivity of the joints, the more realistically one can estimate the stability and deformability of rock bodies in the design of large underground cavities, and the better one may be able to cope with foundation problems in designing bridges, dams, or power plants. In addition, information on the distribution of open and healed or sealed joints is essential for the prediction of fluid flow, and for the reconstruction of migration paths of ore-forming or hydrocarbon fluids. It is the presence of unhealed or only weakly sealed joints and their inherent hydraulic conductivity that poses a major problem in the planning of underground nuclear-waste depositories or the storage of gas and oil in depleted reservoirs. Here again, the stress field is a major factor, since the hydraulic conductivity of barren, or only weakly sealed joints depends very sensitively on the rock pressure that keeps the joints closed.

A further point to mention is that joint-mechanical insight can be very useful in the evaluation of in-situ measurements of rock stresses (by overcoring, flatjack tests, hydraulic fracturing, bore hole break-outs, etc.). Strictly speaking, the measurements merely represent the state of stress at tiny spots in the whole field relevant to the engineering operation. Unfortunately, it is quite common that the rock stresses vary greatly over distances not much larger, or even smaller than the dimensions of the measuring device. These fluctuations are caused by heterogeneities of the rock material (e.g. layering of the rock) and by the presence of discontinuities, such as joints, faults and bedding planes. Thus, the in-situ stress measurements are not much more than "pinpricks" in the relevant stress field. Naturally, this raises the question of how reliable these data are as input in a prognostication of the stresses one may have to cope with as the engineering work progresses. We shall see, later in this course, how the mechanical interpretation of joints, or joint sets and systems may be of assistance in evaluating and weighting in-situ stress data.

As mentioned before, the confusing multiformity in the appearance of rock joints should become simpler, more intelligible and coherent, if jointing is considered as a mechanical process of fracturing. To get started with this approach two basic concepts should be stated. First, rock joints are considered as approximately planar fractures, whose two faces are separated across the fracture plane by a distance that is much smaller than the fracture length. The fracture faces unite at the fracture front. According to the type of displacement three fundamental fracture modes are distinguished, as illustrated in Fig. 1.7. Rock joints are mode I fractures when the relative displacement of the fracture walls is normal to the fracture plane or, alternatively, the joints could be mode II shear fractures (*shear joints*) with a shear displacement of the fracture walls that is "invisible" at the scale of observation.

The second point that should be made clear is that a rock joint is a *brittle* fracture (germ. *Sprödbruch*). What does this mean? The term *brittle* is often used in a somewhat ambiguous way; we should therefore define more precisely what we mean by "brittle". To this aim let us consider how a cylindrical rock sample will fail when axially loaded by a normal stress σ_v that is uniformly distributed over the end faces of the sample. The sample may be laterally unconfined or loaded by a uniform confining pressure σ_c. Disregarding the technical details of the sophisticated testing machineries used in rock mechanics laboratories, Fig. 1.8 summarises how the rock sample may fail.

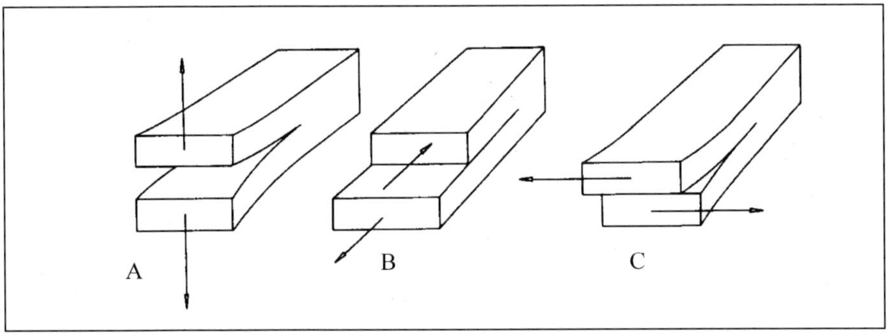

Fig. 1.7. The fundamental fracture modes: A) mode I, opening mode (germ. *Trennbruch*); B) mode II, in-plane shear or sliding mode (germ. *Scherbruch*); C) mode III, anti-plane shear or tearing mode (germ. *Querscherungsbruch*)

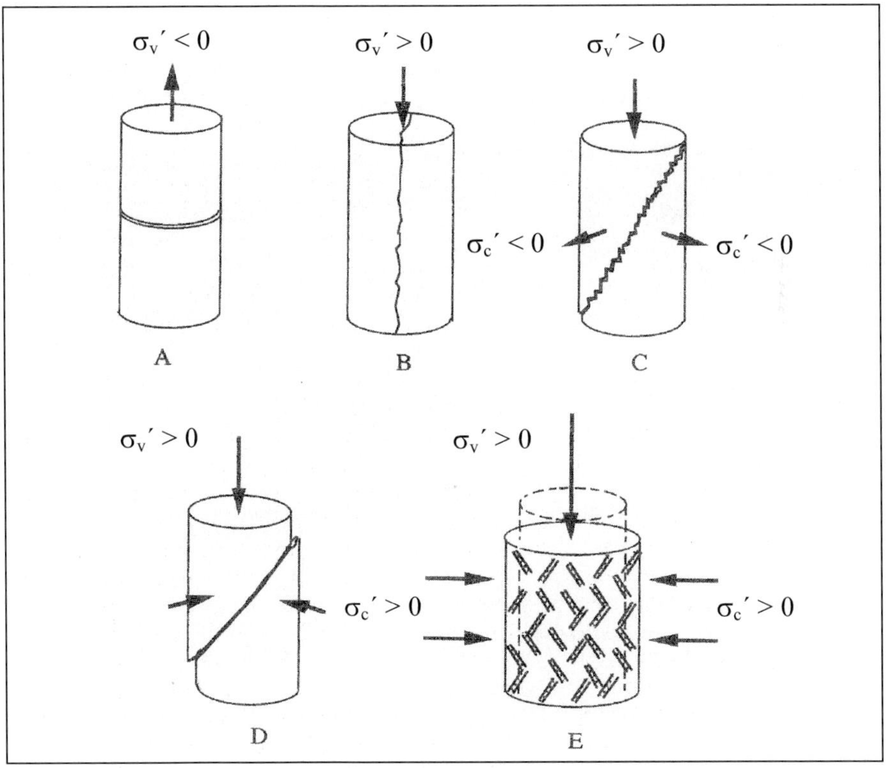

Fig. 1.8. Failure types in cylindrical rock sample under axial-symmetric loading: A) tension fracture (germ. *Zugbruch*); B) extension or "cleavage" fracture (germ. *Spaltbruch*); C) dilational or hybrid extension-shear fracture (germ. *hybrider Dehnungs-Scherbruch*); D) shear fracture (germ. *Scher-* or *Gleitbruch*), E) multi-shear cataclasis (germ. *Multischerungs-Kataklase*)

Now, which of the failure modes A to D in Fig. 1.8, and the associated deformation behaviour of the material should we consider as brittle? And what about the continuous deformation mode in Fig. 1.8E? To answer these questions we have to consider several aspects: First, the stress-strain curves of the loading tests, second, the effect of the rate of loading, and third, the nature of the micro-processes involved. Figure 1.9 shows schematically the type of stress-strain curves that are recorded in axial loading tests.

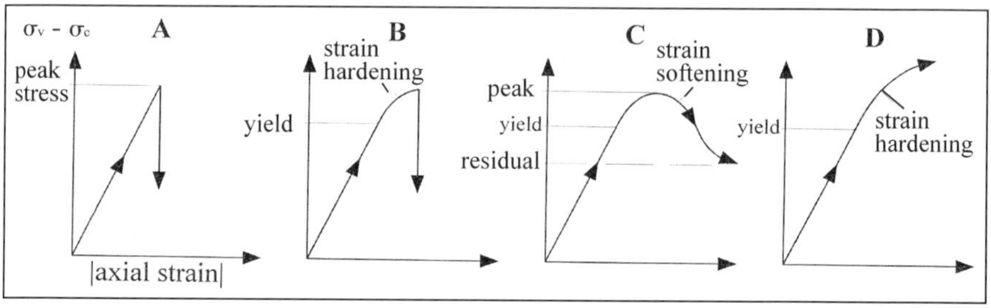

Fig. 1.9. Typical stress-strain responses in axisymmetric loading (Fig. 1.8) (see text for explanation)

Figure 1.9A represents the *ideal* or *very brittle* material behaviour. The material responds to the loading indicated in Figs. 1.8A,B by some small purely elastic extension or shortening, and fails by fracturing at a certain critical stress level, – the "peak stress". The fracturing is accompanied by a sudden stress drop and the release of the elastically stored strain energy, thereby separating the sample into parts which have not undergone permanent deformation and could be fitted together. (Naturally, the elastic strain response need not be strictly proportional to the increase in stress (linear elasticity) as indicated in the figure, and the steepness of the elastic stress-strain curve may somewhat decrease as straining continues.) In reality, rocks do not deform in a purely elastic way up to peak stress, but rather already start deforming inelastically at a lower, somewhat ill-defined stress level – the so-called "yield point" (Fig. 1.9B). This point lies commonly at a level about half the peak stress. The inelastic increase in stress beyond the yield point is referred to as "*strain hardening*". It allows the sample to withstand a further increase in load, until it fractures at peak stress, again with a sudden drop in the axial loading stress σ_v and the complete loss of cohesion along the fracture surface. The stress-strain curves in Figs. 1.9A,B are associated with the loading procedures and failure modes in Figs. 1.8A–C.

Next we consider the situation in Fig. 1.8D, where the rock sample is loaded by axial and lateral compressive stresses, as typical for the compressive stress conditions that generate tectonic faults. The differential stress $\sigma_v - \sigma_c$ reaches a "peak" level under continued elastic/inelastic straining, but decreases thereafter in a more gentle way, as is schematically illustrated in Fig. 1.9C. The post-peak descent of the stress-strain curve implies a gradual reduction in the load-carrying capacity of the material, commonly referred to as "*strain softening*". Under the all boundary compressive stresses, fractures form as shear fractures or narrow shear bands (Fig. 1.8D). On the fracture plane, cohesion may be maintained, at least partly, and a frictional resistance is mobilised by the remaining compressive stresses. This allows the fractured rock to support a certain residual stress difference $\sigma_{v\ res} - \sigma_c$.

The shear failure of the rock under all boundary compression (Fig. 1.8D) is commonly preceded by a continuous inelastic deformation. Depending on the rock type, the magnitude of the inelastic (i.e. permanent) strain will increase as the applied confining pressure σ_c is raised. If raised sufficiently, say up into the kilobar range, strain softening can be completely suppressed (Fig. 1.9D) and pervasive shortening (and lateral extension) of the sample continues under a monotonic increase in axial loading stress σ_v, at least within the range of straining in modern high-pressure testing machines, where shortening seldom exceeds 25%. To illustrate this material deformation behaviour in rocks under compression we insert stress-strain curves from Theodore von Kármán's classical compression tests on cylindrical rock specimens at room temperature (Fig. 1.10). The marble specimens (A) show strain softening at confining pressures up to about 700 bar, but continue to strain-harden at still higher confining pressures; the sandstone (B) behaves similarly. This trend of behaviour has been observed in triaxial testing of a great variety of rocks and loose granular materials all over the world.

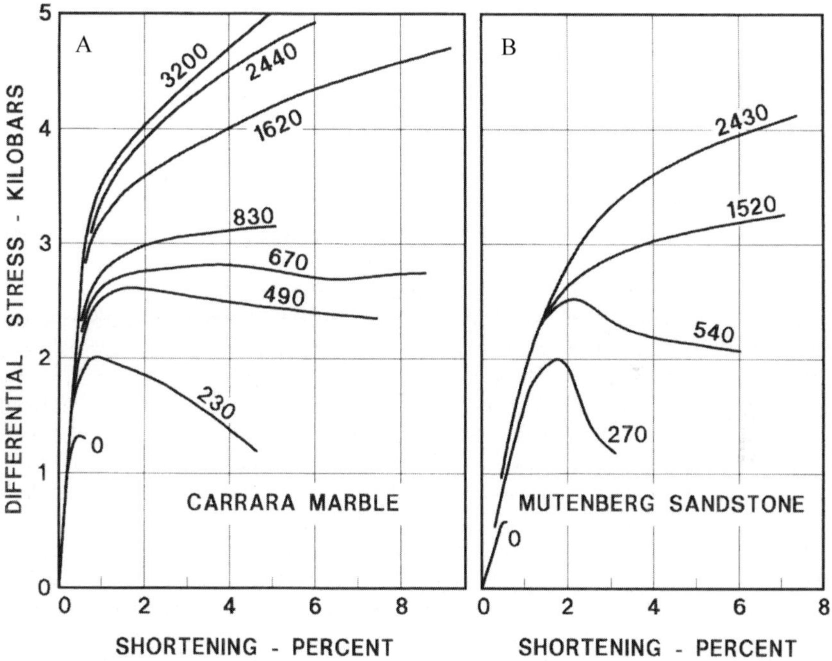

Fig. 1.10. Th. von Kármán's (1911) stress-strain curves for Carrara marble (A) and red Mutenberg sandstone (B) (Differential stress $\sigma_V - \sigma_c$ is plotted against the axial shortening in percent. The numbers on the curves are the confining pressure σ_c in bar.)

Let us now return to the initial question of what we mean by "*brittle*". Which of the different deformation and failure modes considered in Fig. 1.8 together with the associated stress-strain responses in Fig. 1.9 should we consider as brittle? The failure modes A, B, C, D in Fig. 1.8, with associated stress-strain responses A, B, C in Fig. 1.9 have in common the existence of a peak stress and post-peak strain softening. This is the first, but not the only characteristic of "brittleness". The second, equally important constituent of brittle behaviour is the *rate independence* of the pre- and post-peak deformation and fracturing. By that we mean that the inelastic deformation process is not, or only slightly affected by the rate at which the rock is strained or loaded. Naturally, this depends on the deformation processes that operate at the microscale. If the predominant microprocesses are fairly insensitive to the rate of straining or loading, the macroscopic deformation will be so too. Such rate-insensitive processes are the breakage of intergranular bonds, the abrasion, breakage or crushing of grains, frictional intergranular sliding, reorientation of grains, the growth of microcracks, and the associated volume dilation. These energy dissipating "cataclastic" processes operate at moderate temperatures, e.g. below 240–400°C in quartz-rich rocks, and below 400–600°C in feldspar-rich rocks. Thus, disregarding some slight rate dependence, we arrive at the following definition of general brittleness:

> *A macroscopic deformation process is "brittle" in a general sense if it is rate-independent and demonstrates strain-softening in the post-peak region.*

Note that in this broad definition, it is of little relevance whether the post-peak softening occurs as a sudden rupturing and complete loss of strength or as a more gradual reduction of the strength of the material to a residual level.

Given the geothermal gradient, the temperature limits stated above for microcataclastic processes also specify the depth range of the so-called "brittle" upper crust of the Earth, where tension and cleavage joints (i.e. the brittle fractures of type A and B in Fig. 1.8) and brittle shear fractures could be generated under suitable stress conditions. Then, assuming an average geothermal gradient of 25°C/km depth, which is typical for continental areas, the temperature would restrict the formation of tension and cleavage joints in quartz-rich rocks to a depth range of around 10 km.

Let us now return to the cases in Figs. 1.9D and 1.10 where inelastic hardening continues beyond the yield stress. This kind of continuous deformation is commonly called "*ductile*". But note that, when used in this way, the term is purely phenomenological and does not refer to the microprocesses involved. Accordingly, the term is considered as the opposite of "*brittle*", if the use of the latter term is restricted to the violent stress drop of Fig. 1.9A.

In contrast, we prefer to reserve the terms "brittle" and "ductile" to a rheological characterisation of the material by taking into account the stress-strain response of the material. Idealising the real rock behaviour, we distinguish between rate-insensitive and rate-sensitive deformation processes. "Brittleness" then comprises all rate-insensitive deformations which terminate in strain-softening; violent rupturing of the rock is then merely the most extreme case in the whole brittleness range. "Ductility", on the other hand, refers to rate-sensitive processes, associated with "*viscous*" flow, creep and stress relaxation phenomena. These deformations result from diffusive transport processes inside the grains, along grain boundaries, or through the pore water.

At this point, we should note that a deformation such as the continued hardening in Figs. 1.9D and 1.10, although ductile in the phenomenological sense, need not be ductile in the rheological sense, because the processes that operate at the grain-size scale may be predominantly brittle and thus produce rate-insensitive hardening. On the other hand, the macroscopic hardening may result from the combined operation of cataclastic microprocesses and intra-crystalline gliding; it is then referred to as "*semi-brittle*". The multi-shear cataclastic hardening in Fig. 1.8E may be an example of semi-brittle behaviour in marble and various other rocks under high confining load and moderate temperatures.

Finally, it should be understood that the distinction between rate-sensitive and rate-insensitive behaviour strictly pertains to the deformation process itself and not to the conditions that initiate the deformation. To elucidate this point, consider a block of rock salt; when left alone, subject to its own weight, or put under a gently increased surface load, the block will slowly change its shape by viscous flow. If, however, a heavy weight is dropped onto the block, the block may be made to fracture in a brittle manner. Similarly, a rock specimen which deforms in a brittle manner in triaxial testing at room temperature, can be made to deform in a ductile mode at elevated temperatures. In other words, one and the same material can deform either way, in a brittle or in a ductile mode, depending on the particular deformation conditions.

Obviously, these conditions can change in time. A sufficient reduction in temperature of a rock body, say below about 300°C in quartz-rich rock, will stop a ductile deformation and allow brittle deformation processes to operate. Similarly, an increase in the rate of the imposed straining may no longer allow the diffusion-controlled viscous deformation mechanisms to keep pace with the imposed straining. Hence, ductile stretching or shortening of a rock layer may change into brittle fracturing when the rate of the applied extension or shortening is sufficiently raised. Thus joints may also be found embedded in material that has been deformed by viscous flow.

In general, the viscous flow behaviour is not best described as a Newtonian fluid, which implies a linear relationship between strain rates and stresses, but is more realistically modelled by a non-linear *creep* law. In describing the creep behaviour of a rock under a constant load, one can distinguish three phases of creep, as schematically shown in the strain-time diagram of Fig. 1.11: Under a constant load, the rock specimen begins to deform ("primary" creep I), but the strain rate (represented by the slope of the strain-time curve) decreases. At the end of this phase, the creep flow either comes to a complete halt or, under sufficient load, continues and accelerates eventually until rupture occurs. Between this phase of accelerating unstable creep ("tertiary" creep III) and the decelerating "primary" creep, rocks are often found to flow at an approximately constant rate ("secondary" or steady state creep II).

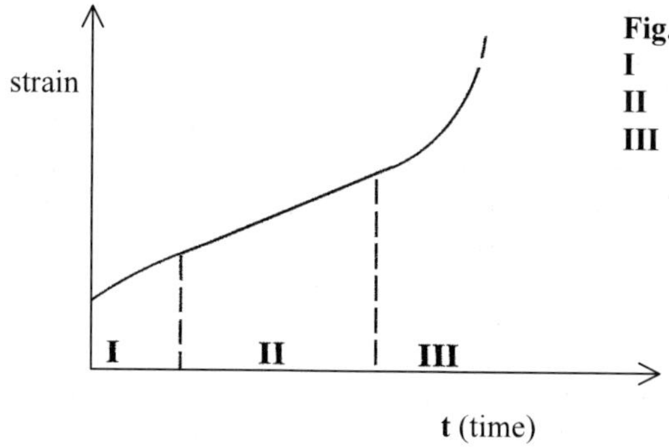

Fig. 1.11. Creep under constant load:
I primary or transient creep;
II secondary or steady state creep;
III tertiary or unstable creep

Occasionally, creep flow may also occur under the pressure and temperature conditions of the brittle upper crust where rocks normally deform in a brittle way. Two typical examples are rock salt and clays or clay-rich sediments. It is well-known that rock salt can flow in steady state (creep phase II) over geological periods. However, clays and clay-rich sediments which have been normally compacted, do not exhibit a steady state flow; they either come to rest after primary creep or, under sufficient load, undergo accelerating tertiary creep that terminates in the formation of discrete fractures.

Chapter 2

Experimental Evidence and Elementary Theory

Strength. In the preceding chapter we reviewed the modes (Fig. 1.8) by which a rock will fail when uniformly stressed beyond a critical state of stress, loosely referred to as the "strength" of the rock. This critical state depends on the rock type and the type of loading. In uniaxial tension (Fig. 1.8A), the critical state is characterised by the maximal tensile stress the rock can sustain ("tensile strength"), in uniaxial compression (Fig. 1.8B) it is the greatest axial compressive stress ("uniaxial compressive strength"), and in compressive shear failure (Fig. 1.8D) it is the maximal differential stress ("shear strength") $\sigma_I - \sigma_{III}$, (σ_I being the greatest, σ_{III} the smallest principal stress) which the rock can withstand.

Strength is not a unique material parameter, even if associated with a certain failure type. It depends to a greater or lesser degree on the confining pressure, as was already illustrated by von Kármán's stress-strain curves in Fig. 1.10, and if a rock sample is laterally confined by two different principal stresses σ_{II} and σ_{III} (Fig. 2.1), the strength may also depend on the intermediate principal stress σ_{II}. In general, the strength decreases when the temperature is substantially raised or when the loading rate is reduced by orders of magnitude. But above all, failure in a porous rock is not affected by the pressure of the pore fluid, as will be seen shortly.

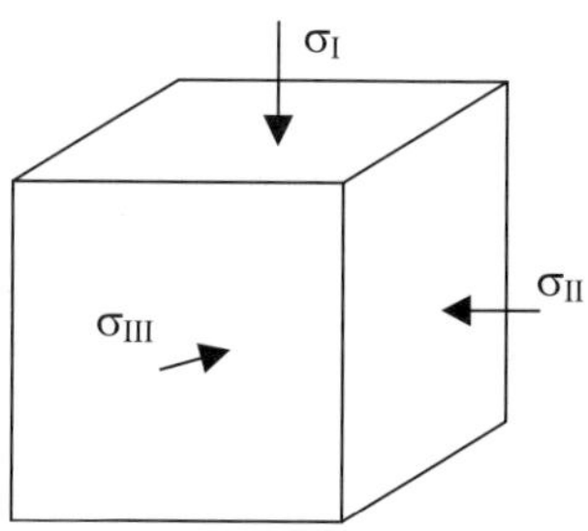

Fig. 2.1. True triaxial loading

Yet, before turning to this important point, we have to draw the readers attention to the sign convention we use when dealing with stresses. Following the common usage in rock and soil mechanics, we ***consider compressive normal stresses as positive***. This choice is motivated by the fact that even in extensional regimes, most normal stresses in the Earth's crust are compressive. To avoid confusion, note that in elasticity theory tensile stresses are counted as positive.

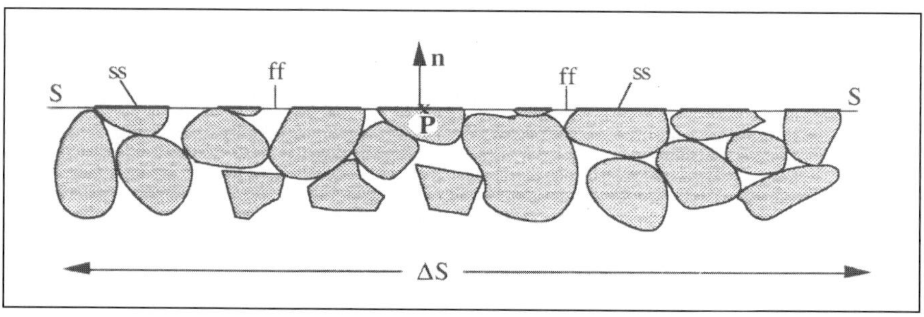

Fig. 2.2. Schematic drawing of the trace of a macroscopic surface element (SS) with centre at P and unit normal **n**. The total area is ΔS, and the intersections of the solid skeleton are indicated by ss, and the pore sections by ff

Pore pressure. So far, we have considered the straining and failure behaviour of *dry* rocks, and that the stresses on test specimens were ***total stresses***; i.e., normal and shear forces transmitted across the total area of a unit cross-section that is very large compared with the dimensions of grains or pores and the like. Thus, this cross-section cuts through the solid skeleton of the rock and pore space alike, as sketched in Fig. 2.2. However, under geological conditions porous rocks are always saturated with *fluids* (water, oil, gas), which differ greatly from the solid components of the rock in both their mechanical and thermal properties, and therefore respond differently to mechanical loading and changes in temperature. In particular, the pore fluid carries a proportion of the total normal stress $\sigma_{\perp}$ on a cross-sectional element inside, or on the boundary of a rock body. Thus it is obvious that in dealing with deformational processes and fracturing in porous rocks (including almost all sedimentary rocks) the *pore pressure* has to be introduced as a separate state variable.

Despite its apparent simplicity, the concept of "pore pressure" conceals a subtlety which has to be pointed out. Note that the concept of stress only makes sense if applied to a material *continuum*. Now, a fluid-saturated rock consists of at least two components, a fluid and a solid part. The pore pressure, p, is simply defined as the sum of all forces that act perpendicularly across the *fluid part* of a large cross-sectional or surface element (SS in Fig. 2.2), divided by the fluid-filled area of the element (force per unit fluid area). The value of the pore pressure is assigned to the centre P of the cross-sectional element. Note that this point may coincide with *any* point of the solid-fluid continuum, irrespective of its position in the fluid or solid part. Hence, the pore pressure field p(P) occupies the whole solid-fluid continuum, although it only results from the pressures that act in the fluid-filled pores of the rock.

Similarly, the *total* normal or tangential stresses that act on the element SS in Fig. 2.1 are simply the sum of all the normal or tangential forces that act across the solid *and* fluid parts of a cross-sectional element SS, divided by the total area of SS, and assigned to its centre P. Naturally, the *state* of total stress at P is defined by the tensorial components of total stress, that act on three mutually orthogonal elements with a common centre at P. Thus we have two stress fields: the field of the total stresses and the field of the pore pressure – both being defined at *any* point of the continuum composed of solid skeleton and pore space.

In the laboratory, fluid-saturated rock specimens are mostly tested in the so-called triaxial-testing apparatus, which is schematically shown in Fig. 2.3. In this apparatus, the pore fluid pressure can be controlled independently from the axially and radially applied compressive stresses. It is also possible to apply pore pressures in excess of the total axial stress σ_V. Naturally, one may expect that the pore pressure will in some way diminish the effect of the total stresses on the failure behaviour of a porous rock. But how to account for this?

During the second half of the last century, an impressive body of careful experimental work on a great variety of porous rocks, and over a wide range of pore pressures, was accumulated. Most remarkably, the data on brittle rock failures convincingly demonstrate that the strength, i.e. the critical state of stress which the material can sustain without failing in extensional or shearing modes shown in Fig. 1.8, is not affected by the pore pressure. That means that brittle failure is not controlled by the total normal stresses σ_i, but by the *effective* stresses

$$\sigma_i' = \sigma_i - p \qquad (i = 1, 2, 3; \mathrm{I, II, III}) \tag{2.1}$$

introduced in soil mechanics by Karl Terzaghi in 1923. Note that shear stresses are not affected by the pore pressure, since the pore fluid cannot transmit net shear stresses across a macroscopic interface.

Hence, an increase in pore pressure, while keeping the total stresses constant, will reduce the strength of the material. Interested readers may find more information on

Terzaghi's "effective stress principle" in G. Mandl, FBR (pp. 88, 114–120, 174–175). Here the following remark may suffice: Although Terzaghi's effective stresses control, rigorously or to a very good approximation, the onset of brittle failure in porous rocks, they are, in general, somewhat different from the stresses that are "effective" in producing the *elastic* pre-failure strains. Linear elastic straining in fluid-saturated rocks is controlled by the effective normal stresses

$$\sigma_i^* = \sigma_i - a.p = \sigma_i - \left(1 - K/K_s\right).p \qquad (i = 1, 2, 3; \mathrm{I}, \mathrm{II}, \mathrm{III}) \qquad (2.2)$$

where $1/K$ is overall compressibility of the porous rock, and $1/K_s$ is the compressibility of the skeletal material (e.g., quartz or calcite). The stresses σ_i* are often referred to as "generalised effective stresses", which is a misleading term, since these stresses only apply to the elastic straining.

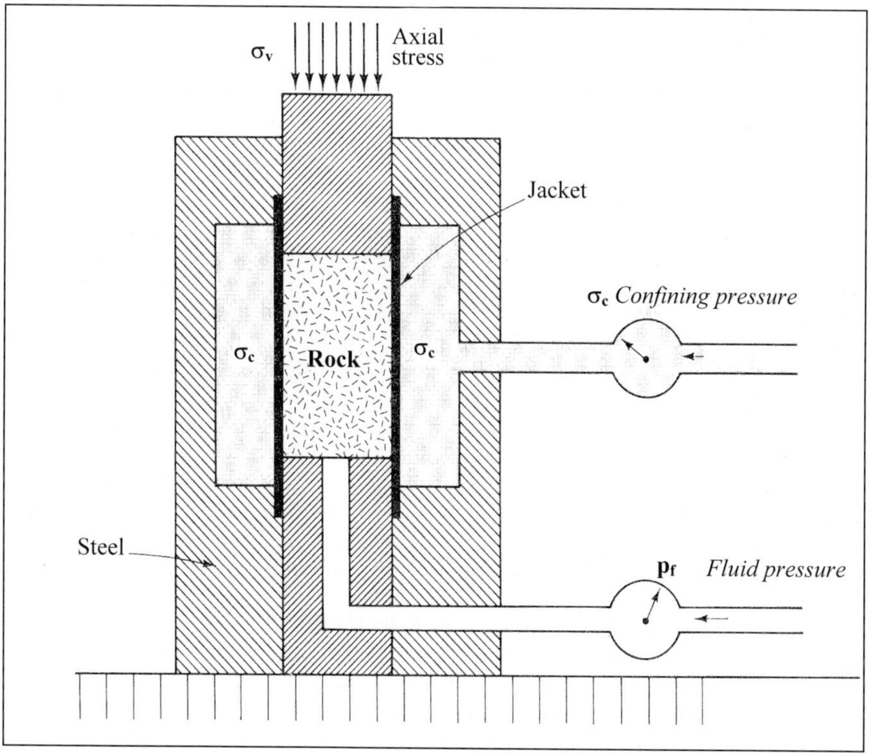

Fig. 2.3. Simplified schematic drawing of triaxial testing apparatus.
The porous rock sample is loaded by an axial stress σ_v and a radially applied confining pressure σ_c (representing two equal principal stresses). The fluid that exerts the confining pressure is separated from the rock by a weak but impermeable jacket. The pore-fluid pressure p_f is controlled independently (from J. Suppe (1985) *Principles of Structural Geology*, Prentice-Hall Int., London)

Tensile fracture (germ. *Zugbruch*). In Fig. 1.8 we have schematically summarized how rock samples may fail by fracturing in a brittle or semi-brittle way. We shall now consider the basic failure modes and the mechanical conditions of their formation in more detail, beginning with the tension fractures. Suppose that a cylindrical rock sample under zero effective confining pressure σ'_c (Fig. 2.4A) is put under an axial tensile load in a way that the axial tensile stress σ_V is fairly uniform across the sample. At a critical value $-T_o$ of the effective tensile stress σ'_v the sample is disrupted along a fracture plane which runs perpendicular to the direction of the applied tensile stress. T_o (> 0) is the ***"tensile strength"*** at zero confining pressure, and the fracture that disrupts the rock sample is a brittle *tension fracture*. It should be mentioned that the experimental determination of uniaxial tensile strength is difficult and produces test data with considerable scatter. Moreover, the measurements are carried out on intact specimens, and therefore are likely to overestimate the actual tensile strength of natural rocks. Typical values of T_o are between 5 and 30 MPa.

The stress state that leads to this failure is represented by the solid circle in the Mohr diagram of Fig. 2.4B. Although the tensile strength was defined for rock specimens under zero confining pressure, tension fracturing will still occur at $\sigma'_v = -T_o$ when the specimen is laterally loaded by a low confining pressure σ'_c. This state is indicated in Fig. 2.4B by a dashed circle. The reader may be reminded that the Mohr circles in Fig. 2.4B represent the normal and tangential stress components on the set of cross-sectional elements which are parallel to the axis of the intermediate principal stress σ_{II}, as illustrated in Fig. 2.4C.

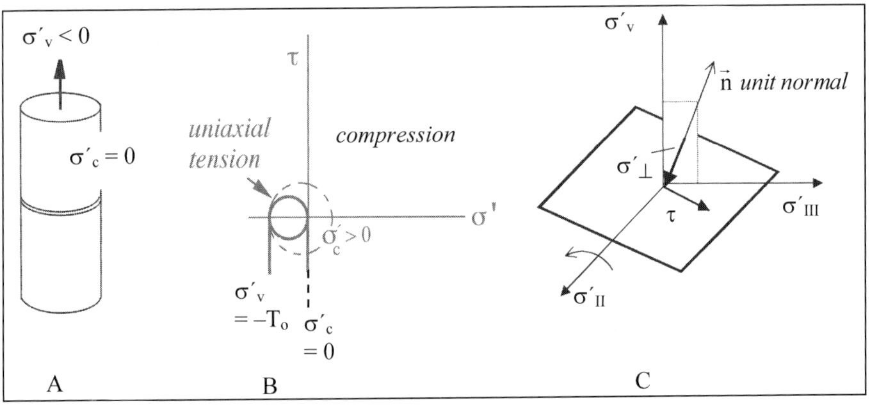

Fig. 2.4. A) Tension fractures under uniaxial extension; B) Mohr circles for tension fracturing; C) normal stress ($\sigma_\perp$) and shear stress (τ) on plane which is parallel to the σ_{II} axis, represent a point on Mohr's stress circle through σ'_I and σ'_{III} on the σ'axis

Some idea of the effective confining pressure that would still allow tensile fractures to develop may be obtained from a failure condition derived by A.A. Griffith (1925) for a flat elliptical crack. According to Griffith's theory, the limiting confining pressure for the formation of tensile fractures is $3T_o$. Applying a tensile load to a specimen under a higher confining pressure would produce failure of a different mode, as will be discussed later, without the tensile stress reaching the tensile strength. Thus the conditions for tensile fracturing are:

$$\sigma'_V \equiv \sigma'_{III} = -T_o, \quad \sigma'_c = \sigma'_I < 3T_o \tag{2.3}$$

where the first condition is purely empirical, and the second purely theoretical and derived for an ideal fracture shape.

In nature, rock fractures which are generated by effective tension are a special class of "joints", which may be called ***"tension joints"*** (germ. *Zugklüfte*) where the word "tension" is used in the sense of *effective* tensile stress. At a given depth in the Earth's crust all total stresses are compressive, and the effective stress condition (Eq. 2.3) for tension joints can only be satisfied when the pore fluid pressure exceeds a total stress component. In fact, high fluid pressures generated inside a rock layer, or supplied from outside, play a key role in the generation of joints. A detailed discussion of this will be given later.

Although the empirical tensile strength T_o is a very useful concept for estimating the geological conditions that promote the formation of tensile fractures, it does not tell us anything about the propagation of fractures or the concentration of tensile stresses near the fracture tip, where decohesion of the material takes place. These problems are the concern of the more sophisticated theory of *linear elastic fracture mechanics (LEFM)*. Our tensile fracture is the "mode I" or "opening mode" fracture shown in Fig. 1.7A of the preceding chapter. Figure 2.5 shows half of such a fracture in two-dimensional view, with a vastly exaggerated aperture. Fracture and stresses are supposed to continue uniformly in the third dimension which makes the cross-sectional plane a deformation plane. Note that the stress components, here referred to a Cartesian coordinate frame, are total stresses. The tensile or compressive "remote" stress σ_2^r acts uniformly at a distance >> L.

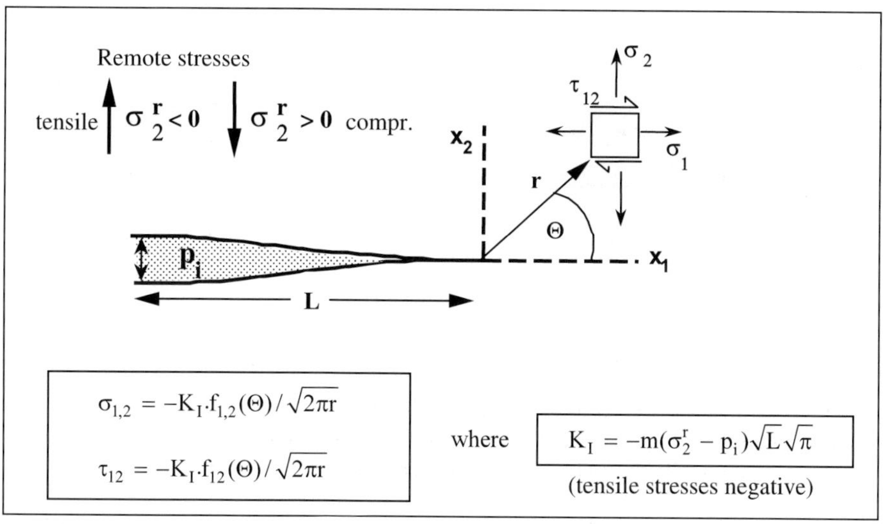

Fig. 2.5. The stress concentration in the near-tip region of a tensile fracture with internal fluid pressure (see text for further explanation)

The formulae given in Fig. 2.5 represent the stress components in a near-tip region whose radius is much smaller than L. The locations in this region are defined by the distance r from the cuspate fracture tip, and the angle Θ. Two separate, well-defined functions describe the dependence of the stresses on r and Θ. According to the formulae, the stresses being proportional to $r^{-1/2}$ increase towards the tip and attain infinite values at the fracture tip. Since such a singularity does not exist in real materials, the formulae do not apply in the very close vicinity of the fracture tip. The parameter K_I in the formulae – the so-called "*stress intensity factor*" – is independent of the coordinates and only determined by the external load system, the geometries of the elastic body and the fractures. For the uniform loading system considered, K_I is given by the expression stated in the figure. Note that the stress intensity increases with the square root of L, which, in a way, expresses the fact that the separated fracture walls exert some leverage on the near-tip region, thereby increasing the magnitude of the local tensile stresses. This is qualitatively illustrated in Fig. 2.6. Assuming that the rock is impermeable, a fluid inside the fracture exerts a pressure p_i on the fracture walls and thus contributes to the leverage action. The term for the "opening" or "driving stress" in the formula for K_I in Fig. 2.5 is therefore $\sigma^r_2 - p_i$, which is the sum of two negative terms, since we consider tensile stresses as negative. If the remote stress were compressive, as also indicated in Fig. 2.5, the pressure of the fracture fluid would have to exceed the remote rock pressure in order to keep the fracture open.

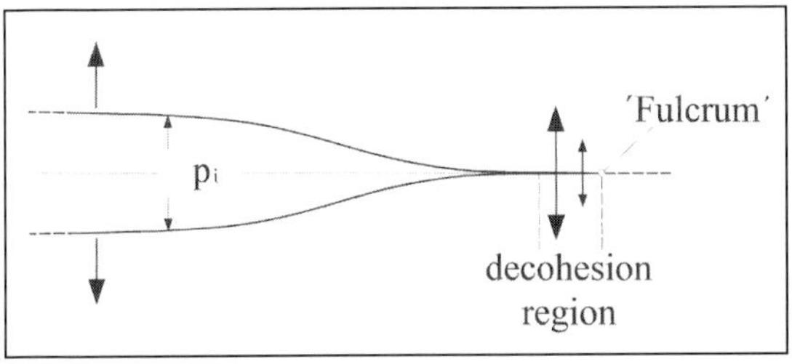

Fig. 2.6. The "leverage" action of fracture walls on the near-tip region (schematically)

Fig. 2.7. Tension joints in a Carboniferous turbidite sequence with high pore pressure (Cornwall, West England). Note that bedding plane irregularities serve as nuclei for fracture initiation

The coefficient **m** in the formula for K_I is a dimensionless "modification factor". It has the value 1 for a straight internal crack at some distance from the remote boundaries of the elastic body. The value of m increases when the ratio L/w of a double-ended fracture contained in an elastic strip of width 2w parallel to the fracture increases. For a fracture that begins open-ended at the edge of a semi-infinite elastic body and ends in a cuspate tip at the distance L from the edge, m attains the value 1.12 (see B.R. Lawn and T.R. Wilshaw 1975, *Fracture of Brittle Solids*). This result is interesting, because it suggests that *tension joints grow more easily from the bedding planes* into a rock bed than from locations inside the bed, provided the fracture nuclei are of comparable shape and orientation. One should expect this tendency to be further enhanced by "notch"-type bedding plane irregularities, as is illustrated in the photograph of Fig. 2.7.

The stress intensity factor provides a refinement of the concept of tensile strength. The net driving stress $\sigma^r_2 - p_i$, will propagate a fracture, when K_I attains a certain critical value K_{Ic} – the so-called ***"fracture toughness"*** (germ. *Bruchzähigkeit*). This is a real material constant, which combines the various material properties that control the decohesion process in the near-tip region of the fracture. From the literature we learn that typical values for K_{Ic} for hard rocks vary between about 1 and 3 $MPa.m^{1/2}$. When we insert K_{Ic} for K_I in the formula for K_I in Fig. 2.5, we obtain the critical tensile "driving stress" for fracture propagation:

$$\left|\sigma^r_\perp - p_i\right|_{crit} = K_{Ic} / m\sqrt{\pi L} \tag{2.4}$$

where $\sigma^r_\perp$ denotes the remote tensile stress ($\sigma^r_2 < 0$ in Fig. 2.5) that acts normal to the fracture plane.

We recall that in deriving this expression the fracture walls were assumed to be impermeable. How should the above formula be modified in order to apply to the more general case of a *tensile fracture in* ***a porous*** *rock*? In brittle fracturing, the build-up of stresses that leads to fracture propagation is of a predominantly linearly elastic nature. The elastic strains in the porous rock with pore pressure p are then controlled by the generalised effective stresses (Eq. 2.2). Hence, when applying Eq. 2.4 to a porous rock, we have to replace the total boundary stresses $\sigma^r_\perp$ and p_i by the effective stresses $\sigma^r_\perp{}^* = \sigma^r_\perp - a.p$ and $p_i^* = p_i - a.p$. Note, however, that because of the minus sign in Eq. 2.4 the value of the left hand side of the equation remains unchanged. Thus, the driving stress that propagates a tensile fracture in a fluid-saturated porous rock is

$$\left|\sigma^{r*}_\perp - (p_i - a.p)\right|_{crit} = \left|\sigma^r_\perp - p_i\right|_{crit} = K_{Ic} / m\sqrt{\pi L} \tag{2.5}$$

Therefore, in terms of the remote total boundary stress and the fluid pressure p_i inside the fracture, the condition for fracture propagation in the fluid saturated permeable rock and in the non-porous rock is the same.

We can draw several important conclusions from this expression: First, since at the onset of fracture growth the fracture pressure p_i is equal to the pore pressure p of the surrounding fluid-saturated rock, the *onset* of fracture growth is controlled by the remote Terzaghi stress (Eq. 2.1)

$$\left|\sigma^{r*}_\perp + a.p - p\right| = \left|\sigma^{r'}_\perp\right|_{crit} = K_{Ic} / m\sqrt{\pi L} \tag{2.6}$$

Secondly, since the driving stress decreases as the fracture grows in length, the fracture growth is accelerated by the increase in the tensile excess load (and by the release of elastic strain energy), if the remote tensile stress $\sigma^r_\perp$ and the pore pressure p are kept constant. In reality, however, the fracture pressure p_i in a propagating double-ended fracture will drop as the volume of the growing fracture increases. This causes an inflow of pore fluid from the surrounding rock at a rate depending on the hydraulic permeability of the rock. At the same time, the decrease in fracture pressure will entail a decrease in the driving stress $|\sigma^r_\perp - p_i|$, which will retard, possibly stabilise or even temporarily stop the propagation of a tensile fracture inside the fluid-saturated rock.

In closing this interlude on elastic fracture mechanics, we recall that LEFM does not provide a theory of fracture initiation, but rather deals with the stress concentration and propagation of *existing* fractures of a certain minimum length. We also have to realize that the concept of the stress intensity factor applies to the tip region of the fracture in a homogeneous material and may break down when macroscopic heterogeneities occur in that region.

After having discussed how the critical tensile driving stress depends on the fracture length, and how this affects the growth of a tensile fracture, we are now confronted with the question of the use and reliability of the "*tensile strength*" T_0 determined by uniaxial tension tests. We first draw attention to the fact that a test specimen which is selected for its homogeneity is still highly heterogeneous on a micro-scale. Fractures of a certain length and favourable orientation may already exist inside the specimen; and, even if this is not the case, microscopic flaws, inclusions, pores of irregular shape, or grains with different elastic properties, etc., will act as local concentrators of tensile stresses and thus serve as nuclei for tensile fractures. Obviously, the tensile stress applied to the test specimen has to be raised to a level that is sufficient to generate suitable fractures and to initiate their growth or to start propagating already existing fractures. This level is the limit $-T_0$ of the tensile stress which the specimen can sustain, since at this stress the propagation of macroscopic fractures, sub-perpendicular to the applied tension, is not only initiated but can continue without any further increase in tensile load, in accordance with Eqs. 2.5 and 2.6. Recall that the tensile load required to propagate the fractures decreases as the fractures grow in length, and that the decrease in the driving stress is even enhanced by the fact that the form factor m increases as the growing fracture approaches the boundary of the rock specimen. Hence, at least in a specimen of dry rock under the critical tensile stress $-T_0$, the growing surplus in net driving force, in combination with the release of elastic strain energy, will cause a rapid and violent rupturing of the test specimen.

Thus in conclusion, we expect the "tensile strength" T_0 to be the remote tensile stress that is required to start the propagation of fractures already present in the unstressed rock or, if such fractures are not present, to generate suitable fractures and to initiate propagation in a direction perpendicular to the remote tensile stress. In this sense, the tensile strength determined in uniaxial tension tests is a material constant only for the *sample of material tested*. But, in general, the rock mass from which a test sample was taken, is more heterogeneous than the sample, and most likely contains more favourably orientated fractures of greater length than the test sample. Therefore, in geological applications, *the tensile strength of rock samples should be considered as an upper bound of the tensile strength of the rock mass from which the samples were taken.*

Extension or "cleavage" ("axial splitting") fracture (germ. *Spaltbruch*). While the tension fractures are fairly well understood, the fracture type, illustrated in Fig. 2.8A, may appear almost paradoxical: A specimen, under zero lateral confining pressure, can be split along one or several axial fracture planes by an axial compressive load $\sigma'_v > 0$. These "cleavage" fractures, which open perpendicular to the direction of the maximum compressive stress in the absence of any tensile boundary stress, are also called "*extension*" fractures; a term, which we exclusively apply to cleavage fractures to distinguish them from the *tensile* fractures, which are generated by a tensile load. Extension fracturing (splitting) of a sample under a uniform uniaxial compression takes place when the maximum effective compressive stress σ'_I (= σ'_v) reaches a critical value $\mathbf{C_0}$ – the "***uniaxial compressive strength***" (germ. *Druckfestigkeit*):

$$\sigma'_I = C_0 \quad \text{and} \quad \sigma'_{III} = \sigma'_c = 0 \tag{2.7}$$

This critical state of stress is represented by the solid circle in the Mohr diagram in Fig. 2.8B. Typical values of C_o reported in the literature lie between 50 and, say, 500 MPa. Note, however, that in lithologically anisotropic rocks, especially in layered rocks, one should expect a certain anisotropy of the uniaxial compressive strength.

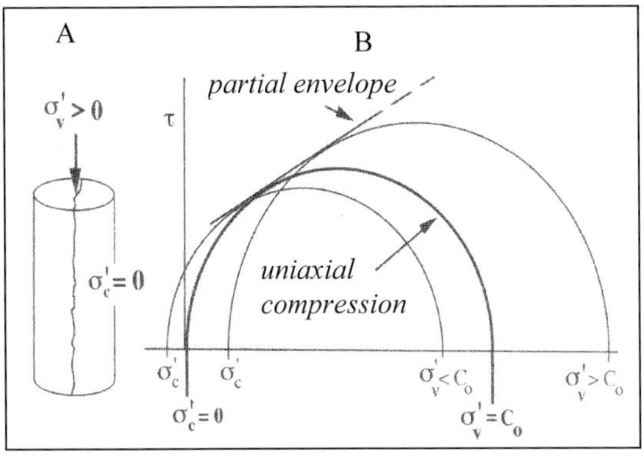

Fig. 2.8. Extension (cleavage) fractures under uniaxial compression: A) Cylindrical specimen under uniform axial loading; B) Mohr circles for extension fracturing (splitting) of a specimen (σ'_c effective confining pressure, σ'_v effective axial stress, C_o "uniaxial compressive strength")

The morphology of these fractures is very similar to that of brittle tensile fractures. It should, therefore, not be surprising that the explanation for this fracture phenomenon has been sought in radial tensile stresses which could have been induced at the end faces of the test specimens by friction between the specimen and the loading pistons of the testing machine. However, in the past, great care was taken to eliminate such end effects in producing axial splitting. Moreover, the same type of fracturing is obtained when a cylindrical specimen is loaded by high fluid pressure on the curved surface while the end faces are kept stress free, or are even loaded by some small compressive axial stress (Fig. 2.9). Under a sufficiently high fluid pressure, the loaded cylinder is "explosively" disrupted by one or several extension fractures perpendicular to the cylinder axis ("disking" in "Bridgman's paradox").

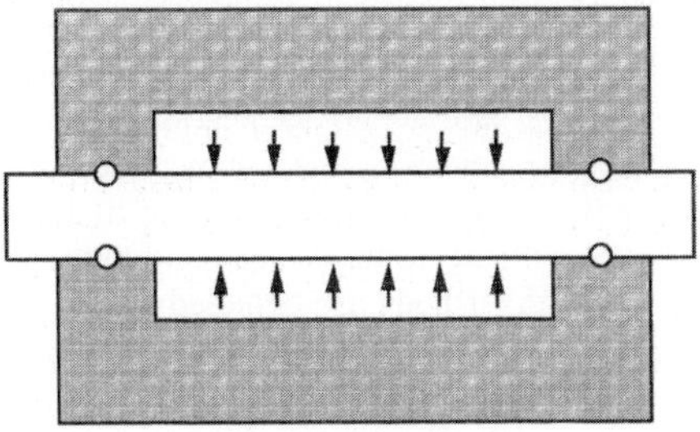

Fig. 2.9. "Pinching-off effect"

Today, the phenomenon is basically understood in terms of compression-induced local tension regions inside the rock specimen. Although the brittle behaviour of rocks in compression and, in particular, that of porous rocks is very complex, it is clear that on a grain or pore scale a variety of mechanisms can give rise to localized tensile stresses inside the rock. In Fig. 2.10, four of these mechanisms are schematically illustrated, all of which have been extensively dealt with in the more recent literature. The mechanism favoured most by theoreticians who attempt to derive the macroscopic deformation and failure behaviour of brittle solids from micro-mechanical processes, is the so-called "wing crack" model (Fig. 2.10A). In the early 1960's W.F. Brace and E.G. Bombolakis, and E. Hoek and Z.T. Bieniawski, demonstrated that cracks which are inclined with respect to the remote maximum compressive stress σ'^r_1 will not propagate as shear cracks in their own planes when the compressive stress is raised, but will have tensile "wing cracks" branching off. It is easily seen (Fig. 2.10A) that an inclined parent crack is under shear which causes extensions of the crack walls at the receding sides near the crack tips. At these locations, the wall-parallel tensile stress attains its highest value and can produce tensile fractures when the remote compressive stress σ^r_I is sufficiently raised. When the walls of the parent crack are not in frictional contact, the "wing cracks" start perpendicularly from the inclined parent crack. If there is frictional contact between the fracture walls, the direction of a wing crack will be inclined towards the tip of the parent crack. The wing cracks grow stably, requiring a progressively larger σ^r_I, and tend to align themselves parallel to the remote maximum principal stress.

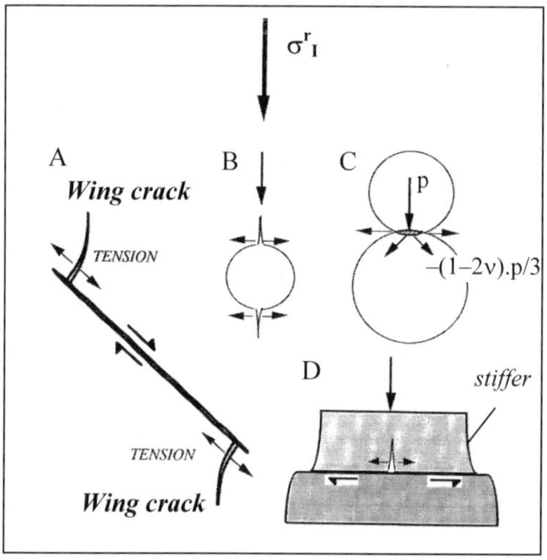

Fig. 2.10. Schematised brittle micro-mechanisms that induce localised tensile stresses (see text for explanation)

A second important mechanism that gives rise to local tensile stresses, is illustrated schematically in Fig. 2.10B: The presence of a spherical or elliptical hole in an elastic continuum under remote uniaxial compressive loading will cause wallparallel tension to concentrate at two points of the cavity wall. Again, tensile cracks will start growing in a stable manner when the remote uniaxial compressive stress is progressively raised. A third mechanism operates at the contacts between grains. As shown in Fig. 2.10C, the maximum tensile stresses act in a radial direction at the circular boundary of the contact area of two elastic spheres pressed together by a compressive load. The contact pressure attains its maximum (p) at the centre of the contact area and produces the maximum tensile stress indicated in the figure. The fourth cause of local concentrations of tensile fracture stresses is associated with an elastic mismatch of particles in compressive contact. In Fig. 2.10D a stiffer and a weaker body are depicted in contact along some finite area; bonds or contact friction prevent slippage when a uniaxial compressive load is applied. The compressive load must then induce shear stresses near the contact because, under the applied load, the weaker body has the tendency to extend further in the lateral direction than the stiffer body. These shear stresses will put the central part of the contact area under tension, and a progressive increase in the compressive stress will eventually produce a stably growing tensile fracture.

There are still other mechanisms, such as the bending of elongated grains, or the indentation of sharp grain corners into neighbour grains which locally induce tensile stresses that propagate small tensile fractures. All these fractures grow in a stable way under a progressively increasing compressive load, and align – in a statistical sense – along the direction of the uniaxial compressive stress. At some level of this stress, some fractures become interlinked, and one may expect the longer fractures to grow more easily than the shorter ones. This will localise the progress of fracturing into macroscopic extension fractures which will eventually split the rock sample in the axial direction

Whatever the mechanisms are by which tensile micro-fractures are induced by uniaxial compression, the growth of the fractures is very sensitively affected by a compressive stress normal to the fracture plane. It is, therefore, quite obvious that a confining pressure $\sigma'_{III} = \sigma'_c > 0$ acting on the test specimen in Fig. 2.8A will have an adverse effect on the formation and growth of microfractures and will obstruct the formation of macroscopic extension fractures parallel to the imposed maximum compressive stress. However, the generation of macroscopic extension fractures will not be abruptly stopped by the application of an arbitrarily small confining pressure; hence, there is good reason to expect a range of very moderate confining pressures which still allow macroscopic extension fractures to be formed. Obviously, the range of these low confining pressures will depend on the rock type. Moreover, axial splitting under some small confining pressure will require a higher axial compressive stress than in purely uniaxial compression. Such a state of stress is represented, in a qualitative way, by the largest circle in the Mohr diagram of Fig. 2.8B. Similarly, one should expect that axial splitting can be produced by an axial compression that is smaller than C_o, when the confining stress is slightly tensile ($\sigma'_{III} < 0$), as indicated by the smallest circle in the Mohr diagram of Fig. 2.8B. Thus, by applying different confining stresses slightly deviating from zero, a set of critical Mohr circles for the extensional (cleavage) fracturing of a given rock sample can be obtained, which increase in size with increasing maximum compressive stress and are enclosed by a pair of inclined envelopes.

Some experimental corroboration of this concept may be found in observations reported by outstanding researchers in rock mechanics in the 1970's, and in recent results of axial loading tests on tuff specimens under very moderate confining pressures (J.M. Kemeny 1993; in: Pasamehmetoglu et al. (eds.) *Assessment and Prevention of Failure Phenomena in Rock Engineering*, Balkema, Rotterdam, pp. 23–33). We must remember that, in experiments of this type, great care has to be taken to avoid friction between the ends of the test specimen and the platens of the testing machine that could promote axial splitting. There is certainly a need for more experimental work in this field.

However, theoretical and experimental results leave no doubt that the propagation of tensile microfractures and, consequently, the formation of macroscopic extension fractures parallel to the imposed compressive boundary stress σ'_I are completely *suppressed* by confining pressures σ'_{III} which amount to more than a few percent of σ'_I.

Geological implications. Figure 2.11 presents examples of joints that have been interpreted as cleavage (extension) fractures. Figure 2.11A shows a fan of extension fractures that formed sub-parallel to a free vertical rock face on which σ'_{III} must vanish. The fractures in Fig. 2.11B,C are most likely cleavage fractures generated in response to unloading by the removal of overburden while the rocks were still under a relatively high bed-parallel residual compressive stress. But what can be stated, more generally, on the geological conditions of jointing under joint-parallel compression?

What can be inferred from the rock mechanical data of cleavage fracturing? First, it may be safely concluded that in nature, the formation of joints is only possible if σ'_{III} is zero, or close to zero. But what can be said about the necessary maximum compressive stress σ'_I? Here we face a serious problem. The uniaxial compression tests in the laboratory determine the axial stress C_o at the instant the test specimen fails by splitting. Yet axial cleavage fractures develop and grow in a stable way with increasing axial load, long before σ'_I has reached the limit value C_o. This is known from experiments (e.g., H. Bock (1980) N. Jb. Geol. Paläont. Abh. *160*, pp. 380–405) and is supported by theories on the development of cleavage fractures, which will be discussed later in this book (see Chapt. 4, pp. 87–94). Unfortunately, there are little data available that would allow a quantification of the onset and growth of cleavage fractures as a function of the applied axial stress σ'_I. Tentatively we may hypothesize that the development of cleavage fractures starts at $\sigma'_I \approx C_o/2$.

Nevertheless, one has to bear in mind that C_o may be substantially reduced by a layer-parallel σ'_{III} which is slightly tensile, though still much less so than the tensile strength of the rock. Furthermore, the collapse mechanisms, (such as buckling,) in unconfined test samples, may differ markedly from the behaviour of a thick horizontal plate under uniform vertical load. And finally, the uniaxial compressive strength also decreases to some degree with increasing volume of the sample and decreasing loading rate (V.S. Vutukuri, R.D. Lama, S.S. Saluja (1974) *Handbook on Mechanical Properties of Rocks*, Vol. I, Transtech Publ.).

Although the incongruities between the rock mechanical experiments and the geological environment forbid an exact quantification of the compressive stress σ'_I that is required for cleavage jointing in the brittle crust, we can draw the following conclusions. First, it is almost certainly erroneous to assume that the necessary effective compressive stress σ'_I for cleavage fracturing has to reach the uniaxial compressive strength C_o of the rock. Secondly, as demonstrated by the Mohr diagram in Fig. 2.12, the smallest value of the maximum effective stress σ'_I that can initiate faulting (of any kind) is, in general, still greater than the compressive stress σ'_I necessary for the formation of cleavage fractures when $\sigma'_{III} = 0$.

Also note that in Fig. 2.12 the regime of cleavage fracturing (see also Fig. 2.8) is separated from the regime of compressional faulting. This means that we consider cleavage fracturing and shear faulting in compression as being two mechanically different processes. Hence, we do not follow the practice of relating the uniaxial compressive strength C_o directly to the shear strength parameters τ_o and φ of the straight Mohr-Coulomb tangent of the stress circles which represent the conditions for the onset of faulting under compressive stresses (Fig. 1.8D).

A

B

C

Fig. 2.11. Extension (cleavage-)jointing approximately parallel to a free surface: A) Fan of subvertical extension joints at a vertical rock boundary (from P. Bankwitz et al. (2000) Z. geol. Wiss. *28(1/2)*); B) buckled exfoliation sheet in granitic rock, Yosemite National Park, Cal., USA (from J. Suppe 1985, *Principles of Structural Geology*, p. 201, Prentice Hall); C) bed-parallel extension joints in limestone, Münster Basin, Germany

Hence, in principle, cleavage fracturing should be possible in all tectonic stress regimes that lead to faulting. However, this does not yet say anything about the aperture and the spacing of parallel cleavage joints; we shall revert to these open questions in Chapt. 4 (pp. 91–94). Here we are content with a few remarks on the existence of cleavage fractures in geological settings.

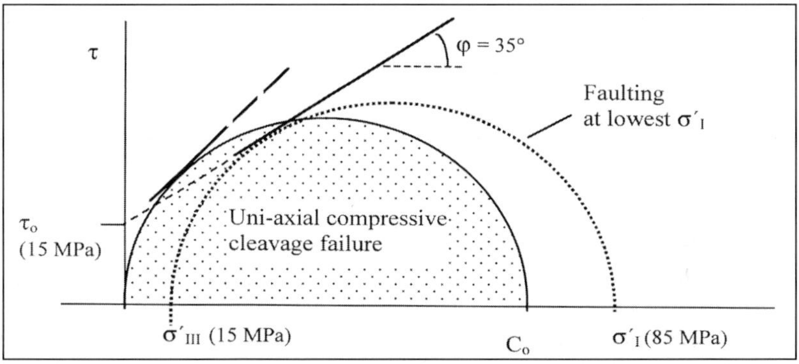

Fig. 2.12. Mohr diagram comparing the smallest σ'_I value in faulting with the uniaxial compressive strength for a typical sedimentary rock (τ_o = 15 MPa, φ = 35°, C_o = 67.5 MPa)

Layer-parallel cleavage joints may form in near-surface rocks under high surface-parallel compression σ'_I. This offers an explanation for the enigmatic *extension joints* (sheet fractures) which were formed *at shallow depth* (at very small σ'_{III}) *sub-parallel to the topographic surface* as it existed at the time of fracturing (see Fig. 2.11B). Note that the cohesive bonds on planes parallel to the layering or bedding of the rock are weaker than in any other direction, and therefore allow easier splitting of the rock. Nevertheless, the layer parallel joints will be restricted to depths of not more than several tens of metres below the surface at the time of fracturing.

Vertical cleavage joints can form in compressive regimes of vertical strike-slip (wrench) faulting, where σ'_I and σ'_{III} are horizontal, and the overburden stress is the intermediate principal stress σ'_{II}. Of course, σ'_{III} has to be reduced to approximately zero. This need not be achieved by horizontal tectonic extension at right angles to the σ'_I trajectories, but may also be the result of abnormally high pore pressures that are generated by the build-up of the high compression σ'_I.

The compression process is described by the Mohr diagram in Fig. 2.13 for a layer of medium stiffness under 2.3 km of overburden. Before the onset of the horizontal compression, the layer is tectonically undisturbed and under the normal hydrostatic pore pressure p° = 23 MPa. The initial effective stresses, represented by the circle I, are changed by the uni-directional tectonic compression. The horizontal stress is raised in the direction of the compression from the original value σ'^{o}_{h} = 15 MPa to a value σ'_H that is only limited by the maximum compressive stress σ'^{limit}_{I} that would initiate wrench faulting. In the absence of horizontal tectonic extensions, the compression also raises the pore pressure by Δp, which in turn reduces the horizontal effective stress normal to the compression direction by –Δp, if we assume that the total stress σ^{o}_{h} remains unchanged (a point to be dealt with at the beginning of the next chapter). In the present case, the pore pressure has to be raised by 15 MPa to reduce σ'^{o}_{h} to zero. At this state, represented by the shaded circle II, the pore pressure is $p^\circ + \Delta p = 38$ MPa, and the ratio of the pore pressure to the *total* overburden stress $\lambda = p/\sigma_v = (p^\circ + \Delta p)/(\sigma'^{\circ}_v + p^\circ) \approx 0.66$. As compared with $\lambda = 0.4$ for normal hydrostatic pore pressures, the λ-factor required for vertical cleavage jointing in the wrench faulting regime represents a moderate overpressure. (In the next chapter, it will be seen that higher overpressures are required if the layers cannot extend horizontally normal to the direction of shortening.)

In the thrusting regime, the effective overburden stress σ'_v is the smallest principal stress. Therefore, λ would have to approach 1 in order to reduce this stress to zero. But this is only possible on a very local scale, since $\lambda = 1$ implies a pore pressure that balances the total overburden weight, and may thus separate bedding planes and produce overburden-carrying fluid "sills". Hence, cleavage fractures would tend to be along the bedding planes.

There remains to be considered the extensional regime that leads to normal faulting. Since in this regime the effective overburden stress is the greatest principal stress ($\sigma'_v = \sigma'_I$), it is this stress that would have to generate vertical cleavage fractures, provided a horizontal effective stress is reduced to approximately zero. Although as mentioned before, we do not know the magnitude of σ'_I that is actually needed for cleavage fracturing, we consider it as rather unlikely that the necessary σ'_v in sandstones and limestones (with C_o values in the rather low range of 50–60 MPa) would be less than 35 MPa, the value assumed in Fig. 2.13. Thus it would seem rather unlikely that vertical cleavage jointing in the extensional regime would take place under less than 2–3 km overburden. With reference to the circles I and II in Fig. 2.13, it will be further seen that the reduction of a horizontal effective stress to zero cannot be achieved by overpressuring, since this would also decrease the effective overburden stress. Hence, the horizontal stress reduction

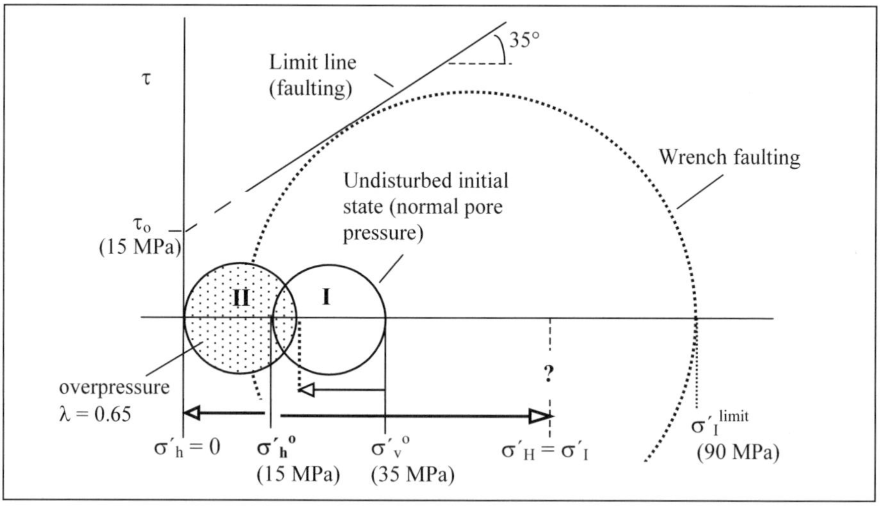

Fig.2.13. Mohr diagram of the overpressuring of the pore fluid in a wrench faulting regime. The layer considered is under 2.3 km of overburden and undisturbed before the onset of the tectonic compression. Initially, the horizontal effective stresses (σ'^{o}_h) are related to the effective overburden stress (σ'^{o}_v) by the empirical relation $\sigma'^{o}_h = K_o.\sigma'^{o}_v$, with $K_o = 0.43$ chosen for a relatively stiff layer (circle I). During horizontal compression, circle I is shifted into position II by the overpressuring, and the greatest principal stress is raised to a value less than σ'^{limit}_I

would have to be produced by a tectonic extension. At a depth below a few kilometres it may therefore become difficult to distinguish between vertical tension joints and cleavage (extension) joints.

Summary. In closing this chapter we can summarise the main points:

- A fluid-saturated porous rock is considered as a ***single continuum*** consisting of a solid and a fluid component. Total stresses and fluid pressure are then defined at any point of the continuum, irrespective of whether the point lies in the solid or fluid part of the rock.

  The formation and propagation of fractures are controlled by ***Terzaghi's effective stresses*** (Eq. 2.1).
- ***Tension joints*** (germ. *Zugklüfte*) (Fig. 2.4) are tensile fractures (mode I) which form and open perpendicular to the direction of the smallest principal stress ($\sigma'_{III} < 0$) when the magnitude of this tension reaches the tensile strength T_o of the material, while the greatest principal stress ($\sigma'_{I} > 0$) remains below a certain limit (Eq. 2.3).

  In general, a relatively undisturbed rock sample has a higher *tensile strength* than the rock mass from which the sample is taken. In a uniform elastic material, a single fracture will start to grow when the remote effective tensile stress normal to the fracture plane produces a critical stress concentration near the fracture tip (Eq. 2.6). In LEFM, the near-tip stress concentration is characterized by the stress intensity factor K_I (Fig. 2.5) which increases with the square root of the fracture length. Hence, the longest fractures will first start growing when the remote effective tensile stress is raised. The critical stress intensity factor K_{Ic} is a real material constant, – the "*fracture toughness*" (germ. *Bruchzähigkeit*) of the material.
- ***Extension or "cleavage" joints*** (germ. *Spaltbrüche*; Fig. 2.8) have the same morphology as tensile fractures and, similar to tensile fractures, form parallel to the σ'_I direction. But contrary to tensile fracturing, extension fractures are "driven" by a high effective compressive σ'_I, while the confining stress σ'_{III} is close to zero. On the micro-scale, *wing cracks* (Fig. 2.10A) are perhaps the most important mechanism in the formation of the compressive extension fractures.

  In the testing of unconfined samples under uniaxial compression, splitting of the samples occurs when the axial stress σ'_I attains a critical value C_o, the "*uniaxial compressive strength*". However, it is important to note that cleavage fractures already form, prior to splitting the sample, at a lower compressive stress $\sigma'_I < C_o$. Unfortunately, the lack of relevant experimental data does not allow a reliable quantification of this stress, and one has to resort to educated guesses.
- ***Geological environment of cleavage joints.*** Cleavage joints *parallel* to flat lying layers can form in near-surface rocks under a high residual compression σ'_I parallel to the surface, and produce exfoliations (Fig. 2.11B). At greater depth, layer-parallel cleavage jointing can only occur on a very local scale, since besides a high layer-parallel effective compression (thrusting regime), the layer-normal effective stress must be close to zero. Thus in flat lying layers, the pore fluid would have to be so heavily over-pressured that it carries the total overburden load, and possibly is forming fluid sills.

  Vertical cleavage joints in flat layers may be expected in regimes of wrench faulting; in particular, when the pore fluid is moderately overpressured by the layer-parallel maximum compressive stress (Fig. 2.13). In an extensional regime, vertical cleavage jointing would seem possible at burial depths of 2–3 km, when a horizontal effective stress is reduced to zero by tectonic extension.

Chapter 3

Hydraulic Fractures

In the preceding chapter we remarked that high fluid pressures play a key role in the generation of tension joints. Since at some depth in the Earth's crust all total stresses are compressive, the fluid pressures are required to push aside the rock walls against the action of the total rock pressure. Similarly in extension (cleavage) jointing ("axial splitting"), high fluid pressures may contribute in reducing σ'_{III} to almost zero. When the fluid pressure is instrumental in fracturing, the fractures are commonly referred to as *hydraulic fractures*.

Such fractures may start from microfractures or flaws inside a bed or from irregularities at bed boundaries, when the pore pressure *inside* the bed is uniformly raised to a sufficient level, or when the bed is ruptured by the "wedging" action of a highly-pressured fluid injected from *outside*. The first fracture type we call *internal hydraulic fractures*, the second type *hydraulic intrusion fractures*. The two fracture types are schematically shown, for horizontal layers, in Fig. 3.1.

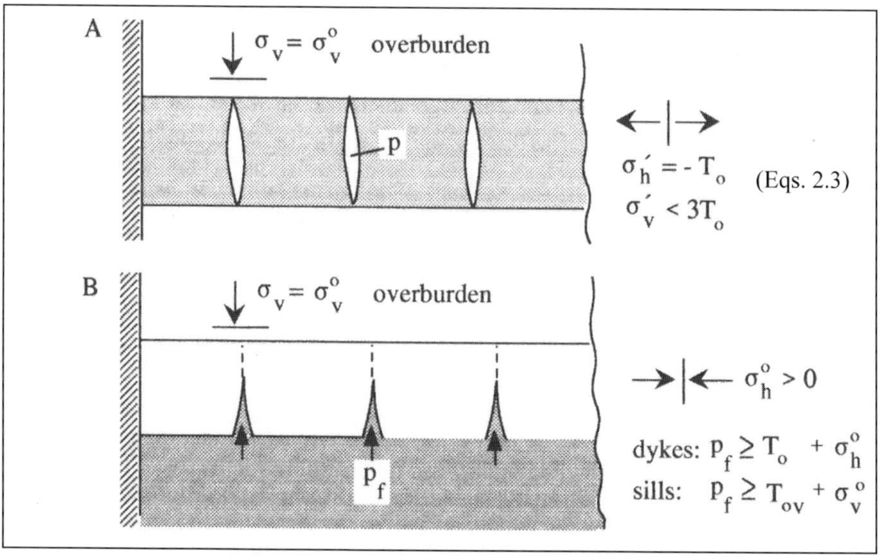

Fig. 3.1. Hydraulic fractures: A) internal fractures, B) intrusion fractures

Internal hydraulic fracturing. In order to allow *tension* fractures to form (Fig. 3.1A), a rise in pore pressure $\Delta p > 0$ is required to reduce the smallest (Terzaghi-) effective stress to the value of the tensile strength $-T_o$ of the material. During the process of overpressuring the rock layer may be allowed to laterally extend, or to remain confined. Hence, we consider two extreme boundary conditions: (*1*) where the interlayer shear resistance is negligibly small, and the layer is completely free to expand in elastic response to a uniform pressure rise Δp, and (*2*) where the layer is confined and not allowed to extend in lateral directions. In both cases, the principal stresses remain vertical and horizontal during the rise in pore pressure, with $\sigma'_I = \sigma_v'$ and $\sigma_{III}' = \sigma_h'$. But the horizontal stress σ_h' behaves differently in the two cases.

When the layer is ***free*** to extend in horizontal directions, both total normal stresses σ_v and σ_h remain unchanged. Hence, all normal effective stresses are reduced by the pore pressure Δp, and the Mohr circle in Fig. 3.2A is shifted towards the tension domain without a change in diameter. In order to satisfy the condition for tensile fracturing (Eq. 2.3), σ_v' has to be smaller than $3T_o$ at the onset of fracturing. Since $\sigma'_I - \sigma'_{III} = \sigma'_v - \sigma'_h$ remains constant during the rise in pore pressure and $\sigma'_h = -T_o$ at the onset of fracturing, the stress difference $\sigma'^o_v - \sigma'^o_h$ must be smaller than $4T_o$ before the rise in pore pressure. In Fig. 3.2 we have also inserted a branch of a Coulomb-Mohr's limit line which is tangential to all stress circles which represent the effective stress states that would induce shear failure (Fig. 1.8D) in a brittle environment of all around compressive effective stresses. The experimentally determined part of the limit line is extended straight (dashed in the figure) to the intersection τ_o with the τ axis, which is commonly referred to as the "cohesive shear strength". Since the radius of the stress circle in Fig. 3.2A is $2T_o$ and, according to rock mechanical data, $2T_o$ is smaller than τ_o, the stress circle cannot touch the Coulomb-Mohr limit line before being shifted into the tension field.

Next we consider the pervasive rise of the fluid pressure in a layer that is laterally completely ***confined***, but is free to deform in the vertical direction. A rise in pore pressure Δp will therefore leave the total overburden load σ_v unchanged, but raise the total horizontal normal stress σ_h, since any lateral extension is suppressed. Therefore, as indicated in Fig. 3.2B, the decrease in horizontal effective stress $\Delta\sigma_h'$ is less than the decrease in vertical effective stress $\Delta\sigma_v' = -\Delta p$:

$$\boxed{\Delta\sigma_h' = -\beta.\Delta p = \beta.\Delta\sigma_v'} \qquad 0 < \beta \leq 1 \tag{3.1}$$

Thus, the stress circle in Fig. 3.2B shrinks while it is shifted towards the tension regime by the increase in pore pressure.

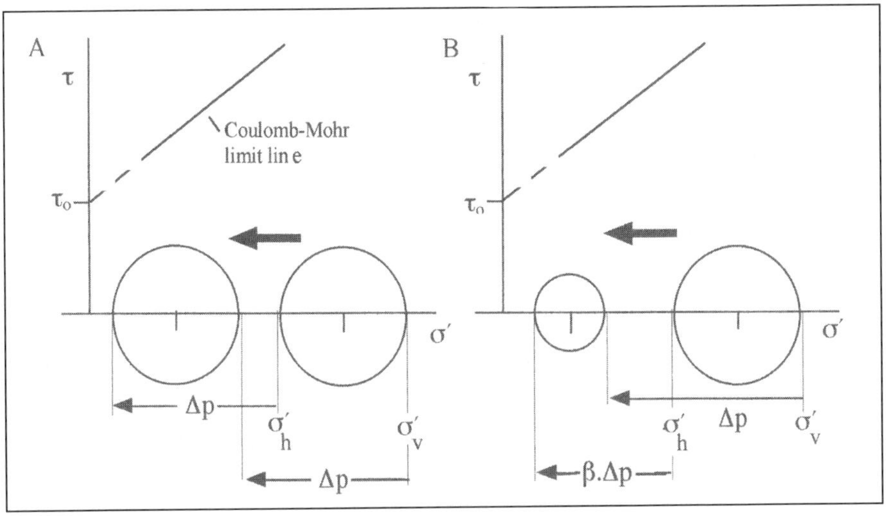

Fig. 3.2. Mohr diagrams of stress changes in horizontal layers due to the rise Δp in pore pressure: A) unrestrained layer, B) laterally confined layer

Noting that the formation of vertical tension fractures requires $\Delta\sigma_h' = -(\sigma_h'^{o} + T_o)$, with $\sigma_h'^{o}$ the effective stress before the pore pressure is raised, the rise in pore pressure required for vertical tensile fracturing follows from Eq. 3.1:

$$\Delta p = (\sigma_h'^{o} + T_o).\beta^{-1} \tag{3.2a}$$

Obviously, this relation reduces to

$$\Delta p = \sigma_h'^{o}.\beta^{-1} \tag{3.2b}$$

for vertical *cleavage* fractures, if both σ_I and σ_{III} are horizontal (wrench faulting regime).

Next we have to determine the factor β, of which we, so far, only know that it is positive and attains its maximum value of 1 when the layer is laterally unconstrained. We assume that the rock is mechanically isotropic in all horizontal directions, and responds in a linear-elastic fashion to the pervasive rise Δp in pore pressure. We recall that the poro-elastic strain increments Δe_i are, in general, not related to the changes in Terzaghi's effective stresses $\Delta\sigma_I'$, but to the changes in the generalized effective stresses $\Delta\sigma_I^*$, as defined by Eq. 2.2 of the preceding chapter. The poro-elastic stress-strain relations for the horizontal x_1, x_2, and the vertical x_3-axis are (see G. Mandl, FBR, pp. 164–175):

$$\begin{aligned} \Delta e_1 &= \frac{1}{E}\left[\Delta\sigma_1^* - \nu(\Delta\sigma_2^* + \Delta\sigma_3^*)\right] \\ \Delta e_2 &= \frac{1}{E}\left[\Delta\sigma_2^* - \nu(\Delta\sigma_3^* + \Delta\sigma_1^*)\right] \\ \Delta e_3 &= \frac{1}{E}\left[\Delta\sigma_3^* - \nu(\Delta\sigma_1^* + \Delta\sigma_2^*)\right] \end{aligned} \tag{3.3}$$

where E is Young's modulus and ν the Poisson ratio.

Because the material is assumed as isotropic in horizontal directions, the pore pressure rise Δp will change all horizontal stresses by the same amount $\Delta\sigma_h^*$. Because of the lateral confinement, $\Delta e_1 = \Delta e_2 = 0$ in the above equations, which thus reduce to

$$\Delta\sigma_h^* = \frac{\nu}{1-\nu}\Delta\sigma_v^* \tag{3.4}$$

and further, since according to Eq. 2.2 $\Delta\sigma_v^* = \Delta\sigma_v - \left(1 - \frac{K}{K_S}\right)\Delta p$ and $\Delta\sigma_v = 0$,

$$\Delta\sigma_h^* = -\left(1 - \frac{K}{K_S}\right)\frac{\nu}{1-\nu}\Delta p \tag{3.5}$$

where K and K_s are the compression moduli of the bulk and the solid material of the porous rock, respectively.

Being concerned with fracturing, we have to rewrite the last result in terms of Terzaghi's effective stress (Eqs. 2.1, 2.2) $\sigma_h' = \sigma_h - p = \sigma_h^* - (K/K_s).p$. Writing Eq. 3.5 in the form of Eq. 3.1, the β-factor becomes

$$\boxed{\beta = 1 - (1 - K/K_s)(1 - 2\nu)(1 - \nu)^{-1} < 1} \tag{3.6}$$

Inserting, for example, for a rock of intermediate stiffness the parameter values ν = 0.25, $K/K_s = 0.3$, one has β = 0.53.

According to Eqs. 3.2a,b, the pore pressure rise necessary for fracturing depends on the horizontal effective stress $\sigma_h'^{\circ}$ prior to the pore pressure rise. Unfortunately, this stress is commonly a subject of speculation, except when the lateral stress can be related to the much better known overburden stress $\sigma_v'^{\circ}$. This is the case in tectonically undisturbed flat strata that often occupy large parts of sedimentary basins. In the absence of horizontal straining in these layers, the stresses $\sigma_h'^{\circ}$ build up in response to the increasing overburden load. But it would be erroneous to describe the development of the horizontal rock stresses by means of Eq. 3.4, since the process is *not* elastic in nature, but predominated by inelastic processes such as compaction and diagenetic consolidation. We therefore prefer to follow soil-mechanics to characterize the final state of $\sigma_h'^{\circ}$, attained in a laterally confined sedimentary rock under a vertical effective stress $\sigma_v'^{\circ}$, by an empirical parameter

$$K_o = \sigma_h'^{\circ}/\sigma_v'^{\circ} < 1 \tag{3.7}$$

or in terms of the ratio of total stresses:

$$\sigma_h^{\circ}/\sigma_v^{\circ} = [\lambda + K_o.(1-\lambda)] \quad \text{with } \lambda = p^{\circ}/\sigma_v^{\circ} \tag{3.8}$$

From soil mechanics and *in situ* stress measurements we infer that typical K_o values for tectonically undisturbed beds will rarely have a value smaller than, say, 0.4 or higher than 0.75.

Inserting from Eq. 3.7, the pore pressure conditions (Eqs. 3.2a,b) are

$$\Delta p = (K_o.\sigma_v'^{o} + T_o).\beta^{-1} \qquad \text{for tension joints}$$

and (3.9)

$$\Delta p = K_o.\sigma_v'^{o}.\beta^{-1} \qquad \text{for cleavage joints}$$

At the end of the preceding chapter (Fig. 2.13) it was shown how vertical *cleavage* jointing in a *wrench faulting regime* is accomplished by a moderate overpressuring ($\lambda = 0.66$) in a laterally unrestrained layer. In accordance with Fig. 3.2A, the Mohr circle was rigidly shifted towards $\sigma_h' = 0$ by the rise in pore pressure. On the other hand, if the layer is laterally completely confined, a markedly higher degree of overpressuring is required to shift the shrinking circle (Fig. 3.2B) towards $\sigma_h' = 0$. Assuming the same initial conditions as in Fig. 2.13, that is $\sigma h'^{\circ} = Ko.\sigma v'^{\circ} = 15$ MPa, $\sigma v^{\circ} = 58$ MPa, $p^{\circ} = 23$ MPa, and inserting the aforementioned value $\beta = 0.53$, the formula in Eq. 3.9 yields $\Delta p = 28.3$ MPa, and a total pore pressure $p = p^{\circ} + \Delta p = 51.3$ MPa. Therefore, the ratio of the total pore pressure to the total overburden $\lambda = p/\sigma_v^{\circ} = 51.3/58 = 0.88$. Thus a "hard" overpressure is needed for the formation of vertical internal hydraulic cleavage joints in laterally completely confined layers.

Turning to the condition of internal hydraulic *tension* jointing under lateral constraints, we have to note that the corresponding relation (Eq. 3.9) is not the only condition, but that the second fracture condition (Eq. 2.3) requires $\sigma_I' < 3T_o$. Since at the onset of vertical tensile fracturing $\sigma_I' = \sigma_v' = \sigma_v'^{\circ} - \Delta p$ and Δp is given by the first Eq. 3.9, the effective overburden stress $\sigma_v'^{\circ}$, prior to the rise in fluid pressure, is restricted by

$$\sigma_v'^{o} < \frac{(3\beta + 1).T_o}{\beta - K_o} \tag{3.10}$$

Obviously, this restricts the depth down to which the tensile hydrofractures can be generated. But this is still not the complete picture. The pervasive rise in pore pressure (Eq. 3.9) is limited, since the pore pressure cannot rise beyond the value that, on a larger scale, reduces σ_v' to zero, and may separate bedding planes and produce overburden carrying fluid "sills".

Since $\sigma_V' = \sigma_V'^{o} - \Delta p$, the vanishing of σ_V' limits the pervasive rise in pore pressure to $\Delta p = \sigma_V'^{o}$, if the usually very small tensile strength across bedding planes is neglected. From (Eq. 3.9) then follows:

$$\sigma_V'^{o} \geq \frac{T_o}{\beta - K_o} \tag{3.11}$$

Combining Eqs. 3.10 and 3.11 we finally arrive at the condition which, prior to the rise in pore pressure, must be satisfied by the effective overburden stress, in order to allow the formation of vertical internal tension type hydrofractures:

$$\frac{T_o}{\beta - K_o} < \sigma_V'^{o} < \frac{(3\beta + 1).T_o}{\beta - K_o} \tag{3.12}$$

What can be inferred from this condition about the depth range of possible tensile internal hydrofracturing? First note that since before a rise in pore pressure the effective overburden stress is compressive ($\sigma_V^{o\prime} > 0$), Eq. 3.12 implies

$$\boxed{K_o < \beta} \tag{3.13}$$

or with β from Eq. 3.6:

$$1 - \frac{K}{K_s} < (1 - K_o).\frac{1 - \nu}{1 - 2\nu} \tag{3.13a}$$

Although Eq. 3.13 is always fulfilled if the layer is laterally unconstrained and tectonically not compressed, since in this case $\beta = 1$ (Fig. 3.2), and $K_o < 1$, the condition need not be fulfilled if the layer is laterally constrained. As noted before, typical K_o values for tectonically undisturbed beds will rarely have a value smaller than, say, 0.4 or higher than 0.75. The weaker the material, the higher its K_o value. On the other hand, Poisson's ratio ν for porous rocks (determined at constant pore pressure) seldom exceeds 0.3. Assuming $\nu = 0.25$, Eq. 3.13a reads $(1 - K/K_S) < 1.5\ (1 - K_o)$ and thus imposes a quite delicate restriction on the admissible K/K_S ratios. When the compressibility of the skeletal material $1/K_S$, (i.e. the reciprocal of the compression modulus) is much smaller than the overall compressibility $1/K$ of the bulk of the porous rock, $1 - K/K_S$ approaches 1 and Eq. 3.12 cannot be satisfied for any reasonable K_o value (>1/3). This is quite possibly the case for highly porous "hard" rocks. But the smaller the porosity of rocks with mechanically similar skeletal materials, the smaller the bulk compressibility $1/K$ in comparison with the compressibility $1/K_s$ of the skeletal material, and the lower the value of the left-hand side in the inequality 3.13a. Thus, the more readily can the inequality be satisfied for reasonable values of K_o and ν.

In rocks whose parameters do not satisfy condition (Eq. 3.13) a rise in pore pressure can generate vertical tension fractures only after K_o is sufficiently reduced by some lateral extensional straining of the layers.

When condition 3.13, 3.13a is satisfied, the inequality 3.12 defines a range in depth z where vertical internal hydrofractures of the tensile mode can be formed in laterally constrained layers. Assuming that prior to the rise in pore pressure the effective overburden stress $\sigma_V'^{o}$ increases by 15 MPa per km depth, the depth range is

$$\frac{T_o(\text{MPa})}{15(\beta - K_o)} < z(\text{km}) < \frac{(3\beta + 1).T_o(\text{MPa})}{15(\beta - K_o)} \tag{3.14}$$

To illustrate the implications of this condition, consider a horizontal, laterally confined layer with $\beta = 0.55$, $K_o = 0.45$, and $T_o = 5$ MPa. The layer would have to lie at a depth of between 3.3 and 8.8 km, in order to be vertically fractured by a rise in pore pressure. According to Eq. 3.9, the pore pressure would have to be raised by 50 MPa at the depth of 3.3 km, and by 117 MPa at 8.8 km. Assuming that the original pore pressure p° increased by 10 MPa per kilometer and the total overburden stress $\sigma_v{}^o$ by 25 MPa, the ratio $\lambda = p/\sigma_v{}^o = (p^\circ + \Delta p)/\sigma_v{}^o$ would vary with depth from 1 at 3.3 km to 0.93 at 8.8 km. This demonstrates the surprisingly high overpressures that are required to hydrofracture a laterally confined layer. Also note, that if the layer were free to expand in lateral directions ($\beta = 1$), the depth range from Eq. 3.14 would lie between only 0.6 and 2.4 km, and the associated λ-factor would vary between 1 and 0.75, again representing "hard" overpressuring.

Aperture of tensile hydraulic fractures. Having established the conditions for the vertical hydrofracturing of a horizontal layer by high internal pore pressures, we face the interesting question of how wide can the fractures be open; or in other words, what is the aperture of the vertical fractures? In answering this question we restrict ourselves to tension joints, since a satisfactory theory for the opening of cleavage joints is not yet available. (Qualitative aspects of the aperture of cleavage joints will be dealt with in connection with the spacing of cleavage joints in Chapt. 4.) We continue with considering the laterally completely constrained layer. As the fractures grow, they open and fill with fluid from the surrounding pore space, and the fracture fluid gradually attains the value of the pore pressure p inside the layer. We assume that, despite the increase in fluid volume due to the new fracture space, the pore pressure is maintained by the fluid contact with adjacent beds, at the level required for fracturing. We further assume that the smallest principal stress $\sigma_{III}{}^o = \sigma_h{}^o$ acts along one horizontal direction only, allowing the formation of a single array of parallel vertical fractures perpendicular to the direction of $\sigma_h{}^o$. At the onset of fracturing $\sigma_h{}' = -T_o$ throughout the whole layer. After formation, the fractures are filled with pore fluid of pressure p, which exerts the total normal stress p upon the fracture walls. Hence, the tensile effective stress $\sigma_h{}' = \sigma_h - p$ vanishes at the location of the fractures, thereby increasing from the negative value $-T_o$ to 0, as is illustrated by the shaded stress circle in Fig. 3.3. In other words, the material of the fracture walls experiences a compressive stress $\Delta\sigma_h{}' = T_o$. As the pore pressure is unchanged by this process, the total stress σ_h and the generalized effective stress $\sigma_h{}^*$ are also raised by T_o. Thus, $\Delta\sigma_h{}' = \Delta\sigma_h = \Delta\sigma_h{}^* = T_o$ inside the wall material.

Now let us consider two adjacent parallel fractures which transect the horizontal layer as schematically shown in Fig. 3.4. The open space of the fractures is produced by elastic compression of the material between the fractures. The maximum opening of the fractures is established when the contraction of the material is not hindered by a cohesive/frictional contact with the adjacent layers. The compression of the inter-fracture material is then produced by the full effective compressive stress $\Delta\sigma_h{}^* = T_o$, and the fracture aperture δd is readily determined by the first two stress/strain equations (Eq. 3.3). Inserting into these equations $\Delta\sigma_1 = \Delta\sigma_h{}^* = T_o$, $\Delta e_2 = 0$ because straining along the strike of the fractures is suppressed, and $\Delta\sigma_3{}^* = 0$ because the effective overburden remains unchanged by the formation of the fractures, the contractive strain Δe_1 becomes

$$\Delta e_1 = \frac{\delta d}{d} = \frac{1-\nu^2}{E}.T_o \tag{3.15}$$

where $\delta d/d$ is the relative change in distance d between the midplanes of the neighbouring fractures.

To illustrate this result, we assume again $T_o = 5$ MPa, $\nu = 0.25$, $E = 10^4$ MPa, and obtain $\delta d/d \approx 0.5 \times 10^{-3}$ from Eq. 3.15. Naturally, the greater the distance between the two fractures, the more fracture opening δd is elastically accommodated. Assuming arbitrarily a "large" distance d = 1 m, the aperture δd is only 0.5 mm. But note that this result most likely overestimates the apertures that occur in real cases, where the contraction of the fractured layer is opposed by the unfractured ambient layers which react by building up shear stresses on the layer contacts (a problem we shall be concerned with in Chapt. 4). We therefore may safely conclude that the fractures in the laterally constrained layer have the appearance of ***"hairline cracks"*** – at least as long as the fracture volume is accommodated by a purely elastic deformation. (An inelastic accommodation of fracture space by the withdrawal of material by pressure solution will be considered in connection with the formation of veinlets in Chapt. 8, p. 192.)

Next, the question arises as to whether the aperture equation (Eq. 3.15) also applies to a laterally *unconstrained* layer. First we recall that, in deriving the formula, we assumed that the pressure drop which, inside the fracture, accompanies the creation of new fracture space, is compensated for by an inflow of high-pressure fluid from ambient beds. Thus, the constant total stress $\sigma_h = p$ is maintained on the fracture walls, or re-established, and exceeds the original total stress σ_h^o by $\Delta\sigma_h = T_o$. Secondly, Eq. 3.15 defines an upper bound of the fracture aperture, since the material between fractures was considered completely free to contract, unhindered by cohesive/frictional contacts with the adjacent layers. Under these conditions, the fracture-bounded segments in a laterally perfectly unconstrained layer would be pushed sideways in an accelerative motion, since the incremental stress $\Delta\sigma_h$ is not balanced by reactive lateral boundary stresses. Thus, as long as the internal fracture pressure is restored after the drop associated with an increase in fracture space, the vertical fractures would widen, and the fractured layer extend, albeit in a discontinuous fashion.

In reality, however, the adjacent layers react to the widening of the fractures by mobilizing an interlayer shearing resistance which hinders the widening of the fractures and the extension of the fractured layer. Although this subject will be taken up in Chapt. 4, a special point should be mentioned here, since it is relevant to the internal fracturing of layers with unconstrained lateral boundaries. Consider the internally fractured brittle layer

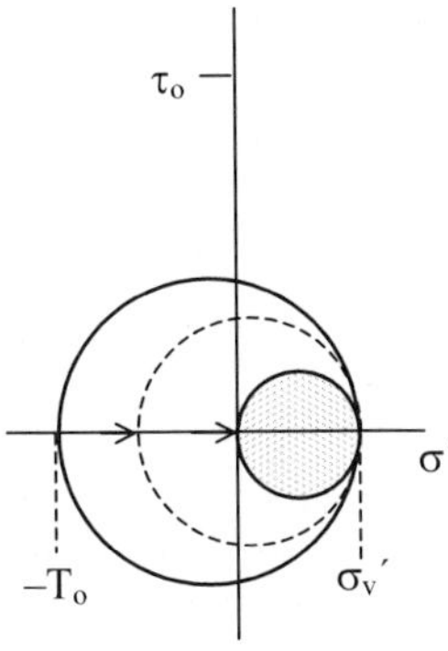

Fig. 3.3. Release of layer-parallel tension by internal hydrofracturing

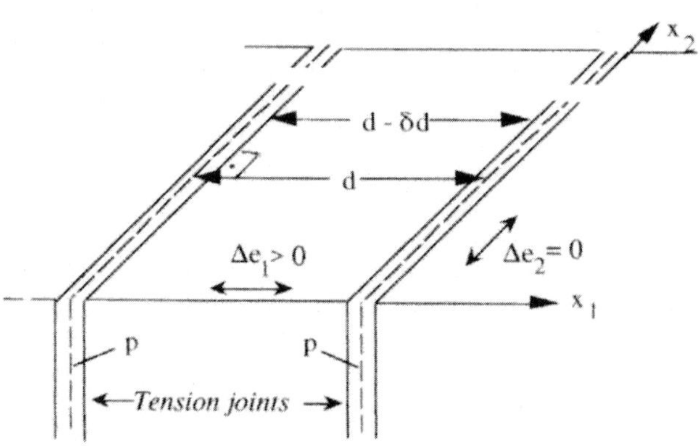

Fig. 3.4. Aperture (δd) of parallel vertical hydrofractures in overpressured, laterally constrained layer

embedded in overpressured beds of a semi-brittle material, which, under shear load, deforms by viscous creep. In this case, the obstructive shear stresses may, at least to some degree, relax in time, and allow for a delayed increase in fracture aperture. Naturally, the process requires that overpressured fluid is supplied from the ambient layers, and/or that semi-brittle bulk material of the ambient layers is squeezed into the fractures. As a likely setting for this process, we envisage a brittle sandstone layer embedded in highly overpressured, normally consolidated shales.

In conclusion, it may be stated that vertical internal tension mode hydrofractures develop as "hairline" cracks, when the opening of the fractures is mainly accommodated by elastic deformation and any layer-parallel extension is suppressed. Conversely, a wider opening of the fractures suggests that the fractured layer was extended in response to a high pore pressure and/or an external force.

Hydraulic intrusion fracturing (Fig. 3.1B). Rock layers under effective compres-sive stress can be fractured by the injection of high-pressured fluids such as water, oil, gas, salt, fluidized sand, or magmas. Intrusion fractures which cut across the bedding (discordant fractures) are termed *dykes* (germ. *bankrechte Gänge*), while hydraulic fractures that intrude the rock parallel (concordant) to its bedding are termed *sills* (germ. *schichtparallele Gänge*). Mostly discussed in the geological literature are magmatic dykes and sills. Here, we are mainly concerned with dyke- and sill-systems that are caused by the injection of water, oil, or gas into sedimentary rocks, as illustrated by the photographs in Fig. 3.5A–E on the next page, and in Fig. 1.5. Notice that in Fig. 3.5A,B the intrusion process is dominated by dyke-type fractures, in Fig. 3.5C,E by sill-type fractures, while in Figs. 3.5D and 1.5 both fracture types contribute about equally to the process of fluid transport. The interplay of the two fracture types in the passage of high-pressured fluids across rock layers of very low or zero permeability is illustrated in the schematic diagram in Fig. 3.6.

While internal hydrofractures form under an effective tensile stress, intrusion fractures may be generated in a layer under effective compression, when the pressure of the injected fluid exceeds the sum of the *total* normal stress and the tensile strength, as already noted in Fig. 3.1B. Considering again, for convenience, horizontal or near-horizontal layers with the smaller layer-parallel total stress $\sigma_h{}^o$ and the total overburden stress $\sigma_v{}^o$ being principal stresses, the formation of intrusion fractures of the dyke or sill type requires that the pressure p_f of the fracture fluid satisfies the basic conditions

dykes: $$p_f \geq \sigma_h^o + T_o \quad (\sigma_h^o = \sigma_{III}^o) \tag{3.16a}$$

sills: $$p_f \geq \sigma_v^o + T_{ov} \quad (\sigma_v^o = \sigma_{III}^o) \tag{3.16b}$$

where T_o is the tensile strength against fracturing perpendicularly across the bedding of a sedimentary rock, and T_{ov} is the tensile strength against bedding-parallel fracturing. Commonly, in sedimentary rocks, the tensile strength is highly anisotropic, with T_{ov} much smaller than T_o, which, therefore may be neglected with respect to the total overburden stress $\sigma_v{}^o$ in Eq. 3.16b.

Again we emphasize that, as discussed in the preceding chapter, the "tensile strength" T_o is not a real material constant, and the conditions (Eqs. 3.16a,b) are little more than crude approximations of reality. In particular, the fracture fluid pressure required to break the cohesive bonds in the tip region of the fracture will decrease as a monotonic function of increasing fracture length. However, in spite of this, the fracture conditions (Eqs. 3.16a,b) may suffice to define the *in situ* stress states that will most likely promote the initiation of intrusion fractures.

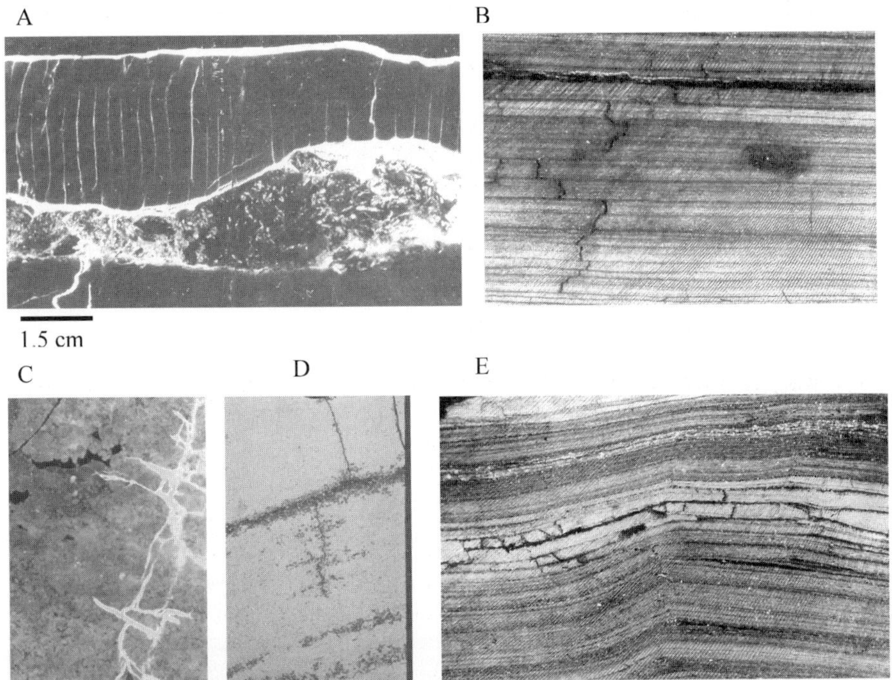

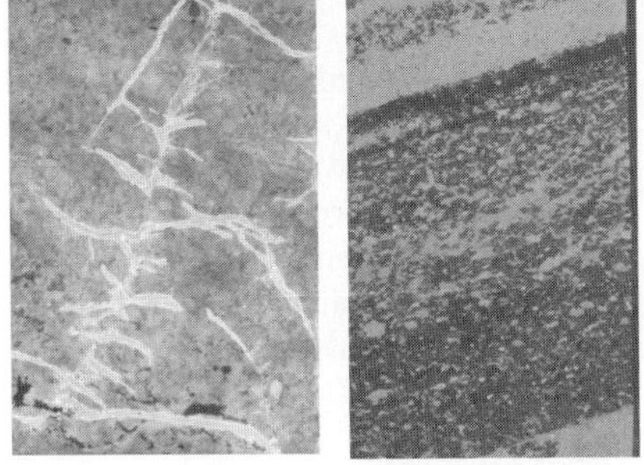

Fig. 3.5. Photographs of fluid dykes and sills in sedimentary rocks:
A) Dykelets formed in lime mud sediments by the injection of fluidised clay from a thin clay band (white). Notice that some fractures terminate already inside the layer (Middle Triassic, Weserbergland, Germany; courtesy D. Meier);
B) Oil dykelets in Carbonate source rock (core sample, Qatar; courtesy Shell Int. Petr. Mij.);
C) Short oil sills connected by dykelets (core sample, Adriatic Sea; courtesy Shell Int. Petr. Mij); D) Oil bearing fractured limestone showing oil intrusion both by flow through per-meable zones and by intrusion fracturing (core sample from 3150 m, Tertiary and Upper Cretaceous, Adriatic Sea; courtesy Shell Int. Petr. Mij.); E) Dyke-sill system in carbonate oil source rock; notice the compressive folding (core sample, Qatar; courtesy Shell Int. Petr. Mij.)

Obviously, the dyke-type hydrofractures are essential in enforcing pathways of fluid migration across beds of very low or zero permeability ("cap rocks"). Hence, we shall focus on this type of hydrofracture. Quite similar to the discussion of parallel vertical internal hydrofractures in Fig. 3.4, we consider in Fig. 3.7 parallel dyke-type hydrofractures. We assume that, before fracturing, the minimum principal stress $\sigma_{III}^{o} = \sigma_h^{o}$ acts along the horizontal x_1 direction only. Under this condition, only a single set of parallel vertical dykes can develop. The bed as a whole is not allowed to extend in any horizontal direction, but the material between the dykes is elastically compressed ($\Delta e_1 = \Delta e_h > 0$) by the extra

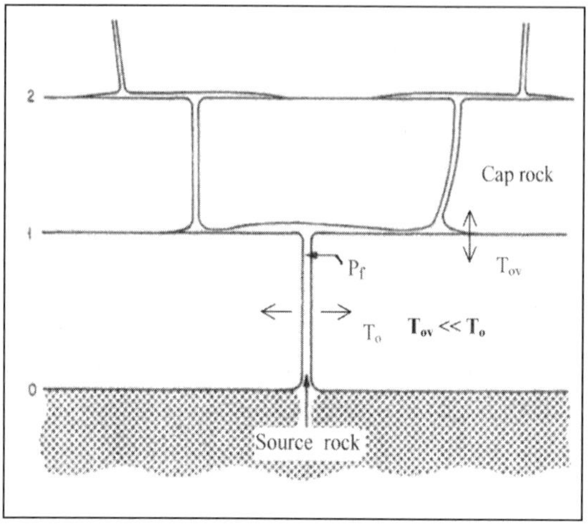

Fig. 3.6. The sill-dyke mechanism in sedimentary rocks (note the anisotropy of the tensile strength)

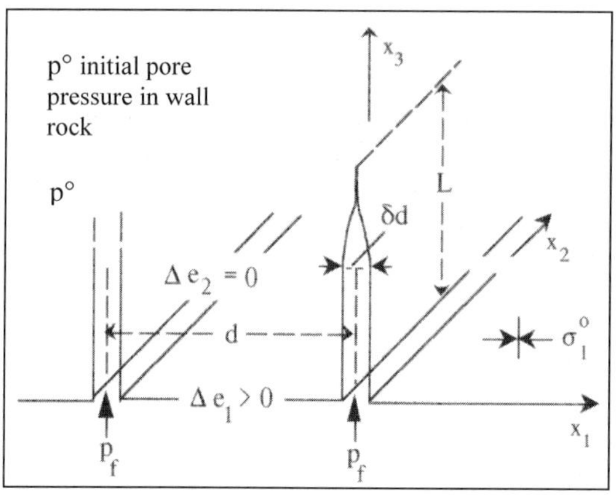

Fig. 3.7. Aperture (δd) of parallel dykes in a horizontal, laterally constrained layer

stress $\Delta\sigma_h = p_f - \sigma_h^o$ exerted by the fracture pressure p_f. Along the strike of the dykes, i.e. in the x_2 direction, straining is suppressed ($\Delta e_2 = 0$). Commonly, sedimentary rocks have some permeability, which allows the fracture pressure and the fluid pressure in the wall rock to interact, and thus to cause a gradual decrease of the fracture stress $\Delta\sigma_h$ *after* the formation of hydrofractures. In due course, we shall deal with this additional complication, but for now we shall concern ourselves with the simpler problem of hydrofracturing of *nonporous* rocks.

The ***aperture*** δd of the parallel dykes is then determined in exactly the same way as the aperture of the parallel vertical internal hydrofractures (Fig. 3.4), which was caused by the extra normal stress $\Delta\sigma_h^* = T_o$. In a non-porous wall rock, the material between two parallel dykes is compressed by $\Delta\sigma_h^* = \Delta\sigma_h = p_f - \sigma_h^o$. Hence, the aperture δd of the parallel dykes is determined by

$$\frac{\delta d}{d} = \frac{1 - \nu^2}{E}(p_f - \sigma_h^o) \qquad (3.17)$$

where d is the distance between the midplanes of neighbouring fractures. Note that the formula would also apply to a porous and permeable wall rock, as long as the pore pressure is maintained at the pre-fracturing level by interaction with an external fluid source.

Another particularly interesting aspect of dyke formation is the ***spacing*** of parallel dykes. Regardless of whether or not the aperture of dyke-type intrusion fractures changes with time, as shall be considered later, the distance between neighbouring parallel dykes is related to the initial aperture. Fracture mechanics (LEFM) provides a relationship between the aperture and the length L of a single hydrofracture, if it is not hindered by neighbouring fractures (e.g., D.D. Pollard, P. Segall (1987); Chapt. 8 in: B.K. Atkinson (ed.) *Fracture Mechanics of Rocks*, London). Referring to Fig. 3.7, where we replace σ_1^o by σ_h^o, the fracture-mechanical relationship between the aperture δd at the injection end ($x_3 = 0$) and the length of the hydrofracture is

$$\delta d(0) = 4\frac{1-v^2}{E}(p_f - \sigma_h^o).L \tag{3.18}$$

The average of the aperture, from the injection end to the fracture tip, is

$$\delta d_{av} = 3.14\frac{1-v^2}{E}(p_f - \sigma_h^o).L \approx \frac{3}{4}\delta d(0) \tag{3.19}$$

In this formula, $\delta d(0)$ is the aperture at the moment the dykes are formed. Since at this instant the interaction between the fracture fluid and the pore fluid has not yet affected the pore pressure, the fracture aperture is determined by Eq. 3.17, which was derived for a non-porous rock, or a rock whose pore pressure was kept constant. Combining Eq. 3.19 with Eq. 3.17, and eliminating δd (= δd_{av}), one arrives at the extremely simple approximation of the fracture spacing d:

$$\boxed{d \approx 3L} \tag{3.20}$$

Note that in deriving this formula it was assumed that neighbouring fractures develop without affecting each other. Hence, the spacing predicted by the formula should be considered as an upper bound of the actual spacing of parallel dyke-type intrusion fractures.

The ***rate*** at which an intrusion fracture grows is mainly controlled by the rate of fluid supply to the fracture tip. If the injection pressure at the injection end is maintained for a sufficient period, an approximate expression for the growth rate of a dyke-type intrusion fracture is obtained by the law for steady, one-dimensional laminar flow of a viscous fluid (viscosity η_f) between parallel plates, separated by the distance δd_{av}. The mean velocity between the vertical plates, which represents the growth rate of the fracture, is

$$\overline{v} = -\frac{\delta d_{av}^2}{12\eta_f}\left(\frac{dp}{ds} + \rho_f g\right) \tag{3.21}$$

where $-dp/ds$ is the upwards driving vertical pressure gradient of the fracture fluid and $\rho_f g$ is the increase in gravitational potential energy of the unit fluid volume per unit vertical distance. We neglect the comparatively small gravity term and approximate the pressure gradient to $-dp/ds = (p_f - \sigma_h^o)/L$, assuming a linear pressure drop from p_f at the injection end to σ_h^o at the closed fracture tip. Inserting for δd_{av} in Eq. 3.21 from Eqs. 3.19 and 3.18, gives a crude estimation of the growth rate of the vertical intrusion fracture:

$$\overline{v} = \frac{3}{4}\frac{(1-v^2)^2}{\eta_f.E^2}(p_f - \sigma_h^o)^3.L \tag{3.22}$$

Thus, the growth rate of the dyke is proportional to the fracture length L and proportional to the cube of the excess pressure in the fracture.

Therefore, the development of sets of hydrocarbon dykelets may be envisaged as sketched in Fig. 3.8. The fractures nucleate, closely spaced, at the base of a bed. Naturally, some fractures will be longer than others and therefore grow faster than the shorter ones.

The shorter fractures are stopped by compression of the rock ahead of them. Thus, the fracture fluid will enter a bed in a diffuse way through many fractures, but while proceeding through the bed, the fractures will decrease in number and increase in width. An indication of this phenomenon may be seen in the photograph of Fig. 3.5A.

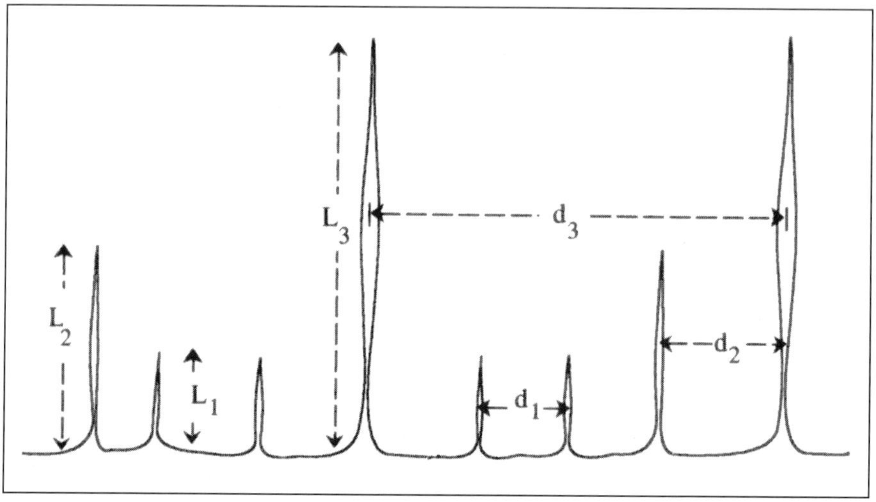

Fig. 3.8. The growth of vertical intrusion fractures (dykes) of different length (after G. Mandl and R.M. Harkness (1987); in: M.E. Jones and R.M.F. Preston (eds.) *Deformation of Sediments and Sedimentary Rocks*, Geol. Soc. Special Publ. *29*, pp. 39–53)

In general, the growth rate of the vertical fractures is high, as can be seen by inserting some typical values in Eq. 3.22, say, $E = 10^4$ MPa, $\nu = 0.2$, $p_f - \sigma_h^o = 1$ MPa, $\eta_f = 10^{-2}$ Poise ($= 10^{-3}$ Pa.s $= 10^{-9}$ MPa.s), and $L = 1$ m. The fracture would advance at a rate of about 7 m/s. If we take into account that high-pressured fracture fluids have a high compression modulus and that, therefore, the discharge of a small volume of fluid from the feeder reservoir will cause a sharp pressure drop, and the restoration of the original fluid pressure in the feeder reservoir will take some time, it is very likely that, temporarily, the injection pressure will drop below the value needed to maintain open fractures at the injection end. Thus, the growth of the fractures will stop until the injection pressure has been sufficiently raised again. In other words, the vertical intrusion fractures grow in a *spasmodic* fashion.

Movement of closed fractures. There is another interesting aspect of the growth of intrusion fractures. Consider a dyke that is continuously fed by high-pressure fluid. In general, the dyke will grow in a regional field of horizontal stress σ_h^o that increases in a downward direction at a greater gradient than the fluid pressure p_f, as indicated in Fig. 3.9A. For convenience, we assume both gradients as constants. When the injection pressure $p_f(0)$ at the base ($z = 0$) is reduced, the width of the dyke at the injection end will decrease, and eventually the dyke will close. But note that closure is not yet achieved when $p_f(0) = \sigma_h^o(0)$, since the average excess pressure $p_{f\,av} - \sigma_{h\,av}^o$ in the fracture exerts a certain leverage on the fracture walls, which tends to keep the lower fracture end open. Hence, complete closure requires that the injection pressure drops to a level less than $\sigma_h^o{}_{(0)}$, as indicated in Fig. 3.9A. If, before closure, the fracture is of sufficient length, the excess pressure $p_f - \sigma_h^o$ in the upper tip region will induce tensile stresses which can break the interparticle bonds of the rock and allow the fracture to propagate upwards.

J. Weertman (1971, J. Geoph. Res. *76*, No. 5) analyzed the problem in two dimensions for water-filled crevasses in glaciers and magmatic dykes by means of a dislocation model. He found that the closed fractures have a slender "teardrop" (or "tadpole") profile because of the stress gradients. He also determined the critical fracture volume that would allow a stable fracture of this shape to migrate vertically through the rock. Using LEFM, D.D. Pollard and O. Müller (1976, J. Geoph. Res. *81*, No. 5) and D.T. Secor, Jr. and D.D. Pollard (1975, Geophys. Res. Letters *2*, No. 11) rederived Weertman's results, and presented the formula

$$L_c = 2\pi^{-1/3}.K_c^{2/3}\left|\frac{d}{dz}(p_f - \sigma_h^o)\right|^{-2/3} \tag{3.23}$$

for the critical length L_c, which an intrusion fracture would have to attain before closing at its lower end in order to allow the fracture to rise, as shown in Fig. 3.9B. The material parameter K_c is the "fracture toughness" K_{Ic}, which was introduced as a measure of tensile strength in Chapt. 2 (Eq. 2.4).

The length of the isolated fracture cannot exceed L_c, because when $L = L_c$ the fracture closes at its base, and the driving fracture pressure cannot rise any further. When

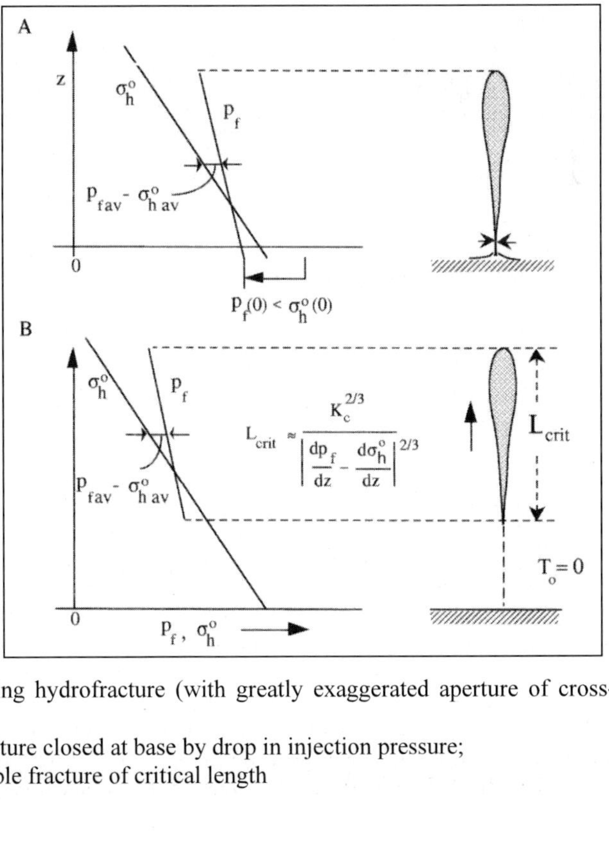

Fig. 3.9. Rising hydrofracture (with greatly exaggerated aperture of cross-sectional tear-drop shape):
A) hydrofracture closed at base by drop in injection pressure;
B) rising stable fracture of critical length

the rock is non-porous, the fracture will rise at constant length through a rock of constant gradient $d\sigma_h^o/dz$. According to the literature, the fracture toughness of hard rocks varies between, say, 1 and 3 MPa $m^{1/2}$, while for soft carbonate rocks, such as chalk, it may be an order of magnitude smaller. Assuming that the fracture fluid is water, and $\sigma_h^o = 0.7.\sigma_v^o$, with σ_v^o increasing in depth by 0.025 MPa per meter, the critical length of the fractures varies between 35 m and 73 m for hard rocks, and drops to a few meters for very soft rocks.

Note that the first hydrofractures propagating through a bed leave traces of practically zero tensile strength, as indicated by a dotted line in Fig. 3.9B. Hence we expect subsequent fluid intrusions to travel along these fracture traces as isolated, very short thin pockets.

The theoretical results were tested experimentally by Akira Takada (1990) who injected various fluids into a block of gelatin. Figure 3.10 shows two photographs from Takada's paper. The cracks were three-dimensional and penny shaped, and detached themselves from the injection needle once they had reached a critical length and started moving. In front view, the tip line of the cracks were elliptical in shape; in side view, they show the typical "frog-larva" shape predicted by Weertman.

Fig. 3.10. Photographs showing formation and propagation of an isolated crack in a gelatin block (0.5 m x 0.3 m x 0.3 m) during and after the injection of silicon oil from the bottom: A) state during fluid injection in front view (left side) and in side view (right side);
B) rising fracture after detachment from injection needle, again in front and side view (from Akira Takada (1990) J. Geoph. Res. *95*, No. B6, pp 8471–8481; reproduced by permission of American Geophysical Union)

The dyke-sill mechanism. So far we have considered dyke-type hydrofractures as growing or rising along a straight path unhindered by barriers imposed by changes in the lithology or the *in-situ* stresses. In dyke-dominated intrusion processes it is frequently observed that the dykes side-step along bedding planes by feeding short sill-type intrusions before continuing their rise through the layered rock. A typical small-scale example are the side-stepping oil dykelets in Fig. 3.5. That a dyke does not directly cross a bedding plane is easily understood when the shear resistance on the bedding plane is low, for instance, because of a low effective overburden stress. When a fracture arrives at the bedding plane, a sharp crack tip cannot be maintained, since, pushed by the fluid pressure p_f of the fracture, the fracture walls slide apart along the lower side of the bedding plane. Hence, the stress concentration that is required to split the rock on the opposite side of the bedding plane is released, and the dyke cannot directly proceed across the bedding plane.

The process was studied by D. Pollard (1973) theoretically and in a series of experiments in which grease was injected into a stack of gelatin layers. Figure 3.11A,B is taken from Pollard's paper, and schematically show an experiment in which a grease sheet was intruded into gelatin at right angles to an interface. When the interface was well lubricated, virtually no shear stress from the advancing intrusion was transmitted across the interface, but the interface was deformed by normal stresses and a small gap was formed as shown in Fig. 3.11A. The formation of the gap is ascribed to the elastic Poisson effect: The aperture of the advancing dyke imposes a horizontal stretching of the material just ahead of the fracture tip; this, in turn, causes a contraction in the vertical direction (Poisson effect), which pulls the lower layer away from the upper layer. The grease then intrudes into the gap along the interface, as sketched in Fig. 3.11B.

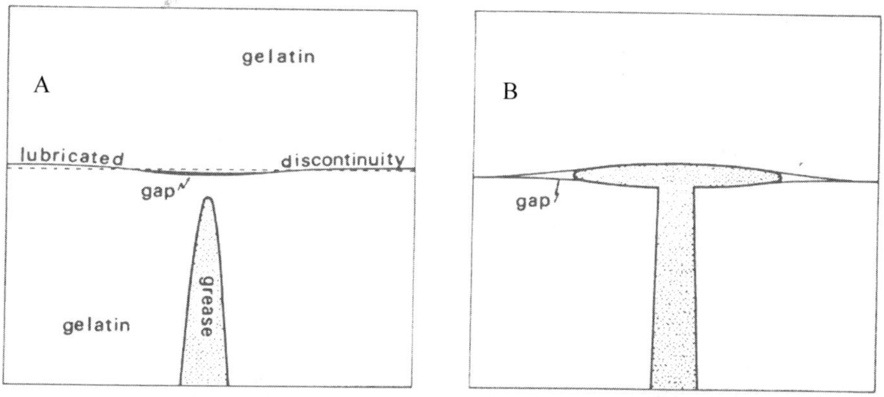

Fig. 3.11. Sketch of David Pollard's experiment in which grease was injected into gelatin perpendicular to a well-lubricated interface. The dyke fluid pressure is smaller than the total overburden stress on the interface (from D.D. Pollard (1973) Tectonophysics *19*, pp 233–269)

Since the process is controlled by very local stress changes induced by the advancing fracture, one would expect the fluid-filled gap to be very short. Nevertheless, the small gap on the interface may extend into a short, though relatively thin sill. This is not easily understood, considering that the pressure p_f of the dyke fluid is smaller than the total overburden stress on the interface. We can suggest two mechanical factors that may contribute to the extension of the fluid-filled gap along the bedding plane: First, when sliding apart, the dyke walls exert shear stresses on the bedding plane in front of the gap ends and these stresses may extend the gap in mode II fashion along the bedding plane of low shear strength. Secondly, and probably more importantly, the fluid pressure of the fracture will abruptly rise due to the inertial forces associated with the sudden arrest of the fluid and the associated vanishing of the driving pressure gradient of the viscous flow.

On the other hand, an advancing hydrofracture will ignore a bedding plane, or any other planar discontinuity in its path, if the high shear strength of the plane prevents the fracture walls from sliding apart. We should also bear in mind that lithological contrasts across bedding planes are often associated with discontinuous changes in the bed-parallel normal stress, as illustrated in Fig. 3.12. Such stress jumps may cause a change in the aperture of a dyke, or even stop the advancing dyke.

In Figs. 1.4 and 3.5E, the fracture system is dominated by sills of oil which is expelled from the layers of a mature hydrocarbon source rock. The sills are interlinked by numerous dykelets which permit the oil to escape from the source rock. We recall that the formation of sills requires (Eq. 3.16b) that the pressure of the injected fluid must be greater than or equal to the overburden stress. This would seem to contradict the generally held view that pore fluid pressures cannot exceed the weight of the overburden column of a unit cross-section, since otherwise the overburden would be in a kind of "floating" state. But this view is not strictly correct. It implies that fluid pressure and overburden stress are averaged over a large area. Locally (that is, in restricted areas) fluid pressures in excess of the overburden column may very well be balanced by the sub-vertical frictional shear resistance which is mobilized in the overburden above the margins of these patches of excessive fluid pressure.

We believe that this mechanism is actually limiting the extent to which a fluid sill can spread out along a bedding plane. Other limiting factors are lithological barriers, or the leakage of fluid from the sill into the cap rock, which is accompanied by a decrease in fluid pressure along the sill.

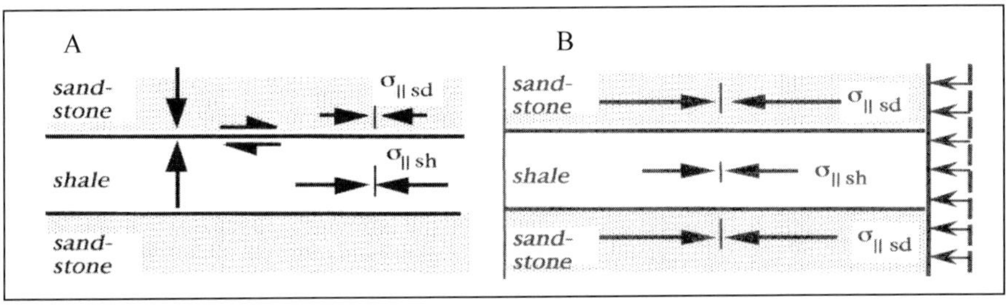

Fig. 3.12. Stress barriers at layer interfaces due to discontinuous changes of layer-parallel normal stresses ($\sigma_{\|}$):
A) tectonically undisturbed layers of different lithologies (the shear couple would be due to an inclination of the layers);
B) tectonic shortening or extension of an alternating sequence of strong and weak beds

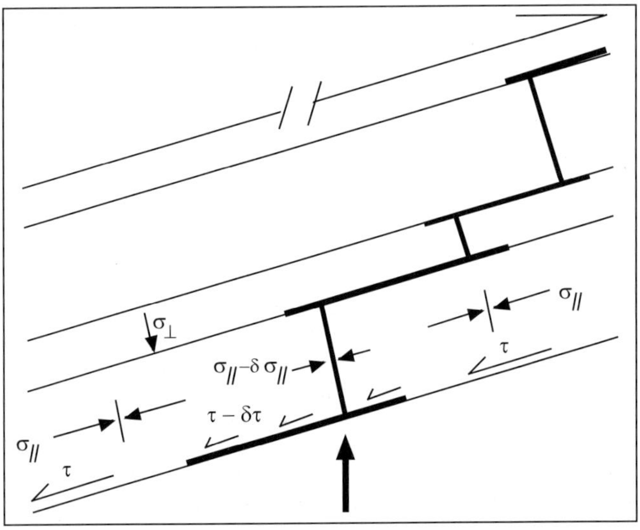

Fig. 3.13. Formation of a sill-dyke mechanism and up-dip fluid migration in the compression regime on a long slope

The next question is how fluid may escape from a sill by breaking through to stratigraphic higher levels and form stacks of sills, as demonstrated by the examples of oil migration in Fig. 1.5 and Fig. 3.5B–E, and schematically illustrated in Fig. 3.6. The detailed dynamics of stepping-up are still not very well known. We must therefore restrict the discussion to a mechanism which may very well be the most important one. A strictly horizontal layering of rocks is rare, and more commonly the beds are somewhat inclined, on a local or larger scale. They therefore experience some bedding-parallel shear stress (τ), exerted by the bed-parallel component of the overburden weight. Consider, for example, gently dipping beds on a long slope (Fig. 3.13). On such slopes the bed-parallel normal stress $\sigma_{//}$ increases downslope, and over a wide distance $\sigma_{//} + T_o$ may exceed the bed-normal stress $\sigma_{\perp}$, thus promoting conditions for sill formation. When a fluid sill forms along such a bedding plane, the effective compressive stress $\sigma_{\perp}'$ across the bedding plane will be locally reduced by the pressure of the sill fluid which, according to Coulomb's friction law, implies a proportional decrease in shear strength. Therefore, at the location of the fluid sill the beds may no longer be able to carry the full shear stress, and the bed-parallel normal stresses $\sigma_{//}$ has to adjust itself in order to preserve static equilibrium. As indicated in the figure, the reduction in shear stress τ will cause a drop in $\sigma_{//}$ in the upslope part of a sill, which is compensated for by an increase of this stress in the down-dip direction to maintain slope stability. The reduction in $\sigma_{//}$ will allow dykes to form and cross beds to feed new sills at stratigraphically higher levels. In this way, a process of upslope sill-dyke migration may be envisaged (Fig. 3.13).

Permeable wall rocks. Before closing this chapter on intrusion fracturing we recall that we have, so far, disregarded (p. 34) the fact that a permeable wall rock may allow the fracture pressure to interact with the pore pressure in the ambient rock. Hence, the pore pressure may be raised by the higher fracture pressure, and consequently reduce, or even close the initial opening of the dyke-type hydrofractures. The reason for the reversal of the compression by the formation of the hydrofractures can be easily explained: The fracture fluid exerts the total stress $\sigma_h = p_f$ upon the porous walls of the vertical fractures. Therefore, as long as the fracture pressure p_f is maintained, the bed-parallel total normal stress σ_h between parallel vertical fractures in a uniform horizontal layer remains equal to p_f. It thus exceeds the total stress σ_h^o, which existed prior to fracturing, by $\Delta\sigma_h = p_f - \sigma_h^o$. In a laterally confined layer, this excess load is balanced by reactive boundary stresses, and is not affected by a rise in pore pressure Δp. If the horizontal layer is of uniform thickness, the total stress σ_h equals p_f throughout the whole layer. In contrast, the effective stress σ_h^* will vary in time and space, because, after an initial rise in pore pressure due to the instantaneous compression of the rock by the formation of the hydrofractures, a further rise in pore pressure proceeds in a diffusion-type process from the fractures into the wall rock. The rise in pore pressure depends on the drainage conditions of the wall rock (permeability, bed thickness, fracture spacing and conditions at the bed surface) and the compressibilities of the porous rock, its solid skeleton, and the fluid.

Our problem is now the generalization of the aperture formula (Eq. 3.17). To simplify the problem, we consider the rock, as before, as being uniform and mechanically isotropic in all horizontal directions, and assume that the mechanical properties of the rock are not affected by the pore pressure changes. Then, within the realm of linear elasticity (Eq. 3.3), we can replace the pore pressure rise Δp (x,t) by its spatial average $\Delta\hat{p}(t)$ between neighbouring vertical fractures. Thus, in accordance with the definition in Eq. 2.2, the spatial average of the transient change in effective stress $\Delta\sigma_h^*(t)$ is

$$\Delta\sigma_h^*(t) = \Delta\sigma_h - a.\Delta\hat{p}(t) = p_f - \sigma_{ho} - a.\Delta\hat{p}(t) \tag{3.24}$$

where $a = 1 - K/K_s$ (Eq. 2.2).

The elastic extension $\Delta e_h < 0$ that is associated with the decrease in effective stress $\Delta\sigma_h^*$, is again determined by the poro-elastic stress-strain relationships (Eq. 3.3), where we insert $\Delta e_2 = 0$ in the second equation and eliminate $\Delta\sigma_2^*$ from the first two equations. With $\Delta\sigma_1^* \equiv \Delta\sigma_h^*$ and $\Delta e_1 \equiv \Delta e_h$, and $\Delta\sigma_3^* \equiv \Delta\sigma_v^* = -a.\Delta\hat{p}$ for the change in the effective overburden stress (the total overburden stress σ_v remaining constant), the poro-elastic equation for $\Delta e_h = \delta d/d$ (Fig. 3.7) can be described as

$$\frac{\delta d}{d} = \frac{1-\nu^2}{E}[(p_f - \sigma_h^o) - (1-\beta).\Delta\hat{p}] \tag{3.25}$$

where β is the material parameter defined by Eq. 3.6.

Hence, in order to maintain some aperture ($\delta d > 0$), the fracture pressure must satisfy the condition

$$p_f > \sigma_h^o + (1-\beta).\Delta\hat{p} \tag{3.26}$$

to ensure a positive right-hand side of Eq. 3.25. If the rock has no porosity, $\Delta\hat{p} = 0$, and the aperture relation (Eq. 3.25) reduces to Eq. 3.17. This would also hold in a fluid-saturated rock, whose pore pressure is maintained at the original level p^o by an ideal hydraulic interaction with ambient rocks. In general, however, the drainage of the fractured beds is rather limited, and the pore pressure will rise ($\Delta\hat{p}(t) > 0$). Part of this rise occurs instantaneously at the onset of hydrofracturing, as the bulk material is compressed by the fracture pressure under virtually undrained conditions. This initial rise in the fluid pressure $\Delta p(0)$ is determined by Eq. 3.32 in the appendix to this chapter.

In the most unfavourable situation, the fractured bed remains completely undrained, as for example, when the fractured bed is enclosed in thick overpressured shales. After the instantaneous rise $\Delta p(0)$ at the point in time $t = 0$, the pore pressures will eventually approach p_f, that is, $\Delta\hat{p} \rightarrow p_f - p^o$. We insert this in Eq. 3.26, and express σ_{ho} and p^o as fractions of the overburden stress σ_v^o. This is done by making use of the K_o-value (Eq. 3.7) and the λ-factor, $\lambda = p^o/\sigma_{vo}$, (Eq. 3.8). Then, with

$$\sigma_h^o = \sigma_h'^o + + \lambda.\sigma_v^o = K_o\,\sigma_v'^o + \lambda.\sigma_v^o = \sigma_v^o\,[(1-\lambda).K_o + \lambda] \tag{3.27}$$

Eq. 3.26 can be written as

$$p_f > \sigma_v^o.\left[\frac{(1-\lambda).K_o}{\beta} + \lambda\right] \tag{3.28}$$

On the other hand, in a dyke-dominated intrusion process one has $p_f < \sigma_v^o + T_{ov}$. In general, T_{ov} may be neglected with respect to σ_v^o, and the bracketed expression in Eq. 3.28 has therefore to be smaller than 1. As one readily verifies, this implies

$$\boxed{K_o < \beta} \tag{3.29}$$

again with $\beta = 1 - (1-2\nu).(1-\nu)^{-1}.(1 - K/K_s)$ as defined by Eq. 3.6.

Not surprisingly, Eq. 3.29 is the same as Eq. 3.13 for open internal hydrofractures in a laterally confined layer under a uniform pore pressure (p. 31). As pointed out before, this condition is not always satisfied. For example, in an undrained low-permeability clay bed, K_o is high, probably > 0.7; furthermore, $0.2 < \nu < 0.4$ and $K/K_s < 0.1$ would be a reasonable guess. This results in $K_o > \beta$, which makes it rather unlikely that water dykelets can remain open in an undrained clay bed. However, we have to concede that during the formation of the dykelets which, as we have seen (p. 38), is a rapid process, the wall rock may be eroded and/or loose material may be dragged into the fracture. Hence after "closure", the intrusion fractures may remain as streaks of higher permeability, and thus act as paths for further fluid migration. In connection with this, also notice in Fig. 3.5A the traces of the fluidized clay (white) that was injected into the dykelets.

So far, both fracture and wall rock were considered as saturated by the *same* fluid. In passing, we draw attention to the practically important case that, instead of water, oil or gas is injected into a water-saturated bed. In general, the pore walls of a water-saturated rock are preferentially "water wet", while gas and most crude oils are "*non-wetting*" regarding the pore walls (Fig. 3.14). The non-wetting fluids only enter the porous wall rock by overcoming a *capillary entry pressure* p_c. The fracture pressure p_f in oil or gas dykelets, therefore, always exceeds the water pressure inside the wall rock by the capillary entry pressure p_c. Hence, the maximum of the average pore pressure rise $\Delta\hat{p}$ in Eq. 3.26 is $\Delta\hat{p}_{max} = p_f - p^o - p_c$. With this expression inserted in Eq. 3.26, and following the argumentation which led to Eq. 3.29, one arrives at the condition for dykes fed by non-wetting fracture fluids to remain open in an undrained water saturated rock:

$$K_o < \beta + \frac{1-\beta}{1-\lambda} \cdot \frac{p_c}{\sigma_v^o} \tag{3.30}$$

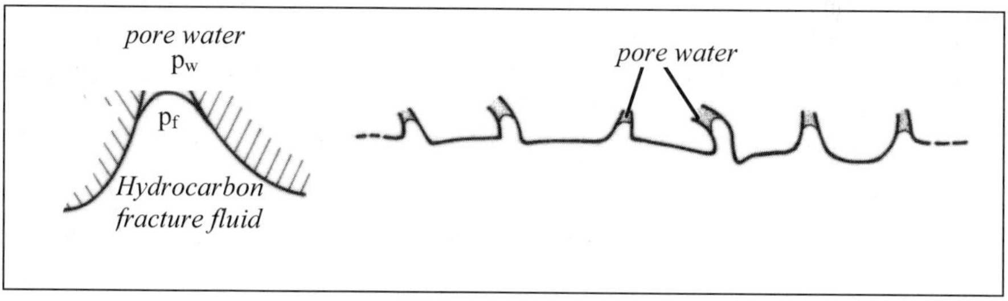

Fig. 3.14. Hydrocarbon/water interfaces at fracture wall (schematic)

We notice that even if Eq. 3.29 is violated ($K_o > \beta$), and therefore water fractures could not remain open in a water-saturated undrained bed, non-wetting fracture fluids may still keep intrusion fractures open, though only down to a certain depth. Interestingly, this limiting depth, given by

$$\sigma_v^o = \frac{(1-\beta).p_c}{(1-\lambda).(K_o - \beta)} \tag{3.30a}$$

increases with the degree of overpressure in the pore fluid ($0.42 \leq \lambda < 1$) prior to fracturing.

Summary. In closing this lengthy chapter on hydrofractures we shall summarise the main points.

- At some depth in the Earth's crust all *total* stresses are compressive. The formation of open fractures at that depth therefore requires fluid pressures in excess of the smallest total normal stress. Fractures of this kind are termed "*hydraulic*" *fractures*. There are two types of hydraulic fractures: *internal* fractures and *intrusion* fractures.

  Internal hydraulic fractures are formed by a pervasive rise in pore pressure. They are of the *tensile* type if the smallest (Terzaghi-)effective stress is reduced by the pore pressure to the negative value of the tensile strength (T_o), while the greatest effective stress remains smaller than approximately three times the tensile strength (Fig. 3.1A).

  The internal fractures are of the *cleavage (extension)* type if the smallest effective stress is reduced to approximately zero, while the greatest (compressive) effective stress has a high value, as already discussed and summarized in Chapt. 2.

  Intrusion fractures are formed by the injection of highly-pressured fluids into a bed from outside (Fig. 3.1B) whereby the injection pressure exceeds the smallest total stress plus the tensile strength.
- ***Vertical internal hydrofracturing*** in a horizontal bed requires that the pore pressure is raised throughout the bed to the value given by Eqs. 3.2a,b for the tensile and the cleavage mode, respectively. Thus the pore pressure rise Δp depends on the value of the smallest bed-parallel effective stress ($\sigma'_h{}^\circ$) prior to the rise in fluid pressure, and is inversely proportional to a dimensionless factor (β).

  If, though it is somewhat unlikely $\beta = 1$, the bed is completely free to expand in response to the pressure rise Δp. In the more realistic case $\beta < 1$, the lateral expansion is hindered by a lateral confinement. If the lateral expansion is completely suppressed, and the material reacts elastically to the rise in pore pressure, β is determined by the elastic rock moduli (Eq. 3.6). (Typical values of β may lie, say, between 0.4 and 0.7.)

  In laterally confined layers, internal tension and cleavage fractures require surprisingly high ("hard") overpressures ($0.8 < \lambda = p/\sigma_v{}^\circ < 1$); and only cleavage type fracture in laterally unconstrained layers may form under moderate overpressures (say, $0.6 < \lambda < 0.7$).

  Besides the high overpressure, *tensile internal hydrofracturing of laterally confined layers* requires (Eq. 3.13 or 3.13a) that $K_o < \beta$, with K_o the ratio of the smallest horizontal effective (Terzaghi-)stress to the overburden stress, both stresses being considered before pore pressure rise. The condition need not be satisfied in highly porous "hard" rocks or overpressured shales, where the formation of internal tension fractures would require some reduction of the K_o-ratio by tectonic extension. But even if the overpressure is sufficiently high and the additional condition $K_o < \beta$ is fulfilled, internal tensile fracturing remains confined to a certain depth range (cf. inequality 3.12 or 3.14).

  If any layer-parallel extension is suppressed, the vertical internal hydrofractures in the tensile mode develop as *"hairline" cracks* (see Eq. 3.15, Fig. 3.4). Conversely, a wider opening of the fractures suggests that the fractured layer was extended by the cooperation of a high pore pressure and extensional external forces, or by tectonic extension alone.
- ***Dyke-dominated intrusion processes.*** Dyke-type fractures serve as migration paths for fluids across rocks of very low or zero permeability ("cap rocks"). The aperture δd (Eqs. 3.18, 3.19), the spacing d (Eq. 3.20), and the growth rate (Eq. 3.22) of parallel dykes are proportional to the fracture length L. Hence, the fracture fluid will enter a

bed in a diffuse way through many closely spaced fractures, but while proceeding upwards through the bed the fractures will decrease in number and increase in width, and thereby stop the shorter fractures (Fig. 3.8).

- ***Rising dykelets.*** In general, the intrusion fractures will grow in a spasmodic fashion, as temporarily, the injection pressure drops below the value needed to keep the fracture open at the injection end. If, before closure, the fracture is of a length $L \geq L_c$, with L_c a critical length determined by Eq. 3.23, an isolated fracture of length L_c will form and rise through the rock driven by the upward decrease in lateral rock stress σ_h^o (Fig. 3.9). The rising fractures have a very slender "teardrop" profile, and a length depending on the fracture toughness K_c of the rock ($L_c \propto K_c^{2/3}$). A rising fracture will keep its critical length as long as no fracture fluid is lost to the wall rock. Otherwise, the rise will come to a halt, until the fluid volume of the fracture is restored by connecting fractures, which ascend along a trail of reduced fracture toughness created by the previous fracture.
- ***Dyke-sill mechanism.*** Even in dyke-dominated intrusion processes, dykes are often forced to make short *side-steps* when arriving at bedding planes of low shear resistance. The dykes are arrested at such bedding planes, because of the release of stress concentration at the fracture tip. Dykes are also arrested by stress barriers associated with abrupt changes in the bed-parallel normal stresses (Fig. 3.12). In either case, the sudden arrest of the fluid flow in the fracture invokes a brief rise in fluid pressure at the fracture tip, which may suffice to "lift off" the overburden around the contact of fracture and bedding plane. If the fracture system is dominated by bedding-plane sills, the fluid may still break through to stratigraphic higher levels, and form stacks of sills with interlinking dykelets (see Figs. 1.5, 3.5B–E and 3.6). This process of fluid migration operates in compressional regimes, in particular on slopes (Fig. 3.13).
- ***Dyke-type fractures in permeable rocks.*** The hydraulic interaction of fracture fluid and pore fluid through permeable fracture walls, tends to reduce the initial pressure difference by increasing the pore pressure of the wall rock. Since this reduces the effective compression between parallel dykes, the original fracture pressure can no longer maintain the initial opening (aperture) of the fractures. In the worst case of a fractured layer that is hydraulically isolated from the adjacent layers, some aperture of the fracture can only be maintained if Eq. 3.29 is satisfied. In view of this condition, it seems unlikely that water dykelets can remain open in an undrained clay bed. The situation is different when oil or gas is injected into a water-saturated bed. Since these fluids are "non-wetting" with respect to the pore walls of the "water wet" rock, they can only penetrate into the wall rock by overcoming a capillary entry pressure. In this case, Eq. 3.29 is replaced by Eq. 3.30, which indicates that non-wetting fracture fluids can still keep intrusion fractures open when water fractures would be closed.

Appendix to Chapter 3

The instantaneous build-up of pore pressure

A part $\Delta p(0)$ of the total pressure build-up $\Delta\hat{p}$ is caused by the undrained isothermal volume change of the rock at the instant of hydrofracturing. It is expressed in terms of the total stress load by the Skempton-Bishop formula (see G. Mandl, FBR, p. 174):

$$\Delta p(0) = B.\frac{\Delta\sigma_1 + \Delta\sigma_2 + \Delta\sigma_3}{3} \tag{3.31}$$

where

$$B = \left[1 + \Phi\frac{K_f^{-1} - K_s^{-1}}{K^{-1} - K_s^{-1}}\right]^{-1} \tag{3.31a}$$

and Φ is the porosity, and the compression moduli are K, for the bulk volume, K_s for the solid skeleton material, and K_f for the pore fluid.

The total stress increments in Eq. 3.31 are: $\Delta\sigma_1 = \Delta\sigma_h = p_f - \sigma_h{}^o$, $\Delta\sigma_3 = \Delta\sigma_v = 0$, and $\Delta\sigma_2 = \Delta\sigma_2^* + a\,\Delta p(0)$, according to the definition of the generalized effective stresses (Eq. 2.2).

The term $\Delta\sigma_2^*$ is obtained by inserting $\Delta e_2 = 0$ into the second equation (Eq. 3.3):

$$\Delta\sigma_2^* = \nu.(\Delta\sigma_h^* + \Delta\sigma_v^*) = \nu.(\Delta\sigma_h^* - a.\Delta p(0))$$

or in terms of total stresses:

$$\Delta\sigma_2 = \nu.\Delta\sigma_h + (1 - 2\nu).a.\Delta p(0)$$

Inserting the total stress increments into Eq. 3.31, the rise in pore pressure at the instant t = 0 of hydrofracturing becomes

$$\Delta p(0) = \frac{(1+\nu)(p_f - \sigma_h^o)}{(3/B) - (1 - 2\nu).a} \tag{3.32}$$

where, as before, $a = 1 - K/K_s$.

As an example, let us insert in Eq. 3.32 the parameter values $\nu = 0.2$, $B = 0.4$, and $a = 0.7$, as may be considered typical for porous rocks of intermediate strength (e.g., G. Mandl, FBR, p. 179). $\Delta p(0)$ is then 17% of the total stress rise $\Delta\sigma_h = p_f - \sigma_h{}^o$ in the undrained rock.

Chapter 4

Termination and Spacing of Tension Joints in Layered Rocks

In the preceding chapter, we discussed the mechanism of natural hydraulic fractures, i.e., internal and intrusion fractures which are mainly produced by high fluid pressures. Here, we shall consider in some detail the behaviour of the mode I fractures that are produced in layered rocks by the effective tension associated with a layer-parallel extension. These tension joints form under the conditions schematically represented below (Fig. 4.1).

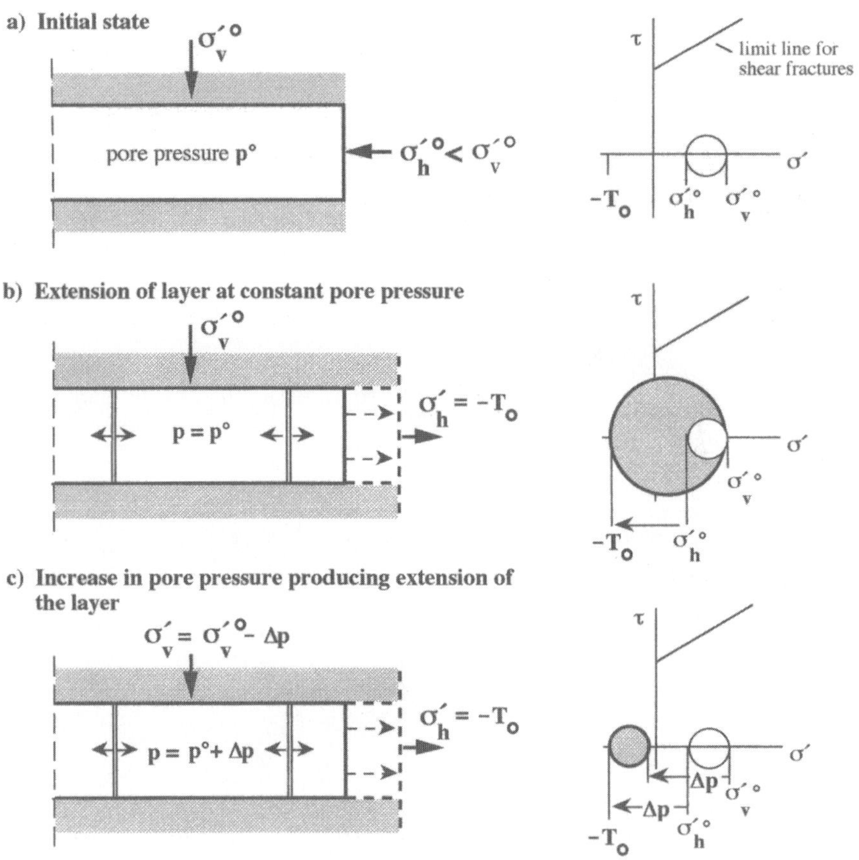

Fig. 4.1. Formation of tension joints in extending layer. Layers are schematised on the left, with the associated Mohr circle diagrams on the right

Termination of tension joints. In layered sedimentary rocks, the tension joints may terminate at bedding contacts, cut straight across contacts, or continue after side-stepping along a layer interface. A typical example of sets of joints confined to single layers was presented in Fig. 1.3, where the joints cut across the limestone beds but not across the interbedded weaker marls. Similar examples of alternating sequences of lithologically contrasting layers are shown in Figs. 4.2 and 4.3. Figure 4.4 represents a case where a systematic joint set (characterised by straight fracture traces on the bedding) cuts straight through a layer interface. Figure 4.5 shows the side-stepping of almost straight

Fig. 4.2. Liassic limestone layers separated by unfractured shales ("Blue Lias" oil source rock), Wessex Basin (South England, Lime Bay)

Fig. 4.3. Flexural slip (left lateral) of folded Carbon., NE rim of Rhenish massif (note the fracture inclination due to the reorientation of the σ_I axis by the slip imposed shear

Fig. 4.4. Orthogonal joint sets in siltstone beds (Cornwall). Wider spaced systematic joints in the thicker bed cut through the bedding plane. Thinner bed is more intensely fractured

vertical fractures on bedding planes, similar to the side-stepping of oil dykelets in Fig. 3.5B of the preceding chapter. These examples may suffice to illustrate the overwhelming evidence of joint termination and propagation in layered sedimentary rocks. And, like in the case of the dyke-sill process in Chapt. 3, we face the problem of what controls the termination, side-stepping or straight propagation of tension joints. The problem has considerable practical and economical implications, that have already been referred to in the introductory Chapt. 1.

Fig. 4.5. A) Profile of vertical joints in an alternating siltstone/shale turbidite sequence. The joint traces are discontinuous at bedding interfaces, as shown by the close-up view (0.5 x 0.5 m) in B) (from D.E. Helgeson and A. Aydin (1991) J. Struct. Geol. *13/8*, pp. 897–911)

Let us first turn to the most striking and ubiquitous phenomenon of sets of parallel tension joints that are confined to single layers, as shown by the examples in Figs. 4.2 and 4.3. In Chapt. 3 ("The dyke-sill mechanism") we mentioned as possible reasons for the arrest of dyke-type mode-I fractures at layer interfaces: (*1*) the weakness of the interface with respect to its shearing resistance and its tensile strength, and/or (*2*) the presence of a stress barrier at the upper side of the interface. Although the tension joints we are concerned with here are not produced by the wedging action of some high-pressured injection fluids, but by layer-parallel extension, one may expect the mechanisms of fracture arrest to be similar in both cases. Consider first, a stack of mechanically very similar layers in direct contact with one another, and let the whole stack be uniformly extended in a layer-parallel direction until the tensile strength is reached. The tension joints that form in the individual layers will then terminate at the bedding interfaces if the interfaces are very weak in tensile strength and shear resistance. In this case, a tension joint approaching an interface cannot maintain the near-tip stress concentration necessary for crossing the interface. As mentioned with regard to dyke-type fractures, the loss in near-tip stress concentration is commonly ascribed to two mechanisms. Firstly, just ahead of an advancing tension fracture, the interface is very slightly opened by the Poisson-type contraction of the material between fracture tip and interface, that accompanies the layer-parallel elastic stretching ahead of the fracture tip due to the aperture of the tension fracture. Secondly, the tip of a tension fracture arriving at a weak interface is blunted and the stress concentration lost, as the fracture walls can slip sideways along the interface. Interestingly, and somewhat unexpectedly, the results of recent numerical modelling experiments suggest that it is the local opening, rather than interfacial sliding, which primarily causes the termination of tension joints at weak interfaces between mechanically very similar layers (M.L. Cooke and Ch.A. Underwood (2001) J. Structural Geology *23*, Nr. 2/3, pp. 223–238).

Although the joints in mechanically similar layers terminate at weak interfaces, and are thus confined to separate layers, the joints in one layer need not develop independently from the joints in an adjacent layer; they may even combine into larger composite joints. A simple, but instructive example of a composite joint is shown in Fig. 4.6, which is taken from the same paper as Fig. 4.5. The figure shows a graphic representation of joints that formed in a stack of siltstone layers. Note that the individual joint segments are aligned in a vertical plane. Also, the initiation points are vertically aligned and are all located at the top of each layer, at the point where the joint from above first arrived. This suggests that the bed contacts still allowed an arriving fracture front to maintain some of its stress concentration which, when added to the tensile stress in the adjacent layer, triggered a new fracture.

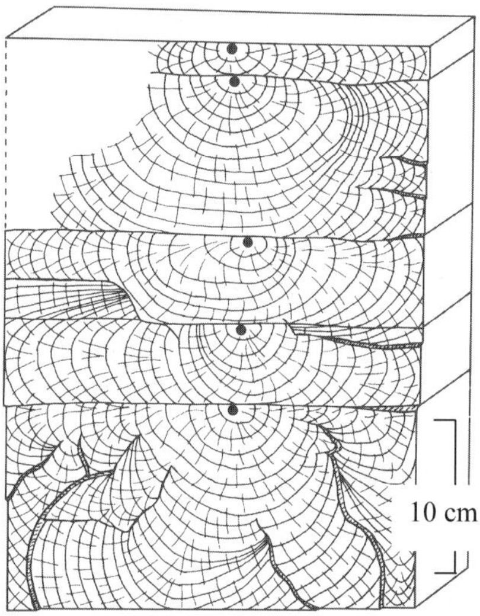

Fig. 4.6. Graphic representation of a composite joint in a stack of siltstone layers, showing the initiation points of the fractures, hackle traces, and reconstructed joint fronts (from D.E. Helgeson and A. Aydin (1991) J. Struct. Geol. *13/8*, pp. 897–911)

Unhealed pre-existing cracks can play a similar role as weak layer interfaces in stopping the propagation of tension joints. There is abundant evidence of younger joint sets abutting a set of earlier joints which have not yet been healed and thus still behave as very weak contact planes. The younger joints are therefore truncated by the older joints. The traces of the older joints may continue straight over wider areas, and are then referred to as "*systematic*" joints (germ. *fundamentale Klüfte* or *Hauptklüfte*). The joints truncated by the systematic joints are called "*non-systematic*" joints (germ. *Nebenklüfte*). The concept of systematic and non-systematic joints is illustrated in Fig. 4.7. A set of systematic joints abutted by non-systematic joints can be seen in Fig. 4.4. Two sets of systematic joints, and truncated non-systematic joints are shown in Fig. 4.8.

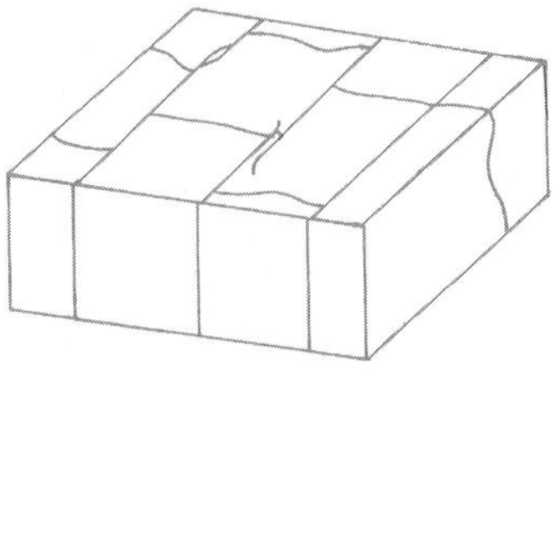

Fig. 4.7. Orthogonal systematic and non-systematic joints (after N.J. Price (1966) *Fault and Joint Development*, Pergamon Press)

The fact that weak layer interfaces or joints can arrest approaching fractures is also exploited by rock engineers in underground excavation operations and in the construction of road cuts. In these operations, pre-existing fractures or artificially created cracks ("pre-splitting") function as barriers to the fractures produced by blasting. This is illustrated by the photograph in Fig. 4.9.

Fig. 4.8. Aerial view of joints in flat-lying Permian sandstone (Canyonlands National Park, Utah). Two sets of systematic joints intersect at an angle of 70°, and are abutted by non-systematic joints under about 90° (photograph by G.E. McGill, from John Suppe (1985) *Principles of Structural Geology*, Prentice-Hall)

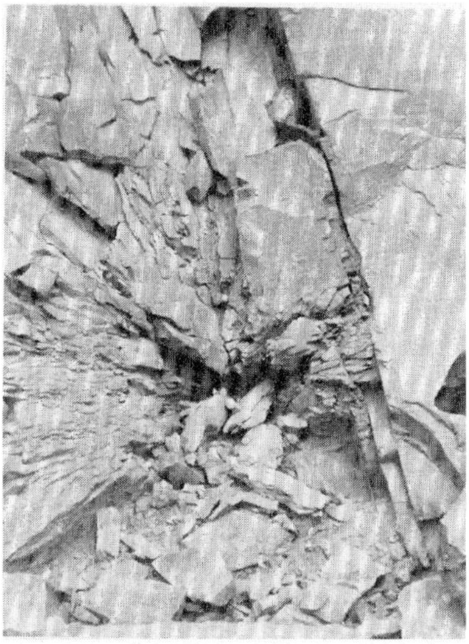

Fig. 4.9. Fractures radiating from a blast hole in a quarry wall. The fractures have been arrested by the pre-existing crack (a natural pre-split) on the right hand side of the photograph (from P.E. Gretener (1983) Bull. Ver. Schweiz. Petroleum-Geol. u. Ing. *49*, Nr. 116, pp. 29–35)

The appearance of the non-systematic joints naturally poses the question as to the stress field that controlled the formation of these tension fractures. Obviously, this stress field cannot be the same as the field that generated the systematic tension joints. We shall deal with this question in Chapt. 5. Here, we conclude the discussion of the joint arrest at weak interfaces between mechanically very similar layers by adding a few remarks on the so-called side-stepping, or step-over, of tension fractures at bedding contacts. Figure 4.5B may serve as an illustration of the phenomenon. Recall that we have already encountered this phenomenon in the context of the dyke-sill mechanism in Chapt. 3 (see Fig. 3.5). There, we expected the intruding high-pressure fluid to play an essential role in producing side-steps at bedding contacts, i.e. short sill-type intrusions, before the dyke-type fracture could continue into the next layer. Here, in the case of tension joints, side-stepping does not involve the intrusion of a high-pressure fluid. Although the side-stepping process is not yet completely understood, and it seems that, in particular, the role of inertial effects in the arrest of fractures has not been taken into account, we may draw a somewhat cursory picture of the mechanical process from recent publications.

Consider in Fig. 4.10 a fracture which propagates in the x_1 direction, and look for the position of the maximal tensile stress $|\sigma_{III}|$ on the line $x_1 = d$, $x_2 > 0$. The position of the maximum (or the maxima) is determined by Eq. 4.1 for the field of the minimum principal stress σ_{III}. Interestingly, the maximum is not attained at $x_1 = d$, $x_2 = 0$, but at $x_2 \approx \pm 0{,}53d$. Hence, for reasons of symmetry, two equal maxima lie symmetrical to the fracture plane at $x_2 \approx \pm 0{,}53d$. The maxima exceed the value of the principal tensile stress at $x_1 = d$, $x_2 = 0$ by about 13%.

$$\sigma_{III} = -\frac{K_I}{\sqrt{2\pi r}}.\cos\frac{\Theta}{2}\left(1+\sin\frac{\Theta}{2}\right) = -\frac{K_I}{\sqrt{2\pi d}}.\sqrt{\cos\Theta}.\cos\frac{\Theta}{2}\left(1+\sin\frac{\Theta}{2}\right) \qquad (4.1)$$

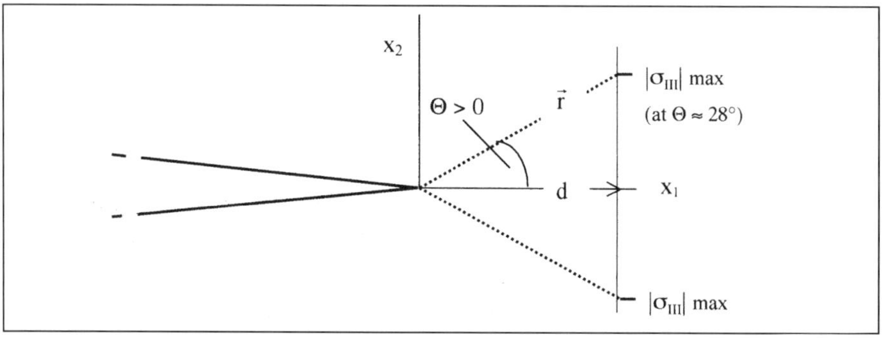

Fig. 4.10. Maxima of tensile stress ahead of mode I fracture

But despite this, the fracture propagates straight ahead in the uniform material along the x_1-direction. This is in accord with LEFM's prediction that the rate at which mechanical energy is released during incremental crack extension ("crack extension force") is maximal when the mode I crack propagates in its own plane. Now, if the fracture approaches a plane of low tensile strength (e.g., a weak bed contact or an unhealed fracture) at $x_1 = d$, the plane will be locally opened by the Poisson effect, and the energy release rate drops and can no longer propagate the fracture. However, tensile stress maxima will still be induced on the plane of weakness at points which lie symmetrically off the x_1 axis, although most likely not exactly at the points indicated in Fig. 4.10. Combined with the overall (far-field) tension on the opposite side of the plane of weakness, the induced tensile stress maxima may trigger new tension fractures to the right and left of the plane of the "parent" fracture.

- In **summary**, a tension joint propagating toward a contact between mechanically similar beds under the same layer-parallel effective tension may either: (*1*) terminate at the contact, (*2*) propagate straight through the contact, (*3*) continue across the contact by triggering a new joint segment on the opposite side of the contact, or (*4*) side-step to the right or left of the bedding contact.
- How the approaching joint behaves at the interface depends mainly on the tensile and shearing strengths of the interface. Fracture termination is favoured by a very low interfacial tensile strength, while very strong interfaces may be ignored by a propagating tension joint. Contacts of moderate strength may cause side-stepping, or the sequential development of joint segments arranged in-plane with each other.
- Since higher effective overburden pressures tend to suppress the small local opening of bedding contacts and increase the frictional strength of the contacts, deep burial may promote the straight propagation of tension joints across bedding planes.

Spacing of tension joints. We now turn to a curious, but rather common phenomenon in layers of sedimentary rocks: the fairly regular spacing (germ. *Kluftabstand*) of parallel tension joints, as shown in Figs. 4.2 and 4.3. What causes a regular spacing of parallel tension joints? Previously, we have seen that joints in adjacent mechanically similar layers may interact with each other. When we wish to understand the reasons for regular spaced tension jointing of a layer, and to define the controlling parameters, it is advisable to first consider a layer which is not affected by jointing in a neighbouring layer. A glance at the jointed layers in Fig. 4.2 tells us that this is best achieved by considering the jointed layers as being separated by thick layers of an "incompetent" material, as for instance, a clayey sediment. By "incompetent" we mean a material which, in comparison, has a relatively low elastic stiffness (Young's modulus E), or even deforms by creep flow under the applied (slow) extension. The extension of the whole sequence then induces sufficient tensile stresses in the stiffer (more competent) layers to generate tension joints, but not so in the lower modulus interlayers, where, in the case of creep flow, the layer-parallel stress may even remain compressive (cf. Fig. 3.12B with reversed arrows). The configuration we are dealing with is schematized in Fig. 4.11.

A number of mechanical models have been developed to explain the regular arrangement of the tension joints in the stiffer layer. Two models deserve special attention, since they bound the range of mechanisms that in reality produce sets of equidistant tension joints. The simpler of the two mechanical models was proposed by N.J. Price (N.J. Price (1966) *Fault and Joint Development in Brittle and Semi-brittle Rocks*, Pergamon Press, Oxford), the more intricate model was presented by D.W. Hobbs (1967; Geol. Magazine *104*, pp. 550–556). In order to obtain a clear picture of how appropriately the models analyze and explain the main features of the real process, we have to discuss the theoretical models and the underlying assumptions (some of which were tacitly made) in detail.

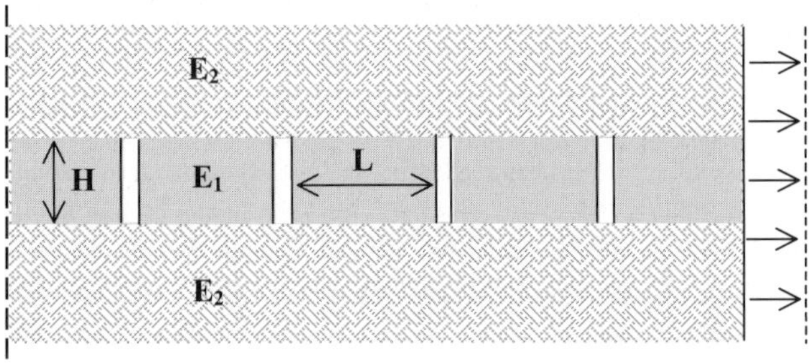

Fig. 4.11. Multilayer of stiff (E_1) and softer layer (E_2) under uniform lateral extension. Extension causing tension fractures in the stiffer bed:

$E_2 < E_1$

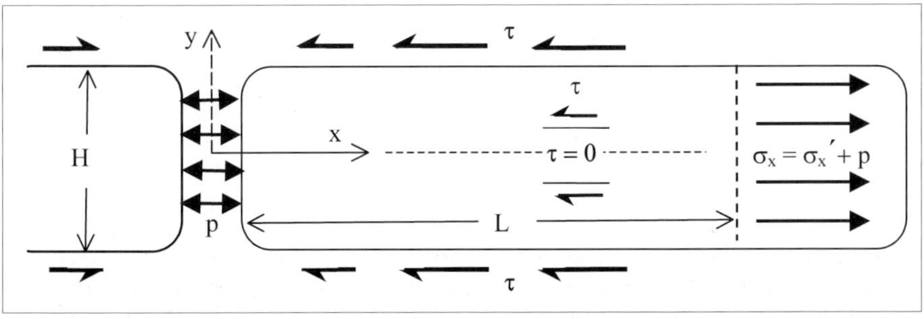

Fig. 4.12. Generation of a single joint (filled by pore fluid of pressure p) in an extended layer; interfacial frictional shear stresses (τ) balance the reduction in tensile stress σ_x'

Price's frictional coupling model. Price's model is basically a static balance of the mechanical forces in a stiffer layer which is in frictional contact with the ambient strata of softer material. The layer is horizontal and uniform, and the whole set of layers is uniformly extended. This induces a uniform tensile stress σ_x' in the stiff layer being considered, while leaving the layer boundaries free of shear loads, and only loaded by the compressive effective overburden stress σ_v'. When the tensile stress reaches the tensile strength ($-T_o$), a first joint is formed normal to the layer, and filled with pore fluid. The state of stress in the vicinity of the fluid-filled tension fracture is schematically shown in Fig. 4.12 for a section parallel to the plane of deformation (plane strain) and representing a slice of unit width.

Assuming a sufficient permeability of the rock, the fluid pressure in the fracture quickly attains the value of the pore pressure p in the layer, and therefore exerts the total stress $\sigma_x = p$ upon the fracture walls. Thus, the effective stress on the vertical fracture walls drops to zero, and the release in tensile stress tends to shorten the separated parts of the layer. If there were no cohesive/frictional resistance along the layer boundaries, the shortening would be uniform and accommodated by the widening of the joint. Further extension of the layer would simply be accommodated by further widening of the joint and slippage of the intact parts of the layer. In reality, however, the slippage of the layer is hindered by interfacial shear stresses (τ) which are mobilized in reaction to the shortening of the layer material.

Let us for the moment assume that we know the average τ_{av} of the induced shear stresses, the total shear resistance on a layer segment of length L then is $-2\tau_{av}.L$. Next, consider in Fig. 4.12 a cross section of height H and unit width at the distance L from the fracture wall. Since the pore pressure part p.H balances the force of the fluid pressure on the fracture wall, the static balance of all forces that act parallel to the layer (per unit width) is simply

$$\hat{\sigma}_x'(L).H = -2\,\tau_{av}.L \qquad (4.2)$$

where $\hat{\sigma}_x'$ denotes the average of σ_x' on the vertical cross section at x.

At this point, we may draw an initial conclusion about the spacing of tension joints. Consider two horizontal layer systems, both linear-elastic, with the same mechanical properties, the same pore pressure, and loaded by the same overburden stress, but differing in geometric scale by a common scaling factor m. Thus, when the thickness of the stiff layer in the first system is H, it is m.H in the second system. As a consequence of the linearity of the laws of linear elasticity, the same horizontal uniform straining induces the same tensile stress σ_x' at corresponding points (x and mx) in the two systems. When a first tension joint is formed in the corresponding layers, the associated unloading ($\sigma_x' = 0$) of the fracture walls causes exactly the same stress perturbations at corresponding points in the two systems. In particular, the same average stress τ_{av} will act upon the two stiff layers considered, and $\hat{\sigma}_x'$ at x = L in the layer of thickness H will be the same as at x = m.L in the layer of thickness m.H. Therefore, if the effective tensile stress generates a second tension joint at a distance L from the first joint in the layer of thickness H, it also generates a second joint at a distance m.L in the layer of thickness m.H. Hence, the ratio L/H in Eq. 4.2 is the same for both layers.

Thus, as was already pointed out by Price, the joint spacing to layer thickness ratio L/H is the same in all sets of a jointed layer and its unfractured embedment, which are geometrically similar (i.e., the bed thicknesses being scaled by a common factor), but do not differ in the (elastic) mechanical properties.

This may explain why in an alternating sequence of, say, limestone beds and shales, the thicker beds may show the same L/H ratio, or nearly so, as the thinner beds. But it does not yet tell us anything about the magnitude and the controlling parameters of the dimensionless spacing ratio. In order to determine L/H from Eq. 4.2 we have to know, or at least estimate, the parameters $\hat{\sigma}'_x$ and τ_{av}.

In Price's model of joint spacing, a new joint can form at a distance L from the first one if the vertical cross-section at x = L is the very end of the zone of the stress perturbation caused by the first joint. At this cross-section, the tensile stress σ_x' is still uniform and equal to the tensile strength. Hence, with $\hat{\sigma}'_x = \sigma_x' = -T_o$ in Eq. 4.2, the distance at which, or beyond which, the next joint can be generated is

$$\boxed{L = H.T_o / 2\tau_{av}} \tag{4.3}$$

If we knew τ_{av} in this formula, the problem of determining the spacing of the joints would be solved. Unfortunately, it is the crux of the whole problem, that we do not know the average shear stress τ_{av}

In Price's "slip" model, it is assumed that the cohesive part of the interfacial shear strength is zero. Hence, the shear stresses induced at the layer boundaries by the formation of the first joint cannot exceed the fully mobilized interfacial Coulomb friction $\tau = \sigma'_\perp \tan\varphi$ (with the friction angle φ of the interface).Although, prior to jointing, the vertical normal stress $\sigma'_\perp$ on the interfaces was the effective overburden stress σ_v', after the formation of a joint it no longer has this constant value. But, as we can show in Appendix A to this chapter, the *average* $\hat{\sigma}'_\perp$ of $\sigma'_\perp$ remains equal to σ_v'.

Since the maximum shear stress, mobilized by the fracture-induced release of tensile stress, cannot exceed the Coulomb limit stress $\tau = \sigma'_\perp \tan\varphi$, the total reactive shear force on an interface cannot exceed $L.\sigma_v'.\tan\varphi$. This value is actually attained if *slip occurs on the whole interfacial segment* from x = 0 to x = L. Making this *"slip" model* assumption, τ_{av} in Eq. 4.3) becomes equal to $\sigma_v' \tan\varphi$ and the spacing to layer thickness ratio is therefore

$$\boxed{L/H = T_o / 2\sigma'_v \tan\varphi} \tag{4.4}$$

How adequately the assumption of complete frictional slippage compares with real situations is a question that will be discussed later in this section (see "Models and reality"). But in any case, the slip assumption provides the greatest possible shear resistance on cohesionless interfaces, and Eq. 4.4 therefore defines a *lower bound* of actual joint spacing to layer thickness ratios.

Also note, that the dimensionless spacing L/H of equidistant joints decreases with increasing depth. This is readily seen by expressing the effective overburden stress σ'_v in Eq. 4.4 in terms of the total stress $\sigma'_v = \sigma_v(1-\lambda)$, (with $\lambda = p/\sigma_v$, the ratio of the pore pressure p to the total overburden stress σ_v), and noting that in sedimentary rocks σ_v increases by about 25 MPa per kilometer depth. Choosing the common value of 0.7 for $\tan\varphi$, we obtain the dimensionless spacing in terms of the thickness z (km) of the overburden at the time of jointing:

$$\left(\frac{L}{H}\right) = \frac{T_o(\text{MPa})}{35(1-\lambda)(\text{MPa/km}))z(\text{km})} \tag{4.4a}$$

For example,if the pore water is under normal hydrostatic pressure, i.e. $\lambda = 0.4$, and the layer has a tensile strength $T_o = 5$ MPa, joints with L/H = 0.5 would be formed at the depth of only about half a kilometer. Also note, that the slip model would also account for relatively small joint spacings; for instance, joints with L/H = 0.1 could be formed under about 2.5 km of overburden ($\lambda = 0.4$).

It should also be recalled, that Eqs. 4.4 and 4.4a were derived for flat-lying layers. If the layers dip under angles of more than a few degrees, one should expect the formation of joint sets to be affected by the normal stresses and shear stresses which are induced, prior to jointing, by the components of the overburden weight normal and parallel to the interfaces. We shall return to this complication later.

Finally, it is an important implication of Price's slip model that, once a layer is transected by a set of more or less regularly spaced joints, *further extension of the layers cannot generate new joints in between ("infill joints").* This may be seen as follows.

By the formation of the second joint the tensile stress $\hat{\sigma}'_x$ between the joints is further reduced, and in reaction, the interfacial shear stresses are reduced as well. But the shear stresses are raised again when the whole set of layers is extended further. The reason for this lies in the different straining of the layers, as illustrated in Fig. 4.12A. Although all layers are extended by the same amount, the extension in the fractured layer is mainly accommodated *by the widening* of the joints, which leaves the joint-bounded segments nearly undeformed, whereas the extension of the adjacent material is accommodated by continuous straining. Hence, the layer-parallel displacements decrease across the interfaces of the joint-bounded segments, which results in a layer-parallel shear stress τ_{av}, directly proportional to the displacement gradient. The shear stresses increase with increasing extension of the layers until slippage takes place. Therefore, in the slip model, the shear stress cannot exceed the Coulomb-frictional shear stress ($\tau_{av} = \sigma_v' \tan\varphi$). In the case of complete interfacial slippage, the *balance* relation (Eq. 4.2) states that, midway between two joints, $\hat{\sigma}'_x(L/2) = -(L/H)\sigma_v' \tan\varphi$. Or, when compared with Eq. 4.4: $\hat{\sigma}'_x(L/2) = -T_o/2$. Thus, the average effective tensile stress inside a joint-bounded segment cannot exceed half the value of the tensile strength.

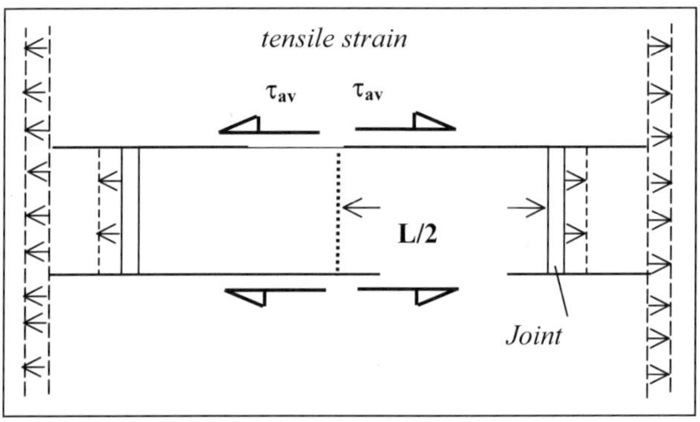

Fig. 4.12A. Accommodation of joined extension of fractured and unfractured layers

The Hobbs model*.* Somewhat curiously, the spacing formulae (Eqs. 4.3, 4.4) do not contain elastic material parameters, such as the Young moduli of the layers. The reason for this is, that the stiff and weak layers in the model are decoupled by an interfacial shearing resistance (τ_{max}) that is independent of the actual stiffnesses of the layers. This is different in the model, introduced by D.W. Hobbs (1967; Geol. Magazine *104*, pp. 550–556) which shall be discussed next.

Consider in Fig. 4.13 a stiff layer (1) and ambient weaker (low modulus) layers (2) that are "welded" together by strong cohesive bonds which exclude any interfacial slippage. It is assumed that the weaker layers are considerably thicker than the stiff layer of thickness H. The layers are supposed to respond elastically to extensional straining. When prior to jointing the three-layer unit is subjected to a uniform extensional displacement of its vertical boundaries, the induced layer-parallel displacement u° has a uniform gradient du°/dx throughout the whole three-layer unit. Hence, so far, the straining has not yet induced a shear stress on the contact planes, although the tensile stresses differ between the layers due to the difference in elastic moduli ($E_1 > E_2$).

As the uniform straining increases, the higher tensile stress in layer 1 eventually reaches the tensile strength, and a first joint is formed (Fig. 4.13). From this moment on, the displacement fields in layer 1 and layers 2 develop differently into the (differentiable) displacement fields $u_1 = u_1$ (x,y) and $u_2 = u_2$ (x,y). On the interfaces the displacements are the same on both sides ($u_1 = u_2$) because slippage is excluded, but the displacement gradients $\partial u/\partial y$ will differ because of the difference in shear moduli ($G_1 \neq G_2$). However, the shear stress τ_{xy} associated with the displacement gradient will vary continuously across the interfaces:

$$\tau_c = \tau_{xy}(x, 0) = G_1.(\partial u_1/\partial y)_{y=0} = G_2.(\partial u_2/\partial y)_{y=0} \tag{4.5}$$

Obviously, the value of the shear stress τ_{xy} (x,y) will decrease monotonically with increasing distance from the interface. Thus $\partial u_{1,2}/\partial y \neq 0$, at least within a certain distance from the interfaces. But this entails in static equilibrium ($\partial\sigma_x/\partial x + \partial\tau_{xy}/\partial y = 0$) that the layer-parallel normal stress σ_x varies in the x-direction. Consequently, the longitudinal strain e_x which, prior to the fracturing in layer 1, was uniform throughout the three-layer system, becomes non-uniform by the formation of the first joint. Thus, while the average strain in the layers 2 remains

$$e_{x2}^{av} = U/W \tag{4.6}$$

with W being the half-length of the layered system, and u = U the uniform boundary displacement (see Fig. 4.13), the local strain e_{x2} in the unfractured embedment is altered by the onset of fracturing in the stiff layer 1. This is an important point to be kept in mind during the exploration of the assumptions and implications of Hobbs' "welded layer" model.

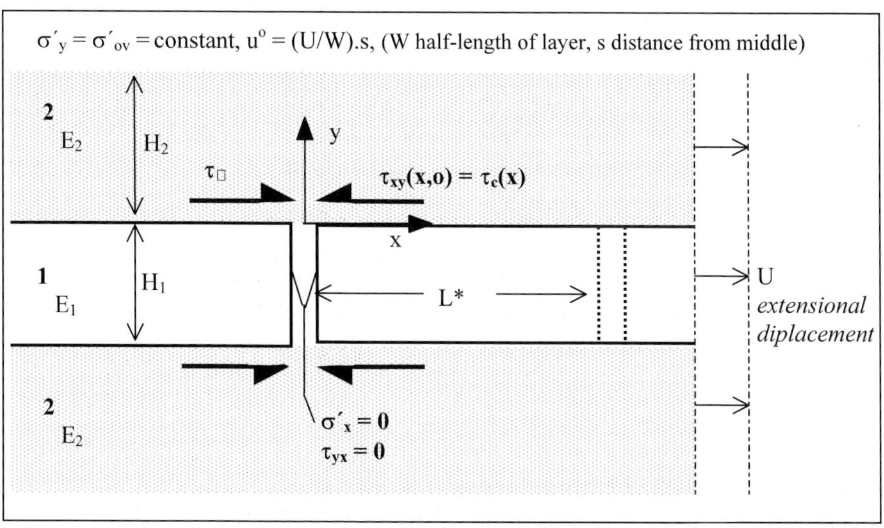

Fig. 4.13. D.W. Hobbs' elastic model of joint spacing in a stiff layer (1) "welded" to the ambient low-modulus layers (2) (see text for explanation)

The aim of the model is to determine the distribution of the tensile stress in layer 1, in order to establish the closest distance L between neighbouring joints that develop simultaneously or consecutively. In tackling the formidable problem of analysing the stress and strain history in the three-layer unit, Hobbs resorted to various simplifying assumptions concerning the relationship between the stresses and strains. Two of the assumptions may seem rather drastic and will be introduced after the uncontroversial part of the theory.

To start with, Hobbs does not consider the fields of stresses $\sigma_{x1}(x,y)$ and displacements $u_1(x,y)$ in layer 1, but the cross-sectional averages $\hat{\sigma}_{x1}(x)$ and $\hat{u}_1(x)$. Furthermore, he disregards pore pressure effects, apparently considering only non-porous rocks. We shall include the pore pressure (p) by following Hobbs' approach in terms of Terzaghi's effective stresses, assuming that the pore pressure p in layer 1 is uniform and remains unchanged during the extension of the layers. Reasonably, the layers are assumed to deform in plane strain in the x,y-plane, while the effective overburden stress σ'_y remains unchanged. The poro-elastic stress-strain relations (Eq. 3.3) then relate the increments of the local average strain $\Delta\hat{e}_{x1}(x)$ to the change in local average stress $\Delta\hat{\sigma}'_{x1}(x)$, where, for simplicity, the generalized effective stresses σ^* (Eq. 2.2) are replaced by Terzaghi's effective stresses σ'(Eq. 2.1) (implying that the compression modulus K_s of the solid matrix material is much higher than the compression modulus K of the bulk of the porous rock):

$$\Delta\hat{\sigma}'_{x1} = E_1.(1-\nu^2)^{-1}.\Delta\hat{e}_{x1} \tag{4.7}$$

In exception to our usual notation practice, we follow here Hobbs in counting *tensile strains and stresses as positive.* Also note, that in Hobbs' paper the ν^2-term is omitted, which, however, is of little relevance to the further derivations.

Further, it is important to recall that strains and stresses in Eq. 4.7 are referred to the same initial state of layer 1. Since the tensile stress $\Delta\hat{\sigma}'_{x1}(x)$ is built up from the *unstressed* state $(\sigma'_{x1}(x,y)=0)$, the extensional strain $\Delta\hat{e}_{x1}(x)$ is also referred to this state. Commonly, however, part of the total extensional strain in layer 1 is already expended in releasing a compressive pre-stress. Recall that, even in tectonically undeformed sedimentary rocks, layer-parallel compressive pre-stresses are built up during the deposition. To release the pre-stress in layer 1 the "welded" layer system had to be subjected to a uniform tensile strain e_o in x_1-direction. Therefore, the extensional strain $\Delta\hat{e}_{x1}(x)$ in Eq. 4.7 is the total local tensile strain $d\hat{u}_1/dx$ minus the pre-strain e_o. Henceforth denoting the total tensile stress in layer 1 by $\hat{\sigma}'_{x1}(x)$, Eq. 4.7 becomes

$$\Delta\hat{\sigma}'_{x1} = E_1.(1-\nu^2)^{-1}.\left(d\hat{u}_1/dx - e_o\right) \tag{4.7a}$$

It is this stress that has to vanish at the location of tension joints in layer 1:

$$\hat{\sigma}'_{x1}(x=0) = \hat{\sigma}'_{x1}(x=L) = 0 \tag{4.7b}$$

At the instant a first tension joint is formed at x = 0, the unloading and lateral shrinkage of the material in layer 1 is counteracted by reactive shear stresses $\tau_{xy} = \tau_c(x)$ at the contact planes (y = 0, –H). Hence, in static equilibrium and at uniform pore pressure, the gradient of the effective tensile stress $\hat{\sigma}'_{x1}$ is balanced by the interfacial shear stresses:

$$d\hat{\sigma}'_{x1}/dx = -2\tau_c(x)/H \tag{4.8}$$

So far, the derivation is still in line with the general theory of linear elasticity. Now Hobbs' first assumption is introduced concerning the gradient of the force $H.\hat{\sigma}'_{x1}$ in the fractured layer 1. The value of $H.d\hat{\sigma}'_{x1}/dx$ at x is assumed to be proportional to the difference between the local average displacement $\hat{u}_l(x)$ *after* the formation of the joint, and the displacement $u^o(x)$ at the same location just prior to the formation of the joint. Thus

$$H.d\hat{\sigma}'_{x1}/dx = 2C.[\hat{u}_l(x) - u^o(x)] \tag{4.9}$$

This equation is differentiated with respect to x:

$$\frac{d^2\hat{\sigma}'_{x1}}{dx^2} = \frac{2C}{H}.\left[\frac{d\hat{u}_1}{dx} - \frac{du^o}{dx}\right] \tag{4.10}$$

(Note that the contribution of the displacements associated with the release of a compressive pre-stress in layer 1 cancel out in the equation.)

Equation 4.10 is turned into an ordinary differential equation of $\hat{\sigma}'_{x1}$ by employing Eq. 4.7a to express $d\hat{u}_l/dx$ in terms of $\hat{\sigma}'_{x1}$, and recalling that $du^o/dx - e_o$ is the uniform pre-fracture straining "e" of the three-layer unit *after* a compressive pre-stress in layer 1 has been released. The differential equation for the tensile stress then reads:

$$\frac{d^2\hat{\sigma}'_{x1}}{dx^2} = k^2\hat{\sigma}'_{x1} - \frac{2C}{H}e \tag{4.11a}$$

where

$$k^2 = \frac{2C.(1-\nu^2)}{H.E_1} \tag{4.11b}$$

and the uniform layer-parallel strain e of the three-layer unit is attained just *prior* to the formation of a joint in layer 1, and is referred to the stress-free state $(\sigma'_{x1}(x,y) = 0)$.

The general solution to Eq. 4.11a is the stress distribution

$$\hat{\sigma}'_{x1}(x) = E_1.(1-\nu^2)^{-1}.e + A.\sinh kx + B.\cosh kx \tag{4.12}$$

with the constants A and B are determined by inserting $\hat{\sigma}'_{x1} = 0$ at the joint positions $x = 0$ and $x = L$.

Since the formation of a first joint in layer 1 is accompanied by the strain "e" switching from a uniform to a non-uniform state, while the boundary displacement U (Fig. 4.13) remains fixed, the strain "e" in Eq. 4.12 may be interpreted in two ways: as the uniform strain just prior to jointing, or as the average e_x^{av} (Eq. 4.6) of the strain in the layers 2 just after the formation of the first joints in layer 1.

The interested reader may wish to follow more closely the derivation of the stress equation 4.12. We substitute in Eq. 4.11a:

$$\hat{\sigma}'_{x1} = Y + E_1.(1-\nu^2)^{-1}.e \tag{4.11c}$$

which transforms the differential equation into the homogeneous form

$$\frac{d^2Y}{dx^2} = k^2.Y \tag{4.11d}$$

The reader may then readily verify by insertion, that

$$Y(x) = A.\sinh kx + B.\cosh kx \tag{4.11e}$$

is the general solution to Eq. 4.11d.

Before determining the constants in Eq. 4.12 we have to decide on the sequence of fracturing events in layer 1. Hobbs assumed that a single joint is first formed at a critical (uniform) strain e, and that a further increase in (average) strain $e_x^{av} > e$ in layers 2 is needed to produce a second joint in layer 1. Without disputing the adequacy of this assumption, we take the liberty of deviating from Hobbs' conception by assuming that a first set of parallel joints, separated by a distance L, is simultaneously formed when the layer-parallel strain reaches the critical value e (again interpreted as uniform pre-jointing strain just prior to jointing, or as average strain just after joint formation). The constants A and B are then determined by inserting $\hat{\sigma}'_{x1}(x = 0) = \hat{\sigma}'_{x1}(x = L) = 0$ in Eq. 4.12:

$$B = -E_1.(1-\nu^2)^{-1}.e \quad \text{for } x = 0, \text{ and } A = -B.\sinh\left(k\frac{L}{2}\right).\left(\cosh\left(k\frac{L}{2}\right)\right)^{-1} \quad \text{for } x = L \qquad (4.13)$$

With these expressions inserted, and after some manipulation with relations of hyperbolic functions, Eq. 4.12 yields the stress distribution between two joints which are separated by the distance L and are produced by the uniform external strain e (again referred to the initial state $\sigma'_{x1}(x,y) = 0$ of layer 1):

$$\hat{\sigma}'_{x1}(x) = \frac{E_1}{1-\nu^2}.e.\left[1 - \cosh\left(k\frac{L}{2} - kx\right).\left(\cosh\left(k\frac{L}{2}\right)\right)^{-1}\right] \qquad (4.14)$$

Note that the post-jointing tensile stress $\hat{\sigma}'_{x1}$ in Eq. 4.14 vanishes at x = 0 and x = L, as required, and attains its maximum $\hat{\sigma}'^{\,max}_{x1}$ midway between the neighbouring joints at x = L/2:

$$\hat{\sigma}'^{\,max}_{x1} = \frac{E_1}{1-\nu^2}.e.\left[1 - \left(\cosh\left(k\frac{L}{2}\right)\right)^{-1}\right] \qquad (4.14a)$$

The stress distribution (Eq. 4.14) is a function of the applied extensional strain and the separation of the joints. We eliminate the extensional strain by dividing Eq. 4.14 by Eq. 4.14a, which gives the dimensionless stress as a function of the single parameter L/H:

$$\frac{\hat{\sigma}'_{x1}(x)}{\hat{\sigma}'^{\,max}_{x1}} = \frac{\cosh\left(\kappa\frac{L}{H}\right) - \cosh\left(\kappa\frac{L}{H} - \kappa\frac{2x}{H}\right)}{\cosh\left(\kappa\frac{L}{H}\right) - 1} \qquad (4.15)$$

with

$$\kappa = k.H/2 = \sqrt{\frac{C.H}{2}.\frac{(1-\nu^2)}{E_1}} \qquad (4.16)$$

A graphical presentation of the distribution of the dimensionless tensile stress between neighbouring joints is shown in Fig. 4.14. Interestingly, numerical differences in $\hat{\sigma}'_{x1}/\hat{\sigma}'^{\,max}_{x1}$ do not appear until the third decimal, although the parameter κ.L/H in Eq. 4.15 was varied over the range from 0.1 to 1.2, corresponding with an interval $0.1 \le E_2/E_1 \le 1$ of the stiffness ratio, and $0.25 \le L/H \le 2$ of the fracture spacing to layer thickness ratio; in the graph, these variations merely show as variation in the thickness of the curve.

The external strain e = e* that has to be applied in order to generate tension joints at x = L/2 is obtained by setting $\hat{\sigma}'^{\,max}_{x1} = T_o$ in Eq. 4.14a:

$$e^* = \frac{T_o.(1-\nu^2)}{E_1}.\cosh\left(\kappa\frac{L}{H}\right).\left[\cosh\left(\kappa\frac{L}{H}\right)-1\right]^{-1} \tag{4.17}$$

or in terms of the inverse hyperbolic function $\cosh^{-1}$:

$$\frac{L}{H} = \kappa^{-1}.\cosh^{-1}\left(\frac{e^*}{e^* - T_o(1-\nu^2)E_1^{-1}}\right) \tag{4.17a}$$

with κ defined by Eq. 4.16.

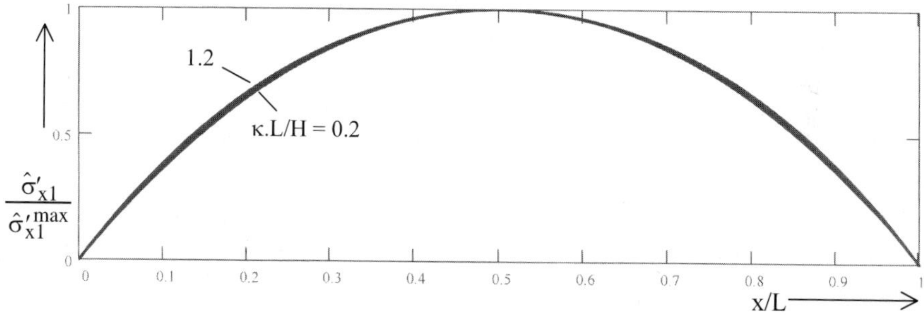

Fig. 4.14. Graphical representation of the stress solution to Eq. 4.15 showing the variation of the normalised tensile stress with the position x/L between neighbouring tension joints at a spacing L. Variations in the parameter κ.L/H of Eq. 4.15 hardly affect the shape of the curve (courtesy of Norbert Tschierske)

Notice that the strain e* in Eqs. 4.17 and 4.17a is not the extensional strain e that produced the first joints at the spacing ratio L/H, but the somewhat higher (average) strain that is required to raise the midway stress $\hat{\sigma}'^{\,max}_{x1}$ to the highest possible value T_o. Thus the strain e* in Eqs. 4.17 and 4.17a is the external strain (referred to the initial state $\sigma'_{x1}(x,y) = 0$ of layer 1) required to produce joints at L/2H. Further, note that because of the strain releases in layer 1 due to jointing, the strains in the joint-bounded segments will be smaller than the external strain e*. Since, according to Eq. 4.7, the local maximum of the strain at x = L/2 in the jointed layer is $T_o.(1 - \nu^2)/E_1$, the denominator on the right hand side of Eq. 4.17a remains positive.

At this point it may be good to summarize the procedure followed so far. By virtue of the assumption regarding the stress gradient (Eq. 4.9), which may well be considered as basic to Hobbs' theoretical model, the local stresses $\hat{\sigma}'_{x1}$ (averaged over vertical cross-sections) were determined without regard to the stress and strain perturbations inside the weak embedment of the fractured layer. The penalty for taking this short cut lurks in the, as yet undetermined constant C of the material parameter κ (Eq. 4.16).

But, remarkably despite this incompleteness, the Eqs. 4.17 or 4.17a already describe the effect of the *tectonic extension* e* on the spacing of the joints. For the purpose of illustration, let us specify the term $T_o.(1 - \nu_{(1)}{}^2)/E_1$ in Eq. 4.17a. Assuming $\nu_{(1)} = 0.2$, T_o = 5 MPa, and $E_1 = 2.4\text{x}10^4$ MPa, as reasonable values for, say, a medium strength sandstone, we have $T_o.(1 - \nu_{(1)}{}^2)/E_1 = 2\text{x}10^{-4}$. With this value inserted in Eq. 4.17a, κL/H attains the values of 0.47, 0.33 and 0.23 for e* = 0.2%, 0.4% and 0.8% respectively. Since, as will be shown a little later on, the parameter κ solely depends on elastic moduli, in our example the spacing/thickness ratio L/H is reduced by a factor of approximately 0.7, when the imposed extensional strain is doubled.

The reduction in the spacing of joint sets by increased tectonic extension is, at least qualitatively, in agreement with what is observed in the field. When the extensional pre-jointing strain varies along a layer, the spacing of joints will vary too. In folded layers, the spacing decreases as the extensional straining increases with increasing curvature. It is, for example, quite common that the joint frequency in a layer increases at the transition from straight fold limbs to hinge zones. Figure 4.15 is included to illustrate this point, although it should be said that jointing in the flexural slip fold is the result of more than just one mechanism. We shall consider this problem in Chapt. 7 when dealing with extensional jointing in compressional folds. An increase in joint frequency is also to be expected in the vicinity of a fault, where layers are bent by the frictional drag along the fault. A somewhat different case is shown in Fig. 4.16, where layers are stretched on the down-thrown side of a normal fault to accommodate the local flattening of the fault.

Fig. 4.15. Tensional jointing in the hinge zone of a small-scale flexural slip fold (Swiss Jura Mount.)

Fig.4.16. Flattening of steep normal fault accommodated by splaying of the fault and closely spaced jointing. (Bristol Channel, South England)

In order to complete the derivation of Hobbs' formula for L/H, the constant C has to be specified. To this end, Hobbs introduced additional assumptions concerning the shear stress τ_{xy} that is induced in layers 2 by the formation of a fracture in layer 1 (Fig. 4.13). Obviously, this stress perturbation in layer 2 decreases with the vertical distance y from the interface, and may be considered to effectively vanish at a distance "d". Assuming that the decrease of τ_{xy} from a value τ_c at the interface y = 0 is *linear*, i.e., $\tau_{xy} = \tau_c.(1 - y/d)$, Hobbs determines the constant C as

$$C = 2G_2/d \tag{4.18}$$

where G_2 (= $E_2.(1 + \nu_{(2)})^{-1}/2$ is the elastic shear modulus of the layers 2.

The interested reader may wish to retrace the derivation of this expression. First, compare relation 4.9 with the equilibrium condition (Eq. 4.8), and find C related to the contact shear stress τ_c:

$$C = \tau_c/(u^o - \hat{u}_1) \tag{4.18a}$$

In this expression, both numerator and denominator are functions of x only. Recall that the denominator is the difference between the displacement u^o of the three-layer unit just prior to the first fracturing, and the displacement $\hat{u}_1$ (averaged over vertical sections) in layer 1 after fracturing. As to the displacements u_2 in layers 2, which prior to fracturing were independent of y (i.e., $u_2(x) = u^o(x)$), it is assumed that the perturbation by the fracturing in layer 1 dies out at a distance "d" from the interfaces, where u_2 attains the value u^o of the undisturbed displacement. Since the displacement gradient $\partial u_2/\partial y$ is equal to the layer-parallel shear stress $\tau_c.d/2$ and Eq. 4.18b divided by the shear modulus G_2, one has

$$u^o(x) - u_2(x, y = 0) = \int_0^d \frac{\partial u_2}{\partial y} dy = \frac{1}{G_2} \int_0^d \tau_{xy}(x,y)\, dy \tag{4.18b}$$

Inserting $\tau_{xy} = \tau_c.(1 - y/d)$ in the integral on the right, according to the assumption of a linear decrease of τ_{xy}, the integral attains the value $\tau_c.d/2$, and Eq. 4.18b turns into

$$u^o(x) - u_2(x,y = 0) = \tau_c.d/2G_2 \tag{4.18c}$$

Finally, when the actual displacement $u_2\,(x,y = 0) = u_1(x,y = 0)$ at the interface $y = 0$ is replaced in Eq. 4.18c by the average displacement $\hat{u}_1$ of the fractured layer, comparison with Eq. 4.18a immediately leads to relation 4.18 for the constant C.

Naturally, the substitution of $\hat{u}_1(x)$ for $u_1(x,y = 0)$ involves a further approximation, which somewhat underestimates the shear stress τ_c, since $\hat{u}_1(x) > u_1(x,y = 0)$, as will be seen later.

Unfortunately, the formula 4.18 for C still contains the unknown depth "d" of the effective penetration of the fracture-induced shear stress τ_{xy} into the embedding layers 2. Hobbs remedied this uncertainty by assuming, somewhat daringly, that d = H. Thus, Eq. 4.18 becomes $C = 2G_2/H$, and Eq. 4.16 turns into

$$\kappa = \sqrt{\frac{G_2.(1-\nu_{(1)}^2)}{E_1}} \tag{4.19}$$

With this expression for κ, Eq. 4.17a finally attains the form

$$\frac{L}{H} = \sqrt{\frac{E_1}{G_2.(1-\nu_{(1)}^2)}}.\cosh^{-1}\left(\frac{e^*}{e^* - T_o(1-\nu_{(1)}^2).E_1^{-1}}\right) \tag{4.20a}$$

where E_1 and $\nu_{(1)}$ are the elastic moduli of the fractured layer, and G_2 is the shear modulus of the embedding layers ($G_2 = E_2(1 + \nu_{(2)})^{-1}/2$).

Hobbs' formula (Eq. 4.20a) is instructive in demonstrating the effects which *extensional straining*, *stiffness contrast*, and *thickness* of the fractured layer have on the spacing of vertical tension joints. Due to the special choice of the shear penetration depth "d", the joint spacing (L in our notation) scales with the thickness H of the fractured layer, provided that the layers under comparison have the same tensile strength and elastic moduli, and that the embedding layers have similar elastic properties.

Although the predictions of Eqs. 4.20 and 4.20a are found to be in fair agreement with the field observations, the simplifications in Hobbs' analytical approach to the truly formidable boundary problem of the fractured three-layer unit need to be borne in mind. The boundary conditions of the problem, recapitulated in Fig. 4.13, have to be satisfied by the average stresses and displacements of Hobbs' model. Also, the simplifying assumptions (Eqs. 4.9 and 4.18) on the stress-coupling between the layers must concur with these requirements.

At this point, some questions arise as to the conditions at the outer bedding planes of the three-layer-unit. Hobbs assumed that the embedding layers (2) sufficiently exceed the stiff interlayer (1) in thickness so as to ensure that at the outer bedding planes of the three-layer-unit the uniformity of the strain e (= e_x = U/W), which is imposed by the uniform boundary displacement U, is not affected by the fracturing of the interlayer. Thus, at the outer boundary planes (y = const.) $\partial e_x/\partial x = 0$; this implies $\partial \sigma_x/\partial x = 0$ because of Hooke's law; and further $\partial \tau_{xy}/\partial y = 0$ because of the equilibrium condition ($\partial \sigma_x/\partial x + \partial \tau_{xy}/\partial y = 0$).

The last condition is not satisfied by Hobbs' assumption, implicit in condition 4.18, of a linearly varying shear stress τ_{xy} in $0 \leq y \leq d$, since $\partial \tau_{xy}/\partial y = -\tau_c/d$. Consequently, neither e_x nor σ_x are uniform at y = d. Also note that the non-uniform strain e_x cannot abruptly change across the plane y = d into a uniform strain, since this would imply a discontinuous change in the layer-parallel displacement u_2.

Naturally, one might argue that the local deviations of the extensional strain from the uniform state of strain (e_x = U/W), which alternate between contraction and extension, may remain tolerably small. Nevertheless, the fact remains that Hobbs' assumption (Eq. 4.18) does not comply with the undisturbed boundary displacements u(x), indicated in Fig. 4.13 for the upper boundary plane; instead, it complies with the stress conditions $\sigma_y = \sigma_{ov}$, $\tau_{xy} = 0$ on the planes at a distance "d" from the layer 1. In fact, this stress condition is the adequate boundary condition for the ideal layer system which we shall consider a little later, and where, in our opinion, an amended Hobbs model may seem even more appropriate than in the case of the three-layer-unit above.

Thin weak interlayers. In nature, the unfractured weaker layers (2) are very often much thinner than the fractured layers (1). Examples of this can be seen in this book. Now, what prevents us from applying Hobbs' model to just this particular case?

Naturally, some idealizations are again required. Let us consider a *pile* of identical three-layer units under uniform extension, a few layers of which are illustrated in Fig. 4.17. By some geometrical or mechanical triggering, the first joints in the stiffer layers (1) are vertically aligned. This alignment might be brought about, for example, by swell structures of the weak interlayers, or by a local accumulation of flaws, or by some slight local bending of the beds. But more likely, the formation of a first joint in a layer (1) localizes the position of a second joint just above the first one in a neighbouring layer (1). As illustrated in Fig. 4.17A, we envisage the vertical alignment of the joints as being produced by the transmission of some layer-parallel shear stress across the thin interlayer (2). The situation is similar to the alignment of joint nuclei in the stack of layers in Fig. 4.6. But while that alignment was ascribed to the near-tip tensile stresses of the fracture arriving at the contact of two competent layers, it is the maximum interfacial shear stress τ_{max} induced by the first joint which determines the position of the next higher joint. The thinness of the weak interlayer allows the (strongly) reduced τ_{max} to be felt at the base of the next higher layer (1) where it locally raises the tensile stress of the already stretched layer. Thus, except for some minor side-stepping, the joints would be vertically aligned by the maxima of the interfacial shear stresses.

Once two neighbouring layers (1) are cut by the aligned joints (Fig. 4.17B), the reactive shear stresses τ_{xy} must vanish along the midplane of the weak interlayer (2), and the perturbed stress fields above and below this plane are each others mirror images. Although we do not exclude the possibility of τ_{xy} vanishing (or nearly so) inside a mid-band rather a midplane, we may assume that the width of this band decreases with decreasing height H_2 of the weak interlayer (2). At the same time, the linear stress function $|\tau_{xy}| = |\tau_c|.|y|/d$ for $0 \le |y| \le d$ may more closely approximate the real non-linear function of y. Therefore, when $d = H_2/2$ is inserted in the constant C (Eq. 4.18), the parameter κ of Eq. 4.19 becomes

$$\kappa^* = \sqrt{\frac{H_1}{H_2/2}.\frac{G_2}{E_1}.(1-\nu_{(1)}^2)} = \kappa.\sqrt{\frac{H_1}{H_2/2}} \tag{4.19a}$$

Then, with κ in Hobbs' formula (Eq. 4.20a) on page 63 replaced by κ*, the spacing formula for thin weak interlayers ($H_2 << H_1$) becomes:

$$\frac{L}{H} = \sqrt{\frac{H_2/2}{H_1}}.\kappa^{-1}.\cosh^{-1}\left(\frac{e^*}{e^* - T_o(1-\nu^2)E_1^{-1}}\right) \tag{4.20b}$$

Hence we expect weak interlayers which are considerably thinner than the fractured layers, to drastically reduce the spacing of regularly spaced joint sets.

Finally, when inspecting the formulae 4.20a,b, it should be clear that although the formulae constitute a continuous relation between the dimensionless spacing L/H and the

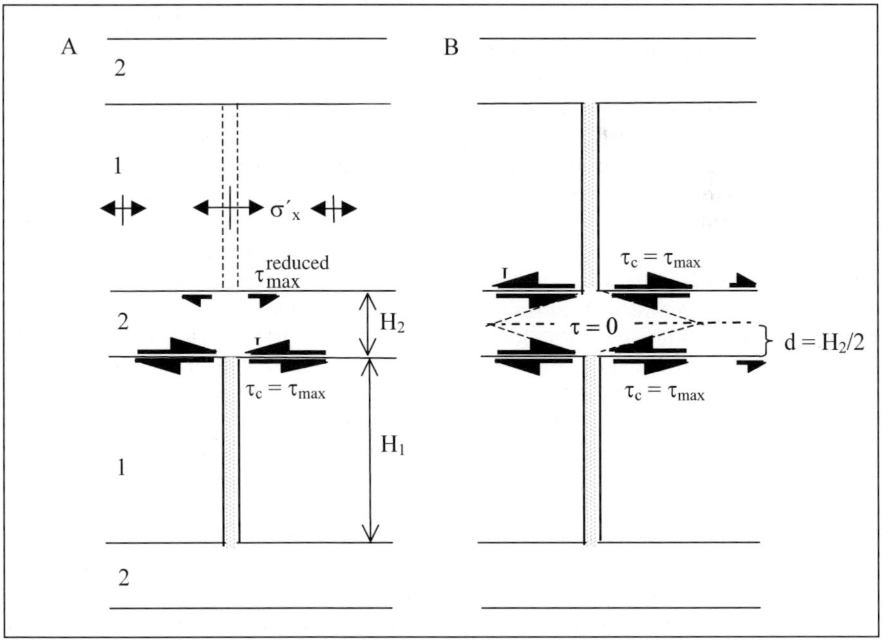

Fig. 4.17. Vertical alignment of tension joints in a stack of identical three-layer units with thin weak layers ($H_2 << H_1$):
A) Maxima of reactive shear stresses in weak interlayer (2) on top of a first tension joint in layer (1). Nucleation of a second tension joint (stippled).
B) Depth of shear penetration "d" in thin interlayer (2) when $H_2 << H_1$

extensional strain e*, this does not imply that *every* change in e* results in a change in L/H. In fact, the formation of new joints, midway between the already existing ones, is a discontinuous process which requires quite specific increases of the external strain e*. This process of "infill jointing" between already existing, more or less regularly spaced joints, is our next issue.

Infill jointing. Price's and Hobbs' models of joint spacing differ greatly in the mechanism of stress transfer between the intact embedment and the fractured interlayer. Although, in both models, the fracturing of the stiff layer is preceded by a uniform extension of the whole multilayer, in the slip model the strain release by fracturing is counteracted by the fully mobilized interfacial Coulomb friction, whereas Hobbs' "welded-layer" model allows for arbitrarily high reactive shear stresses on the interfaces. As a consequence of this difference, the slip model allows for the formation of only one set of regularly spaced tension joints, while in Hobbs' model the continued straining of the intact ambient layers can raise the tensile stress in the segments of the fractured layer until the tensile strength is reached and new joints are formed.

If, in the case of infill jointing, the spacing ratio of the very first joint set was L, the dimensionless stress distribution after the nth infill jointing is determined by Eq. 4.15, when $L/(2^n)$ is inserted for the spacing of the nth infill set, since the spacing L of the first joint set has been halved n-times. This may give the impression that infill jointing in Hobbs' model can be made to go on "*ad infinitum*", as was criticized by some authors.

In fact, the infill process is severely limited by the onset of fracturing or faulting in the embedding layers which were considered as linear elastic. Let us illustrate this limitation by assuming $E_2/E_1 = 0.2$, $E_2 = 0.5 \times 10^4$, $\nu_{(1)} = 0.2$, $\nu_{(2)} = 0.4$, $\kappa^{-1} = 3.82$, and, as before (p. 63), $T_o(1 - \nu_{(1)}^2).E_1^{-1} = 2 \times 10^{-4}$ for a medium strong sandstone. Prior to the extension, the layers were undisturbed, under normal pore pressure, and at a depth of about 1.5 km; the compressive stresses in layer 1 and 2 are assumed as $\sigma'_{x1}{}^{\circ} = 9$ MPa and $\sigma'_{x2}{}^{\circ} =$ 15 MPa, respectively. We recall, that the strain e* in formula 4.20a is referred to the state $\sigma'_{x1} = 0$; hence, prior to the application of e*, a reduction of the compressive stresses $\sigma'_{x1}{}^{\circ}$ to zero is accompanied by a reduction of $\sigma'_{x2}{}^{\circ}$ to 13.2 MPa. If the layers 2 fracture at a tensile stress of, say, –5 MPa, the maximum strain e* supported by the intact layers would, according to Hobbs' formula (Eq. 4.20a), be $e^* = 36.4 \times 10^{-4}$, which accounts L/H = 1.3 in the stiff layer. But, as mentioned before, the parameter L/H in Hobbs' formula refers to neighbouring joints, whose midway tensile stress has just reached the tensile strength. Hence, in our example, the smallest spacing ratio possible would be L/H = 0.65.

Of course instead of fracturing, the embedding layers may also deform in a viscous or frictional-plastic mode; a process, to which Hobbs' model does not apply.

Improvements* and *"fracture saturation". In Hobbs' model of joint spacing we considered the tensile stresses in the fractured layer as stresses ($\hat{\sigma}'_x$) averaged over orthogonal cross sections of the stiff interlayer. Up to the formation of a first set of tension joints, the bonded layers were equally stretched and the tensile stress in the uniform interlayer was a strictly uniform principal stress, since the contacts between equally stretched layers are free of shear stresses whatever the stiffness difference of the layers may be. Thus, there was no need to distinguish between the average stresses and the actual local stresses in applying the fracturing condition ($\hat{\sigma}'_{x1} = \sigma'_{x1} = T_o$) to the first generation of joints. But, the stresses between neighbouring joints of the first, or any later generation, are highly non-uniform, as was indicated by the distribution of the average tensile stress in Fig. 4.14. Thus it seems possible, or even likely, that the actual tensile stresses σ'_{x1} in the fractured layer not only vary in the x-direction, but also in the y-direction.

Then, if $\sigma'_{x1}(x,y)$ strongly deviates from $\hat{\sigma}'_{x1}$, the condition $\hat{\sigma}'_{x1} = T_o$ in Hobbs' model would be a rather crude criterion for the formation of infill joints. Since by definition of the average stress, a difference between the two stresses at x implies that $\sigma'_x(x,y)$ exceeds $\hat{\sigma}'_x(x)$ on a part of the cross section at x, and may there generate a tension fracture, even though $\hat{\sigma}'_{x1} < T_o$. Only if such an "infill fracture" is stopped before cutting across the whole layer, the existing set of fully developed joints will have the smallest spacing possible under the given mechanical and geometrical conditions. In other words, the joint set would be "*saturated*".

We should, therefore, understand how the stresses σ'_x vary in a fracture-bounded segment of the stiff interlayer. To this aim, we start again from the fact that the vanishing of the effective stress σ'_x on the walls of a fluid-filled fracture causes a release of tensile stress in the interlayer material, while the associated shortening is gradually reduced by the reactive shear stresses at the interfaces. If the walls of the joint would accommodate the shortening of the adjacent material like frictionless rigid pistons, the shortening would not vary in the y-direction, and σ'_x would only depend on x. But the fracture walls are flexible, as sketched in Fig. 4.18A, and the widening of the joint will be greatest in the central part, and least at the firmly bonded interfaces where the joint can open up only as far as allowed by the stretching of the ambient layers. Therefore, the tensile stress will maintain its highest value at the interfaces, where the material is stretched in unison with the adjacent intact units, and will decrease towards the mid-plane of the fractured layer.

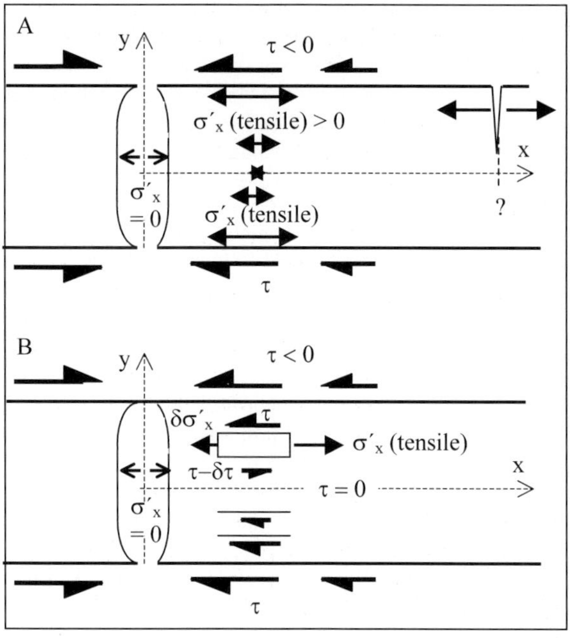

Fig. 4.18. Joints reduce the tensile stress (considered as positive) in the fractured layer, and induce shear stresses.

A) The layer-parallel tensile stress component decreases from a maximum at the interfaces to a minimum at the mid-plane (y = 0), in accord with the deformation of the stress-free fracture walls ($\sigma'_x = 0$).
B) The shear stress (τ) decreases in magnitude from the maximum at the interfaces to zero at the mid-plane, causing the tensile stress to increase with increasing distance from the joint

As sketched in Fig. 4.18B, the reactive shear stresses τ on the interfaces are transmitted into the layer while decreasing in magnitude towards the mid-plane y = 0. For reasons of symmetry, τ vanishes at the mid-plane, as was already indicated in Fig. 4.12. Considering the static balance of the layer-parallel forces on a rectangular volume element in Fig. 4.18B, one notices that the vertical increase in $|\tau|$ must be balanced by a horizontal increase in the tensile stress σ'_x (here counted as positive).

In differential form, this balance is expressed by the equilibrium equation

$$\frac{\partial \sigma'_x}{\partial x} = -\frac{\partial \tau}{\partial y} = \frac{\partial |\tau|}{\partial y} > 0 \tag{4.21}$$

where, with reference to Fig. 4.18B, both sides are positive.

When differentiated again with respect to y, the left hand side of Eq. 4.21 remains positive, since the (positive) σ'_x increases in the positive y-direction. Therefore $\partial^2|\tau|/\partial y^2 \geq 0$, with the equality sign applying if $\partial\sigma'_x/\partial y = 0$. This means, as sketched in Fig. 4.19A, that as long as σ'_x increases on a cross section x = const. (dashed) from mid-plane towards interfaces, the curve $|\tau| = |\tau|(y)$ steepens in the positive y-direction. If in a vertical interval Δy the tensile stress σ'_x remains constant ($\partial\sigma'_x/\partial y = 0$), the shear stress gradient $\partial|\tau|/\partial y$ must be constant as well, and τ will either vary linearly with y, or remain constant, and possibly even zero. We shall shortly specify this in more detail.

Thus, in a broad sense, the tensile stress σ'_x increases with increasing distance from the joint, and decreases from its maximum value at the interfaces towards a minimum at the mid-plane of the fractured layer. The tensile stress σ'_x will therefore exceed the average stress $\hat{\sigma}'_x$ near either interface, and drop below the average stress in the middle part of the layer. Hence, layer-orthogonal fractures can nucleate near the interfaces and start growing in the vertical direction before the average stress has reached the "critical" value T_o, as is suggested on the right of Fig. 4.18A. This tendency is further enhanced by the fracture mechanical result (see Chapt. 2, p. 16) that, even under uniform tension, tension fractures grow more easily from the bedding planes into a rock bed, than from locations inside the bed. We also know from Chapt. 2 (Eq. 2.4) that the critical tensile "driving stress" for fracture propagation decreases with increasing square root of the fracture length. (Although this result was derived for fractures under homogeneous tensile loading, the fracture length may safely be expected to facilitate the fracture growth also under the inhomogeneous stress conditions of our case.)

To developing into a joint proper, the fracture has to cut across a central lamina, where the tensile stresses are lowest, or even fully released. The latter may be the case if the opening joints are very slender and thus allow a flat central part of the fracture wall to be uniformly displaced. Consequently, as illustrated with some exaggeration in Fig. 4.19A, the displacement will also be uniform, or nearly so, in a central strip (shaded in the figure). Since in this region the displacement gradient in the y-direction vanishes, or nearly vanishes, the vanishing shear stress τ (= $G.\partial u_x/\partial y$) cannot suppress the release in tensile stress. We therefore envisage a seam, not necessarily of uniform width, around the mid-plane y = 0 of the jointed layer, in which $\sigma'_x \cong 0$. This central seam will thus form an obstacle to the propagation of a fracture.

However, the central zone of zero tension cannot maintain the blockade of infill fractures when the extension of the ambient layers is sufficiently increased. This is seen by reference to the uniqueness theorem of linear elasticity: stresses and strains in an elastic body are uniquely determined by the tractions or displacements that act on the whole boundary of the body. This implies, for our problem, that by increasing the extensional strain (and hence the displacements u_x) in the ambient layers by a constant factor "m", all shear and tensile stresses in the joint-bounded segment of the interlayer are uniformly increased by this factor. Note that the boundary conditions are multiplied by "m" on the *whole* boundary of the joint-bounded segment considered, because the effective pressure on the joint walls remains zero. Therefore, the tensile "driving force" of fracture propagation is increased by the factor "m". Hence, if the tensile straining of the ambient layers is sufficiently raised, new fractures are propagated across the interlayer, and become new infill joints, essentially in accordance with Hobbs' model.

So far we have tacitly assumed that two neighbouring joints were so far apart (Fig. 4.19A) that the perturbations of the stress field by either joint did not interfere with each other. What happens when the joints are too close together? It stands to reason that the interaction of the two joints will restrain the shortening between the joints. As sketched in Fig. 4.19B, the layer-parallel displacement u_x decreases from the fracture walls towards the cross section midway between the two joints. Thus, the material of the middle strip is now under layer-parallel contractive strain. Hence, with the exception of the close vicinity of the fracture walls, where $\sigma'_x = 0$, a compressive stress is superimposed on the original tension. If the neighbouring joints are sufficiently close, the contractive straining of the middle strip renders the stress σ'_x negative (here again, compression is considered as negative); and the closer the neighbouring joints, the higher are the values that the compressive σ'_x will attain along the x-axis. In addition, one should expect $|\sigma'_x|$ to increase between x = 0 and x = L/2, since it remains zero at the fracture walls.

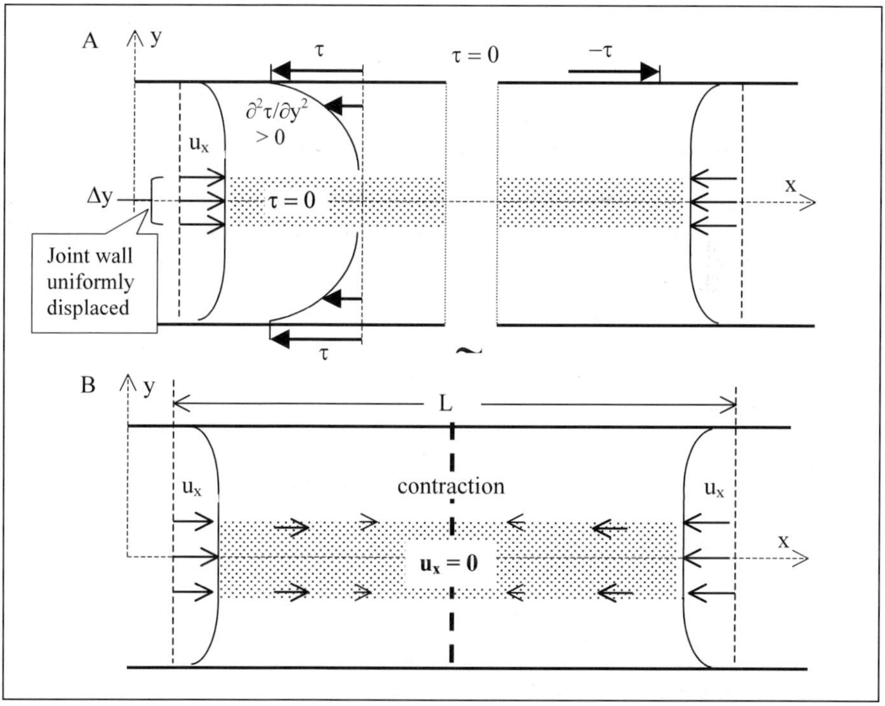

Fig. 4.19. Displacement effects of the opening of tension joints (strongly schematized): A) Widely spaced tension joints: the opening of the joints causes the vertical displacement gradient ($\partial u_x/\partial y$) and the associated shear stress ($\tau = G.\partial u_x/\partial y$) to vanish, or nearly vanish, in a central zone (shaded), allowing the tensile stress there to be fully, or nearly, released ($\sigma'_x = 0$). B) Closely spaced tension joints: interference of the displacement fields of neighbouring joints causes contraction and compressive stresses σ'_x in a central zone (shaded)

Our qualitative assessment of the developing stress field in a joint-bounded segment of the fractured layer may help to understand what, at first sight, might appear as a baffling result of a recent numerical analysis of the problem by T. Bai and D.D. Pollard (2000; J. Structural Geology *22*, pp. 1409–1425). The authors applied a two-dimensional finite element code, designed for the study of elastic fracture problems, to the elastic three-layer problem we are dealing with. Accordingly, a uniform layer-parallel displacement U was applied to the vertical boundaries of the three-layer unit, while the horizontal outer boundary planes were kept free of shear stress, but subjected at the upper plane to a uniform vertical downwards displacement in order to simulate the compressive action of an overburden stress. In fact, these are the same boundary conditions as those complied with in Hobbs' model on the planes at distance "d" from either interface. (While $d = H_1$ in Hobbs' model, $d = 1.5H_1$ in the specific configuration chosen for the numerical model.) Thus it is not assumed that the horizontal strain along the horizontal outer boundary planes is uniform, i.e. undisturbed by the fracturing of the interlayer, as was assumed in Fig. 4.13.

Figure 4.20 shows two contour plots of σ_x (which we may also read as effective stress σ'_x) from Fig. 3 of Bai's and Pollard's paper. In plot A, obtained for a joint spacing layer thickness ratio $L/H = 1$, $\sigma_x = 0$ in a central region, as we would have expected for small interference of the neighbouring joints. In plot B, the closer spacing layer thickness ratio $L/H = 0.8$ results in a compressive region that extends across the central area of the fractured layer almost from joint to joint. Note that the stresses obtained for a specific L/H ratio are the same in all geometrically similar layer systems with identical elastic properties and the same external strains.

Admittedly, the tensile stresses in the contour plots of Fig. 4.20 are too high to be sustained by rocks in the field. But the numerical results illustrate the concept of a central compressive belt in a fractured layer under extension. Although at first sight somewhat enigmatic, the existence of such a compression belt may be understood as the result of interfering displacement perturbations of closely spaced joints, as discussed above in connection with Fig. 4.19B.

Obviously, the presence of a central compression belt presents a formidable obstacle to the propagation of a vertical tension fracture that originates from a flaw or a

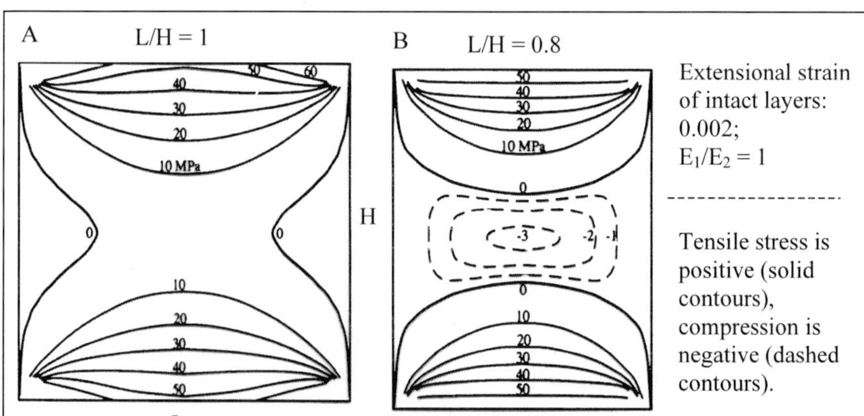

Fig. 4.20. Contour plots of the layer-parallel normal stress (σ_x) between neighbouring vertical joints at different joint spacing to layer thickness ratios (L/H) after T. Bai and D.D. Pollard (2000; J. Structural Geol. *22*, pp. 1409–1425, Fig. 3)

"notch"-type irregularity near an interface. Nevertheless, as A.H. Lachenbruch (1961) has shown in a remarkable paper on "*Depth and spacing of tension joints*" (J. Geophys. Res. *66/12*, pp. 4273–4292), a tension fracture need not be arrested when σ'_x passes from tension to compression, but a fracture may penetrate into or even transect a compression region by exactly halfway between adjacent pre-existing joints. We would rather expect infill joints to form at a shorter distance from the earlier joints, where the compressive stress is lower (see Fig. 4.20B). Hence, we would expect infill jointing at the saturation stage, or shortly before, to somewhat disturb a regular array of earlier joint sets.

Now, what are the conditions that prevent a tension fracture from transecting the central compression zone? The answer lies in an energy balance – the basic concept of elastic fracture mechanics. An actual or even virtual incremental extension of a fracture is "driven" by the release of mechanical energy of the system, and is impeded by the mechanical and/or thermodynamic energy that is consumed during the incremental extension. In our case, the release of mechanical energy involves just the strain energy that is stored in the joint-bounded segments. The energy consumed during the incremental fracture extension consists of two parts: the energy required to rupture and/or plastically deform the material in the vicinity of the advancing fracture tip, and secondly, the energy consumed by increasing the strain energy of the compressive region. In other words, the fracture can only grow if the strain energy release G per unit extension area, i.e. the so-called "strain-energy-release rate" or "crack extension force", balances or exceeds the rate G_c at which energy is consumed per unit area of fracture extension – the "crack resistance". A stationary (stable) growth of the fracture requires $G = G_c$.

It is obvious, that fractures that enter the compressive region from outside stop when the strain-energy-release rate G just equals the rate of strain-energy increase G_{comp} in the compressive region, since no energy would be left to rupture the intact material. Since the compressive strain energy of the central region, and the rate of its increase G_{comp}, increase as L/H decreases, one may readily imagine a strained state of the joint-bounded segment, where the stored tensile strain energy does not allow for a strain release rate $G > G_{comp}$. Hence, when the elastic three-layer model is in a state where $G - G_{comp} = 0$, further infill jointing is certainly impossible, and the process of infill jointing comes to a halt. A state of *joint saturation* is reached.

The state of joint saturation is characterized by a certain critical joint spacing to layer thickness ratio $(L/H)_{cr}$. It is important to note that the state $G - G_{comp} = 0$, if once attained, cannot be altered by further extensional straining of the layer system. This follows from the definition of the strain energy density (referred to an initial zero state), expressed in principal stresses and strains:

$$U_{el} = \frac{1}{2}(\sigma_I e_I + \sigma_{II} e_{II} + \sigma_{III} e_{III}) \tag{4.22}$$

As mentioned before, an increase of the external extensional strain by a factor m entails an increase of the stresses and strains in the linear-elastic material by the same factor. Therefore, the release of strain energy in the tensile region, and the increase in compressive strain energy in the compressive region per unit extension area of the fracture, will both be multiplied by the same factor (m^2). Thus, further increase of the external extensional strain does not affect the state $G - G_{comp} = 0$.

In the aforementioned numerical analysis, Bai and Pollard also analyzed the propagation behaviour of vertical cracks in the three-layer model. The authors calculated the stress intensity factor K_I (Chapt. 2, p. 15) – essentially an equivalent of the strain-energy-release rate G – for vertical tension cracks of varying length and position in the inhomogeneous stress field midway between neighbouring joints. We restrict our interest to fractures that start growing from a location near an interface, since this starting position is more favourable for the propagation of a fracture than a position further inside the fractured layer. In fact, the most favourable starting position at the interface is midway between neighbouring joints, and it is the fracture propagation from this position which is analyzed by Bai and Pollard.

As the fracture grows, the stress intensity factor changes because of the increase in fracture length and the inhomogeneity of the stress field. Naturally, K_I is positive in the high tension region near the interface where the fracture originates. In order to intersect the layer, K_I has to remain positive. If, at a certain L/H ratio K_I drops to zero somewhere on the virtual fracture path, we may safely assume that the fracture cannot cut through the entire layer since it cannot overcome the fracture toughness K_{Ic} of the material. (Note that this is completely analogous to the above condition $G - G_{comp} = 0$.) As noted before, an increase in the externally applied strain cannot change $K_I = 0$ into $K_I > 0$. The limiting joint spacing to layer thickness ratio $(L/H)_{cr}$ that is associated with the occurrence of $K_I = 0$ on the virtual fracture path midway between neighbouring joints, was found by the numerical analysis as $(L/H)_{cr} = 0.55$. If $L/H < (L/H)_{cr}$, infilling fractures can only partially cut the fractured layer. On the other hand, if $L/H > (L/H)_{cr}$, the formation of new infill joints may be promoted by increasing the applied extensional strain.

Again, we may wonder whether the position halfway between neighbouring joints would indeed be the most favourable for infill jointing. True enough, the midway position provides the best starting condition for vertical fracturing from the interface, since it provides the highest tensile stress (and K_I-value), but at the same position the central compression belt offers the greatest resistance to vertical fracturing. As said before, a glance at Fig. 4.20B would have us suspect the state of joint saturation to be achieved by infill jointing away from the midway position between already existing neighbour joints.

Finally, three more remarks may be added concerning the numerical analysis of the infill fracturing. First, the limiting spacing/thickness ratio of 0.55, was determined without taking into account the material resistance against rupturing ("fracture toughness"; see Chapt. 2, p. 17). Hence, $(L/H)_{cr} = 0.55$ is a *lower bound* for the occurrence of infill joints, but will hardly coincide with the actual minimum joint spacing. From the data in the Bai and Pollard paper (especially Fig. 8) we would rather infer a realistic minimum joint spacing $(L/H)_{min}$ of not less than 0.8, below which a central compression zone (Fig. 4.19B) will prevent further infill jointing. Secondly, like in Hobbs' model, the numerical analysis is limited by the assumed linear elastic behaviour of the layers. The numerical analysis by Bai and Pollard was carried out for $E_1/E_2 = 1$, and the applied extensional strain was 0.2%. A more realistic ratio of the Young moduli may raise the external strain, required for minimal joint spacing, close to, or possibly beyond the elastic limit of the rock. And thirdly, the joint saturation is determined for a single configuration of the layers only, and thus does not yet provide information on the influence of the thickness ratio H_2/H_1.

Models and reality. When interpreting field data on joint spacing in terms of the theoretical predictions of the "slip" model or the "welded-layer" model, one should first bear in mind that the models imply drastic, and mutually exclusive, simplifications of the stress coupling between the fractured layer and the intact surroundings, i.e. complete interfacial slip vs. perfect cohesive contact. One consequence of this is that the joint spacings predicted by the two models differ. Secondly, one would wish to know whether a set of equidistant tension joints, observed in nature, can adequately be explained by Price's slip model, or by the "welded-layer" model, – or by neither.

Let us reconsider the theoretical models with regard to aspects of their application. Assume that the inspection of a uniform flat-lying layer, embedded between unfractured layers in nature, reveals a constant spacing to thickness ratio L/H, and that the appearance of the adjacent unfractured layers suggests that fracture interference from other layers can be excluded. Furthermore, suppose that traces of the activity of overpressured pore fluids during fracturing will not escape the attention of the observer.

Then, if there is sufficient evidence of interfacial slip of the joint bounded segments, one may put the slip model to the test and tentatively apply Eq. 4.4a to relate the observed L/H ratio to the depth of the flat lying layer at the time of fracturing. The slip model may then be accepted as a reasonable mechanical explanation of the observed joint set, when after choosing reasonable values of T_o, and the λ-factor (say, between 0.4 and 0.6, to allow for moderate water overpressures during fracturing), Eq. 4.4a predicts values of the overburden thickness z (km) that correlate to the estimates based on the pre-fracturing history of the layer.

If there are no indications of interfacial slip of the joint-bounded segments and, in particular, if there is evidence that several generations of joints constitute the observed joint set, one will turn to the "welded-layer" model. When reasonable values of T_o and of the elastic moduli of the fractured interlayer and the softer embedment are inserted in Eq. 4.20a or 4.20b, the equation predicts the extension of the multilayer that supposedly produced the joint set. The model can then be validated by comparing the predicted extension with the layer extension that follows from a reconstruction of the deformation history.

As already said, both models are based on extreme idealizations of the shear coupling between the fractured layer and its surroundings. It might very well be that neither the assumption of perfect interfacial bonding, nor that of complete interfacial slippage along cohesionless interfaces, are fully realized in the natural process. Instead, to some degree, the actual shear coupling might involve both slip- and non-slip conditions. In particular, near the joint tips the maxima of the interfacial shear stress might exceed the interfacial shear strength and debond the interface material. This very local rupture of the interface may or may not grow and spread out into an interfacial shear fracture that delaminates the fractured layer from the adjacent layers. Regarding the welded-layer model, we therefore face two problems: the first, and probably more difficult one, concerns the possible nucleation of an interfacial shear crack by the stress concentration at a joint tip; the second problem concerns the possible growth of the initial rupture into a fully developed delamination fracture. We turn first to the problem of the nucleation of interfacial shear fracturing by the termination of a tension joint.

Unfortunately, maxima of the interfacial shear stresses were not recorded in the Bai and Pollard paper. Nor do we know of a satisfactory fracture mechanical analysis of the problem. But a first indication of interfacial rupturing at the point of joint termination may be obtained by comparing the interfacial shear stresses of the two "end-member" models of Price and Hobbs. Referring back to the distribution of the dimensionless tensile stress $\hat{\sigma}'_{x1}/\hat{\sigma}'^{max}_{x1}$ in Eq. 4.15 and Fig. 4.14, we insert the highest possible value for the maximum stress $\hat{\sigma}'^{max}_{x1}$ between the neighbouring joints, namely To, and find for the derivative of $\hat{\sigma}'_{x1}$ at the joint $x = 0$:

$$\left(d\hat{\sigma}'_{x1}/dx\right)_{x=0} \cong 4T_o/L \tag{4.23}$$

erting this in the force balance equation (Eq. 4.8) gives the maximum magnitude τ_{max} of Hobbs' interfacial shear stress:

$$\tau_{max\,(Hobbs)} = \frac{H}{2}\left(\frac{d\hat{\sigma}'_{x1}}{dx}\right)_{x=0} \cong 2\frac{H}{L}.T_o \tag{4.24}$$

When compared with the frictional slip stress τ_{av} in Price's force balance (Eq. 4.3 or 4.4), which is the maximum interfacial shear stress of the slip model, Hobbs' maximum interfacial shear stress at a joint position turns out to be about four times the maximum shear stress of Price's slip model:

$$\tau_{max\ (Hobbs)} \cong 4\tau_{max\ (Price)} \tag{4.25}$$

provided the jointing occurs in both the "welded" and the non-cohesive layer at about the same spacing to thickness ratio L/H.

What can be inferred from this relation? Consider two flat-lying layers with the same mechanical properties, both being loaded by the same effective overburden stress σ'_v, and jointing being initiated in both layers at about the same spacing to thickness ratio L/H. In both cases, the frictional part ($\sigma'_v \tan\varphi$) of the interfacial shearing resistance is the same, but while one layer is "welded" to its neighbours ("cohesion" $\tau_o > 0$), the other layer is in purely frictional contact ($\tau_o = 0$). To remain welded during jointing, would then require for the "welded" layer: $\tau_{max\ (Hobbs)} < \tau_o + \sigma'_v \tan\varphi$. This implies (see Eq. 4.25) the *no-rupture condition*

$$\tau_o > 3\sigma'_v \tan\varphi \tag{4.25a}$$

But note that this condition refers to the most critical locations where joints meet the interface. Let us assume again $\tan\varphi = 0.7$, and the effective overburden stress σ'_v increasing by 15 MPa per kilometer, as typical for sedimentary rocks with normal hydrostatic water pressure ($\lambda = 0.4$); condition 4.25a then reads: $\tau_0(\mathrm{MPa}) > 31.5\ z(\mathrm{km})$. Considering that the cohesion τ_o in intact sedimentary rocks is of the order of 10 to 20 MPa, and that these values are very likely higher than the cohesion at layer contacts, which may not amount to more than a few MPa, the condition 4.25a suggests that, below a depth of a few hundred meters, decohesion of the sedimentary interface near the joint ends can hardly be avoided. (Note that this may be different in crystalline rocks where τ_o may reach 50 MPa.) Thus, we consider it highly likely that in most cases of "welded" sedimentary layers jointing generates interfacial rupture nuclei.

We shall, therefore, start the analysis of interfacial shear fracturing in sedimentary rocks from the assumption that *any* point of joint termination at an interface is a potential nucleus for interfacial shear rupture. It remains then to be seen next, under what conditions, and to what extent the shear fractures can spread out into a wider detachment (delamination) zone.

The reader may have already noticed that the assumption of joint-induced nuclei of shear rupture conflicts with the model of joint saturation presented in the preceding section (Improvements and "fracture saturation"). As shown in the numerical analysis by T. Bai and D.D. Pollard (loc. cit.), infill joining is impeded by a central compression zone which forms between too closely spaced joints. In the numerical analysis and in our explanation of the phenomenon (Fig. 4.19, p. 71), the joints remain closed at the tips. If the fracture walls are allowed to slide apart at the fracture tips, the change in fracture shape will affect the stress gradients inside the joint-bounded segment, and therefore also affect the "stress barriers" against infill jointing. The possible effect of this on the minimum spacing L/H is not yet known.

Delamination. It should be possible to obtain an estimate of the range of interfacial detachment by considering the changes in energy during the propagation of a detachment front. We have defined the plane problem in Fig. 4.21. Figure 4.21A shows a vertical slice of a part of the fractured layer of unit width, with the delamination (indicated by dotted lines) caused by a first tension joint. As the delamination front F has advanced by the length δx, it has traversed the frontal volume element S = H.1.δx. Thereby, the interfacial material has been weakened by reduction of the cohesive shear strength τ_o of the interfaces to zero. For simplicity we assume that the weakening is a linear process confined to the frontal segment S, as illustrated in Fig. 4.21B. Taking into account the linear decrease $\delta\tau_c$ of the shear stress from X to X – δX, the shear force on an interfacial area δx.1 of S is reduced by $\delta\tau_c^{av}.\delta x = (\tau_o/2).\delta x$. Adding the shear force reductions on the two interfaces, the balancing force is provided by the decrease $\delta\hat{\sigma}'_X$ of the (cross-sectional) average tensile stress $\hat{\sigma}'_X$ from X – δx to X in layer (1):

$$\delta\hat{\sigma}'^{(1)}_X.H = -\tau_o\,\delta x \tag{4.26}$$

And since the shear stress is linearly decreasing from τ_o to 0 as x increases from X – δx to X, the decrease in average tensile stress is also a linear function of x:

$$d\hat{\sigma}^{(1)}_X(x).H = -2\tau_o\frac{X-x}{\delta x}dx \quad (X-\delta x \le x \le X) \tag{4.27}$$

In the case of horizontal layering, no external forces assist the delamination process; therefore, a delamination fracture (of unit edge width) is solely propagated by the elastic energy release ΔE_{el} in the three-layer unit. For reasons of symmetry we restrict the energy relation to the upper half (H/2) of the jointed layer (1) and to the adjacent weaker layer (2). Propagation of the decohesion over δx then requires that the elastic energy release exceeds, or at least equals, the energy ΔW_f dissipated by the frictional displacement of the detached interlayer segment of length X – δx, plus the energy $2\Delta\Gamma\delta x$ (Γ measured in Joule/m^2 = Pa m) which is expended as surface energy of the newly generated fracture walls and is consumed in producing a frontal damage zone ("process zone") of length δx:

$$\Delta E_{el} \ge \Delta W_f + 2\Delta\Gamma\delta x \tag{4.28}$$

The stress drop $\delta\hat{\sigma}'_X$ in the frontal element S of layer (1) in Fig. 4.21A is accompanied by an elastic shortening of S, which is accommodated by widening of the joint. This

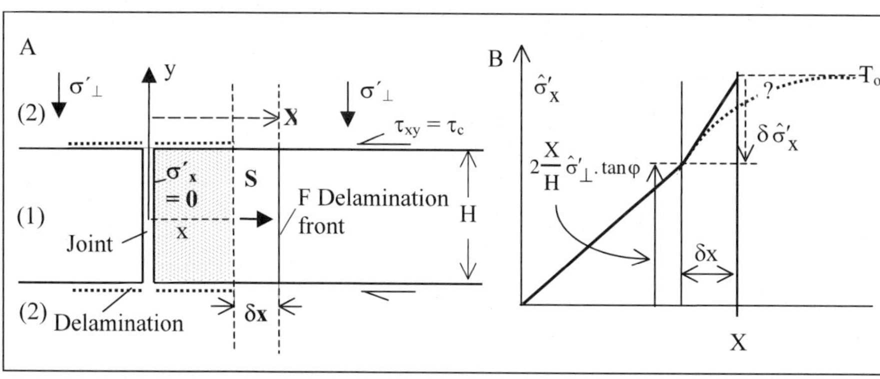

Fig. 4.21A,B. Propagation of decohesion along a jointed horizontal interlayer:
A) Vertical cross section of the decohesion zone (dotted lines) and frontal element S;
B) Hypothetical decrease of tensile stress $\hat{\sigma}'_X$ inside the frontal element S

allows the detached part ($0 \le x \le X{-}\delta x$) of the layer (shaded in the figure) to slip. The elastic plane-strain shortening of S in layer (1) is given by

$$\frac{\delta x - \delta x^*}{\delta x} = \left(1 - \nu_{(1)}^2\right).E_{(1)}^{-1}.\frac{\delta\,\hat{\sigma}'^{(1)}_x\,|_{x=X-\delta x}}{2} = -\left(1 - \nu_{(1)}^2\right).E_{(1)}^{-1}.\frac{\tau_o}{2H}\delta x \qquad (4.29)$$

with δx^* the length of the shortened frontal element. The relation follows from the stress/strain relations (Eq. 3.3) where again the generalised effective stresses σ^* are replaced by Terzaghi's effective stresses σ', and from Eq. 4.27. (Note that the shortening and the change in tensile stress over the interval δx have the same sign.)

According to Eq. 4.29 the (positive) displacement $\delta u^{(1)} = \delta x - \delta x^*$ of the point $X - \delta x$ towards the decohesion front at rest is

$$\delta u^{(1)} = \left(1 - \nu_{(1)}^2\right).E_{(1)}^{-1}.\frac{\tau_o}{2H}\delta x^2 \qquad (4.30)$$

As the debonded segment $X - \delta x$ of layer (1) slips over the distance $\delta u^{(1)}$, it consumes at each interface the energy

$$\Delta W_f = \delta u^{(1)}.(X - \delta x).\hat{\sigma}'_\perp.\tan\varphi = \frac{1 - \nu_{(1)}^2}{2E_{(1)}}.\frac{\tau_o}{H}.(X - \delta x).\hat{\sigma}'_\perp.\tan\varphi.\delta x^2 \qquad (4.31)$$

per advancement δx of the detachment front. Note that the normal stress $\hat{\sigma}'_\perp$ is the average of the normal stress $\sigma'_\perp$ on the debonded part of the interface.

Next we have to evaluate the release of elastic strain energy ΔE_{el} in the three-layer system per advancement δx of the delamination front. Since the change in the density δE_{el} of the elastic strain energy is given by

$$\delta E_{el} = \frac{1}{2}\delta e_x.\delta\sigma'_x + \delta\tau_{xy}.\delta e_{xy} = \frac{1 - \nu^2}{2E}(\delta\sigma'_x)^2 + \frac{(\delta\tau_{xy})^2}{2G} \qquad (4.32)$$

a rigorous determination of the total energy release ΔE_{el} per advancement δx of the delamination front, while the outer boundaries remain fixed, would require detailed knowledge of the field of stress perturbations caused by the interfacial unloading. We circumvent this formidable task by a somewhat cruder estimation of ΔE_{el} based on some simplifying assumptions.

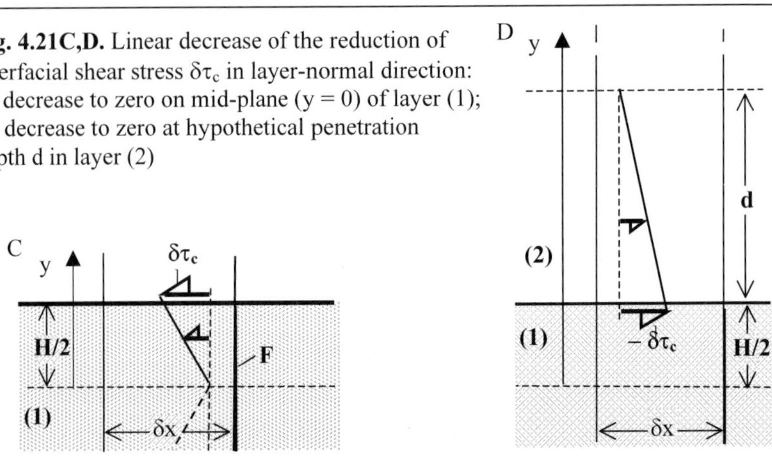

Fig. 4.21C,D. Linear decrease of the reduction of interfacial shear stress $\delta\tau_c$ in layer-normal direction: C) decrease to zero on mid-plane ($y = 0$) of layer (1); D) decrease to zero at hypothetical penetration depth d in layer (2)

First, we consider the whole energy release as concentrated in the frontal strip with the basal area $\delta x.1$, which extends through the upper half layer (1) and the adjacent layer (2). Secondly, we assume, as in Hobbs' model, that the magnitude $|\delta\tau(X.y)|$ of the reduction in layer-parallel shear stress τ_{xy} decreases linearly with the distance from the interface; this is illustrated in Fig. 4.21C,D.

Since the layer-parallel shear stress must vanish at the mid-plane ($y = 0$) of layer (1), we have in the interval $0 \le y \le H/2$ of layer (1):

$$\left|\delta\tau_{xy}^{(1)}(x,y)\right| = \tau_o \frac{X-x}{\delta x} \cdot \frac{y}{H/2} \tag{4.33a}$$

and in the interval $H/2 \le y \le H/2 + d$ of layer (2):

$$\left|\delta\tau_{xy}^{(2)}(x,y)\right| = \tau_o \cdot \frac{X-x}{\delta x} \cdot \left(1 - \frac{y - H/2}{d}\right) \tag{4.33b}$$

where "d" again denotes a hypothetical depth to which the unloading perturbation effectively penetrates.

As a direct consequence of the assumed linear variation of $\delta\tau_{xy}$ along the y-coordinate, the associated change in tensile stress $\delta\sigma'_x$ is independent of y. (This is immediately seen by differentiating the static equilibrium condition $\partial\sigma_x/\partial x + \partial\tau_{xy}/\partial y = 0$ with respect to y.) Hence, we have from Eq. 4.27 for the stress change in layer (1) over the interval δx:

$$\delta\sigma_x'^{(1)}(x,y) = \delta\hat{\sigma}_x'^{(1)}(x) = -\tau_o \frac{X-x}{\delta x} \cdot \frac{\delta x}{H} \tag{4.34a}$$

and in analogy for layer (2):

$$\delta\sigma_x'^{(2)}(x,y) = \delta\hat{\sigma}_x'^{(2)}(x) = -\tau_o \cdot \frac{X-x}{\delta x} \cdot \frac{\delta x}{d} = \delta\sigma_x'^{(1)} \cdot \frac{H}{d} \tag{4.34b}$$

With the stress changes (Eqs. 4.33a,b and 4.34a,b) inserted in the energy expression (Eq. 4.32), and integrating the energy density $\delta E_{el}^{(1)}$ over the frontal half-strip S/2 in layer (1), i.e. from $x = X - \delta x$ to X and from $y = 0$ to $y = H/2$, one obtains

$$\begin{aligned}\Delta E_{el}^{(1)} &= \frac{1-\nu_{(1)}^2}{2E_{(1)}} \tau_o^2 \int_{X-\delta x}^{X} \int_0^{H/2} \left(\frac{X-x}{\delta x}\right)^2 \cdot \left[\left(\frac{\delta x}{H}\right)^2 + \frac{2}{1-\nu_{(1)}} \cdot \left(\frac{y}{H/2}\right)^2\right] dx\, dy \\ &= \frac{1-\nu_{(1)}^2}{2E_{(1)}} \tau_o^2 \cdot \frac{\delta x^2}{3} \left[\frac{\delta x}{2H} + \frac{2}{3(1-\nu_{(1)})} \cdot \frac{H/2}{\delta x}\right]\end{aligned} \tag{4.35a}$$

where G in (4.32) was expressed in $E_{(1)}$ and $\nu_{(1)}$. Likewise, integrating the energy density $\delta E_{el}^{(2)}$ in layer (2) from $x = X - \delta x$ to X and from $y = H/2$ to $y = d$ gives

$$\Delta E_{el}^{(2)} = \frac{1-\nu_{(2)}^2}{2E_{(2)}} \tau_o^2 \cdot \frac{\delta x^2}{3} \left[\frac{\delta x}{d} + \frac{2}{3(1-\nu_{(2)})} \cdot \frac{d}{\delta x}\right] \tag{4.35b}$$

Inserting $\Delta E_{el} = \Delta E_{el}^{(1)} + \Delta E_{el}^{(2)}$ from Eq. 4.35a,b, and ΔW_f from Eq. 4.31 in the basic energy relation (Eq. 4.28) for the upper half of the three-layer unit, we arrive at the inequality

$$a\,\tau_o^2 - \xi\tau_o - 12\frac{E_{(1)}}{1-\nu_{(1)}^2}.\frac{\Delta\Gamma}{\delta x} \geq 0 \tag{4.36}$$

with

$$a=\left(\frac{\delta x}{2H}+\frac{2}{3(1-\nu_{(1)})}.\frac{H/2}{\delta x}\right)+\frac{E_{(1)}}{E_{(2)}}.\frac{1-\nu_{(2)}^2}{1-\nu_{(1)}^2}\left(\frac{\delta x}{d}+\frac{2}{3(1-\nu_{(2)})}\frac{d}{\delta x}\right) \tag{4.37a}$$

and

$$\xi = 3\frac{X-\delta x}{H}.\hat{\sigma}'_{\perp}\tan\varphi \approx 3\frac{X-\delta x}{H}.\sigma'_{V}\tan\varphi \tag{4.37b}$$

In the last equation, the average normal stress $\hat{\sigma}'_{\perp}$ on the debonded horizontal interface is replaced by the effective overburden stress σ'_V. This slightly underestimates the length of the decohesion zone, since σ'_V is somewhat greater than $\hat{\sigma}'_{\perp}$, as is shown in Appendix B to this chapter. Also, note that all terms in the inequality 4.36 have the same physical dimension of a stress squared.

In the following we use the simplified form of the expression 4.37a:

$$a = \frac{2E_{(1)}}{3(1-\nu_{(1)}^2).\delta x}.\left[\frac{(1+\nu_{(1)})}{E_{(1)}}.\frac{H}{2}+\frac{(1+\nu_{(2)})}{E_{(2)}}.d\right] \tag{4.37c}$$

by making the reasonable assumption that the length δx of the damage zone ("process zone") ahead of the interfacial fracture is much smaller than the thickness of the jointed interlayer and the depth, d, of the shear penetration into the adjacent layer ($\delta x \ll H, d$).

Since the delamination condition (Eq. 4.36) is a quadratic inequality in τ_o, it implies

$$a\,\tau_o \geq \xi/2 \overset{+}{\underset{(-)}{}} \sqrt{(\xi/2)^2 + \frac{12E_{(1)}}{1-\nu_{(1)}^2}.\frac{\Delta\Gamma}{\delta x}.a} \tag{4.38}$$

Hence, to start the decohesion process at $\xi = 0$ ($X = \delta x$) requires

$$\frac{12E_{(1)}(1-\nu_{(1)}^2)^{-1}.\Delta\Gamma/\delta x}{\tau_o^2} \leq a \tag{4.39}$$

If this condition is satisfied, the maximum length of the decohesion zone is obtained by choosing the equality sign in Eq. 4.38 and solving for ξ:

$$\xi_{max} = \tau_o.\left[a - \frac{12E_{(1)}(1-\nu_{(1)}^2)^{-1}\Delta\Gamma/\delta x}{\tau_o^2}\right] \tag{4.40}$$

With Eq. 4.37b the dimensionless maximum length of the interfacial slip zone $s_{max}/H = (X_{max} - \delta x)/H$ finally becomes:

$$\boxed{\frac{s_{max}}{H} = \frac{\tau_o}{3\sigma'_v\tan\varphi}\left[a - \frac{12E_{(1)}(1-\nu_{(1)}^2)^{-1}.\Delta\Gamma/\delta x}{\tau_o^2}\right]} \tag{4.41}$$

This equation contains the energy term $\Delta\Gamma$, which in fracture mechanics is also referred to as "fracture resistance". It is associated with the highly complex process of the growth of a shear fracture in a cohesive granular rock. Therefore the fracture path is tortuous, and interrupting intergranular contacts and/or cutting through weaker grains and loosening material ahead of the crack tip. Hence, it would seem unfeasible to attempt an estimation of the magnitude of $\Delta\Gamma$ on micromechanical grounds. Experimental data of $\Delta\Gamma$ for sedimentary rocks, and particularly for interfacial fracturing, are scarce and should only be applied with caution. Although the reported $\Delta\Gamma$ data were obtained under mode I fracturing conditions, we assume that $\Delta\Gamma$ is of the same order of magnitude if the fractures are of mode II. We then infer from the data collected by B.K. Atkinson and Ph.G. Meredith (B.K. Atkinson (ed., 1987) *Fracture Mechanics of Rocks*, Academic Press, Table 11.2) that $2\,\Delta\Gamma$ for sandstones and limestones rarely exceeds 10^2 J.m^{-2} (= 10^2 Pa.m = 10^{-4} MPa.m) in order of magnitude. Considering the reported data, $2\Delta\Gamma = 2 \times 10^2$ J.m^{-2} = 2×10^{-4} MPa would seem a reasonable assumption when quantifying the conclusions which we can draw from Eq. 4.41 in combination with Eq. 4.37c.

Obviously, the sign of the bracketed expression in Eq. 4.41 determines whether decohesion takes place (positive sign) or not (negative sign). For decohesion to occur, the (positive) "a"-term must exceed the second (positive) term in the bracketed difference. Since both bracket terms contain the length δx of the frontal "process zone" in the denominator, δx does not affect the sign of the bracket expression, and the decohesion condition may be written as

$$\frac{H_1}{2}.\frac{1+\nu_{(1)}}{E_{(1)}} + d.\frac{1+\nu_{(2)}}{E_{(2)}} \geq 9\,\frac{2\,\Delta\Gamma}{\tau_o^2} \tag{4.42}$$

While the left side of the inequality contains the elastic moduli and geometrical parameters of the system, the right side exclusively contains the parameters $2\Delta\Gamma$ and τ_o which both characterize the shear fracture resistance of the intact interface, although they stem from quite different theoretical models. Hence, given an alternating sequence of two different sedimentary rocks with well-defined mechanical properties, it will only depend on the thicknesses of adjacent beds as to whether jointing will initiate interfacial delamination or not. In itself this result is almost obvious, since the thicker a bed, the more energy will be released by the jointing to "drive" a delamination process. But one would wish to have an idea of the range of bed thicknesses which in nature would either promote or impede interfacial decohesion.

For this purpose, we compare the orders of magnitude of the two sides of the inequality by inserting reasonable values for the parameters; first for the case of $H_2 > H_1$ ($\equiv$ H), where we may follow Hobbs in assuming $d = H_1$(m). (Recall that in Bai and Pollards numerical analysis $d = 1.5H_1$.) With $E_{(1)} = 2 \times 10^4$ MPa, $E_{(2)} = 0.15\ E_{(1)}$, $\nu_{(1)} = 0.2$, $\nu_{(2)} = 0.4$, and $\tau_o = 3$ MPa, the decohesion condition (Eq. 4.42) then reads

$$\frac{H_1}{2}.10^{-3} > 2\Delta\Gamma\,(\text{MPa.m}) \tag{4.42a}$$

Inserting for $2\,\Delta\Gamma$ the aforementioned value of 2×10^{-4} MPa.m then leads to the conclusion that interfacial decohesion is initiated if $H_1 > 40$ cm.

Next, consider the case of very thin weak interlayers ($H_2 \ll H_1$) that was discussed previously (pp. 66–67). In this type of multilayer the depth "d" of shear penetration approaches half the thickness of the weak layers, $d = H_2/2 \ll H_1/2$, and the decohesion condition (Eq. 4.42) approximates to

$$\frac{H_1}{2}.\frac{1+\nu_{(1)}}{E_{(1)}} \geq 9\,\frac{2\,\Delta\Gamma}{\tau_o^2} \tag{4.42b}$$

Inserting the same values for the material parameters as before, we find that the fractured layers would need a thickness $H_1 > 8$ m to provide the energy release needed for initiating interfacial decohesion.

Naturally, these threshold values of H_1 strongly depend on the mechanical parameters, in particular on the value of $2\Delta\Gamma$. For example, one might easily envisage a sedimentary sequence with a ten times lower interfacial fracture resistance $2\Delta\Gamma$; if all the other mechanical factors are kept equal, delamination would then only require one tenth the thickness (H_1) of the jointed layers.

Further, it should be noted that the Coulomb friction term $\sigma'_v \tan\varphi$ in Eq. 4.41, and thus the effective overburden, has no explicit effect on the *onset* of delamination, but it will affect the delamination *range.* This brings us back to the original question of how far the interfacial friction will allow the delamination to procede once it has been triggered near the upper tip of a joint in layer (1).

To obtain an idea of the magnitude of the decohesion range s_{max}/H_1 we again choose the above material parameters and, in addition, the typical frictional parameter $\tan\varphi = 0.7$ and the effective overburden stress $\sigma'_V = 15$ MPa.z(km). If we again assume $d = H_1$, the expression 4.41 in combination with Eq. 4.37c yields

$$\frac{s_{max}}{H_1} = \frac{12.7}{z(km).\delta x(m)}\left[\frac{H_1}{2} - \frac{2\Delta\Gamma}{10^{-3}}\right] \qquad (4.41a)$$

Assuming $\delta x = 0.01$ m and $2\Delta\Gamma = 2 \times 10^{-4}$ MPa.m, the dimensionless delamination distance becomes $s_{max}/H_1 = 2.6/z(km)$, if the critical H_1 value of 0.4 m is raised, say, to 0.44. Note that the length of the delamination decreases in inverse proportion to the depth of burial z(km), reflecting the increase in frictional resistance with increasing effective overburden.

We consider the above estimates of the delamination length as rather conservative, since the elastic energy release is very likely underestimated by the calculation, the value of the surface energy $2\Delta\Gamma$ of fractures along an interface is probably smaller than the value reported for interior fractures, and inertial forces generated by the elastic surplus energy may drive the delamination further than predicted by the quasi-static model.

The reader may readily convince himself, by varying the values of δx, $2\Delta\Gamma$ and τ_o within reasonable limits in Eq. 4.41, that the decohesion will not remain confined to a narrow region surrounding the initiation point, but the decohesion length will at least exceed the thickness of the jointed layer. This is an important result, since the joint spacings predicted by the slip- and "welded" layer models rarely exceed the bed thickness. Therefore, in general, if interfacial decohesion is triggered by a joint near its ends, the decohesion will reduce the maximum of the reactive interfacial shear stress to the Coulomb friction, the most basic feature of the slip model.

In the above example related to condition 4.42a, the thickness (H_1) of the beds that remained welded to the weaker interlayers during jointing was less than 40 cm. Again, this threshold value depends on the choice of the parameter values; in particular, on the value of the interfacial cohesion τ_o. If, for example, instead of $\tau_o = 3$ MPa, as chosen above, τ_o is reduced to 1.5 MPa, the lower bound of H_1 for interfacial decohesion is raised to 1.6 m; and it is lowered to 0.1 m, if $\tau_o = 6$ MPa. It may appear as somewhat odd that a higher shear strength of the bonded interface should allow easier decohesion. The reason for this seeming paradox lies in the higher energy release rate (Eq. 35a,b).

One further point to be mentioned concerns the role of the effective overburden stress σ'_V. As noted previously, this stress does not explicitly affect the onset of decohesion, since it does not explicitly occur in the decohesion formula (Eq. 4.42). However, it seems very likely that the fracture surface energy $2\Delta\Gamma$ of a macroscopic shear fracture is an *implicit* function of σ'_V, because work in the process zone of the fracture is also per-formed in intergranular sliding and dilation of the material; both processes being affected by the effective compressive stress normal to the macroscopic fracture plane. The lower this stress, the less work will be needed to perform the intergranular sliding and dilation required in propagating the macroscopic shear fracture. Hence, although no experimental data are available, it is reasonable to expect $2\Delta\Gamma$ to decrease (non-linearly) with decreasing σ'_V. This implies that a jointed layer whose thickness does not satisfy the decohesion condition (Eq. 4.42) at a depth of a few kilometers, and thus is perfectly "welded" to the adjacent beds, might become debonded when uplifted to a shallow depth under the concomitant removal of overburden. (The same effect may be achieved by heavy overpressuring of the pore fluid.)

The delamination of the uplifted layer would thus prevent extension of the layer to produce the tensile stresses needed for the formation of infill joints. Instead, the extension of the layer would be accommodated by the opening of older joints that were formed at greater depth. Thus, in the uplifted state "the strata would be saturated with joints", to quote W. Narr and J. Suppe (1991, J. Structural Geol. *13/9*, pp. 1037–1048).

The main conclusions of the lengthy analysis of the delamination process may be summarized as follows:

(1) The maximum interfacial shear stress induced in sedimentary rocks by jointing, is about four times higher in the "welded"-layer model than in the slip model. This suggests that the joint terminations at cohesive interfaces are potential nuclei for interfacial shear rupture (with the possible exception of layers under very low effective overburden).
(2) The shear ruptures grow into delamination zones which extend beyond the distance between neighbouring joints if the jointed layers are of sufficient thickness to provide the release of elastic energy needed for "driving" the decohesion process. As far as may be inferred from the "guesstimated" range of interfacial strength data, the minimum layer thickness needed for delamination lies between a few decimeters and a few meters at most. This implies that, in sedimentary sequences of alternatingly stiff and weak rocks (e.g. limestone beds interspersed with clayey layers), with the stiff rock beds differing in thickness, jointing in the thicker beds may cause delamination, while it may leave the contacts of the thinner beds firmly bonded.
(3) It would therefore appear that application of the "welded" layer model is restricted to the thinner beds, while the slip model would more aptly describe the joint spacing of thicker beds, even if they were strongly bonded prior to jointing.
(4) High effective overburdens reduce the delamination range.

Inclined layers. So far we have examined the models of joint spacing with respect to flat-lying layers. However, in nature, layers are commonly not completely horizontal. If the dip angle (δ) is not more than a few degrees it is of little concern, but greater dip angles may affect the formation of joint sets. So far, this point has received little attention in the literature, and we are therefore limited to a cursory discussion of a few aspects of jointing in inclined layers. Consider in Fig. 4.22A a sequence of parallel, uniform sedimentary layers on a long ideal slope. The slope surface and the layers dip at a constant angle, δ. If the density of the rock does not change with depth, the effective normal stress $\sigma'_\perp$ and the shear stress $\tau_{||}$ exerted on the bedding planes by the overburden weight are simply determined by the condition of static equilibrium (see e.g. G. Mandl, FBR, pp. 19–20, 235–239).

The simple results are stated in the insert of Fig. 4.22 for the case of a long submarine slope. (Note that the coordinate axis z* in the figure is orthogonal to the slope, and thus deviates from the vertical axis z by the dip angle δ.) The state of stress, just before the

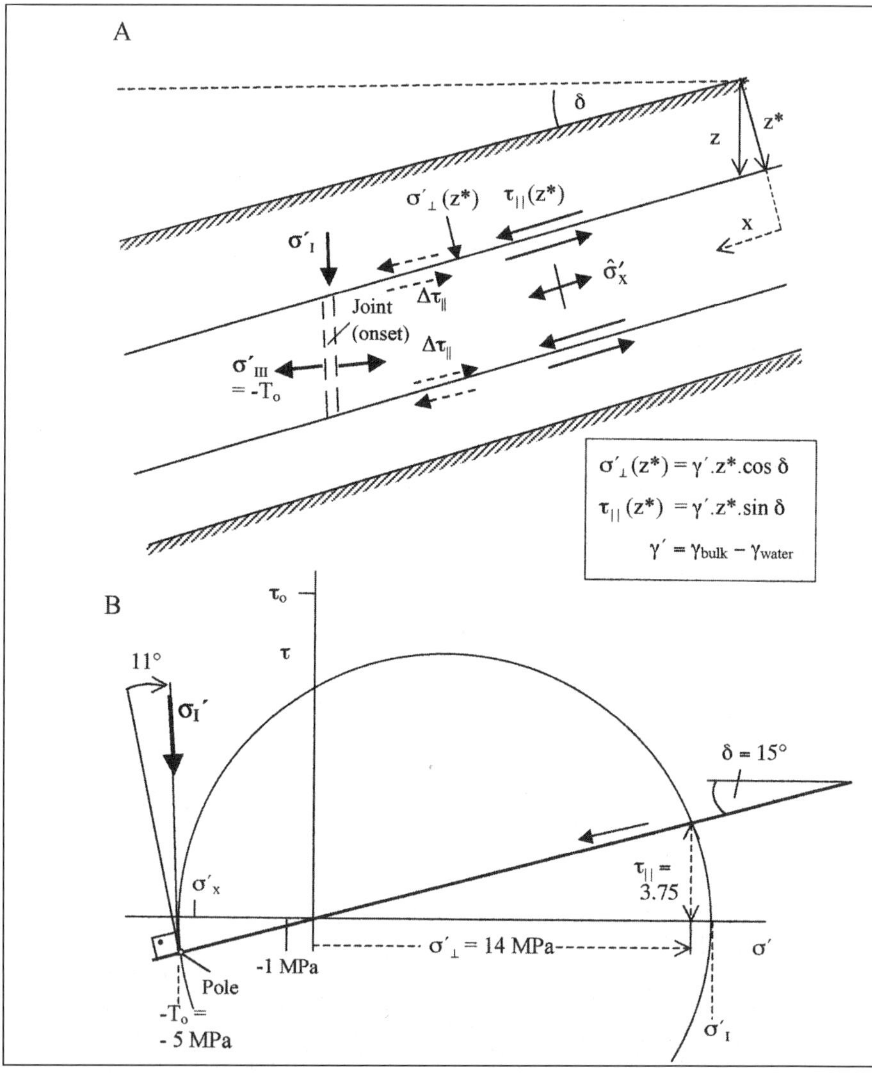

Fig. 4.22. Jointing on a submarine slope:
A) The weight-induced effective normal stress $\sigma'_\perp$ and the shear stress $\tau_{||}$ on the bedding are determined by the formulae given in the insert, with γ' the "submerged" weight of the solid component of the porous rock. The common direction of maximum compressive stress σ'_I and tension joints is inclined with respect to the normal to the bedding planes. The joint-induced shear stress couples $\Delta\tau_{||}$ are indicated by broken arrows.
B) Mohr diagram of the stress state at the instant of jointing at the depth z = 1 km ($\sigma'_v = \sigma'_z$ = 15 MPa.z(km), T_o = 5 MPa, and slope angle δ = 15°); the σ'_I direction is graphically determined by the pole method (see the Appendix to this book)

formation of a first joint, is represented by the Mohr circle in Fig. 4.22B. For illustration purposes, it is assumed that the layer, dipping at the relatively high angle $\delta = 15°$, lies at the depth z = 1 km (z* = 0.97 km), and has a tensile strength of 5 MPa. The Mohr circle is then fully determined by the point $-T_o$, 0 and the stress point $\sigma'_{\perp} = 14$ MPa, $\tau_{||} = 3.75$ MPa on the straight line through (0, 0) which is inclined at $\delta = 15°$ to the horizontal. This slope-parallel line also defines the "pole" of the stress circle (see the Appendix at the end of this book) whose straight connection with the point $-T_o$, 0 has the direction of the maximum compressive stress σ'_I with respect to the sloping layer.

Now note that the angle of the direction of σ'_I with the interfaces is less than 90° (79° in the case considered in the figure). Accordingly, the joints are no longer orthogonal to the layering, a fact already illustrated in Fig. 4.3. Secondly, consider the sense of the shear couples $\Delta\tau_{||}$ (broken arrows in figure A) which are induced upslope by a first joint. Whereas on the upper interface, the joint-induced shear couple $\Delta\tau_{||}$ and the gravitational shear couple $\tau_{||}$ (solid arrows) have the same sense of action, they oppose each other on the lower interface. Therefore, the upslope propagating delamination will now be asymmetric, being facilitated on the upper interface and hampered on the lower one. Conversely, downslope delamination slip will be facilitated on the lower interface.

The asymmetry in delamination may also affect the energy balance (Eq. 4.28) which, so far, has not had to account for work performed by the surroundings of the jointed interlayer during interfacial slip. However, on the slope the gravitational shear stresses perform such work. Although the work expended per unit slip area on one interface is cancelled out by the work consumed on the other interface, the incremental upslope growth of the slip area on the upper interface exceeds that on the lower interface, thus giving rise to the expenditure of net work. This work should be added to the "driving" elastic energy release $\Delta E_{el}.H$ in the balance (Eq. 4.28), and will thus contribute to upslope slip on the upper interface (or to downslope slip on the lower interface).

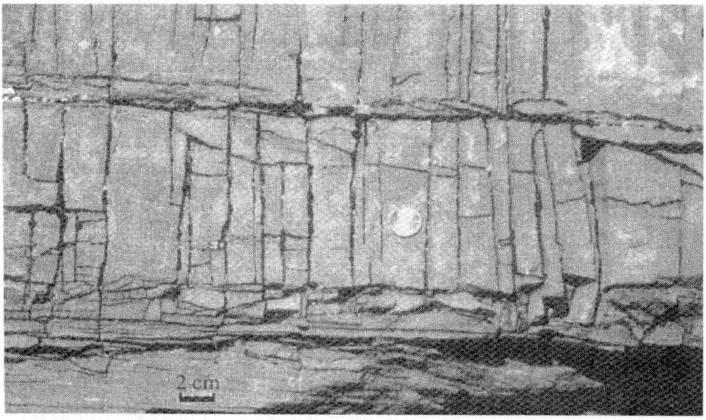

Fig. 4.23. Closely spaced joints in limestone layers (Chimney Rock, Utah) (from T. Bay and D.D. Pollard (2000) J. Structural Geology *22*, pp. 1409–1425)

Irregular spacing and closely spaced joints. Field observations often reveal strong deviations from a regular spacing and, in particular, from the proportionality between spacing and layer thickness even in layers of mechanically identical or very similar rocks under similar overburdens. As to the reasons for the observed deviations from a regular spacing of tension joints, we first note that, in general, a layer contains some (more or less) randomly distributed weaker spots, such as flaws, layer necks, or notch-type irregularities of the interfaces (see Fig. 2.7), which may act as nuclei for the formation of the first tension joints. These precursor joints are therefore likely to be irregularly spaced. But, where neighbouring precursor joints are sufficiently far apart, sets of equidistant infill-joints can then be generated in between, when the tensile stress is sufficiently raised.

Secondly, we have to bear in mind that thin incompetent interlayers may allow tensile fractures of one layer to intrude into, or intersect, an adjacent stiff layer, as was illustrated in Fig. 4.4. Conversely, if fractures are arrested at a layer interface, as in Figs. 4.5 and 4.6, they may induce sufficient additional tensile stresses at the opposite side of the interface to trigger new fractures at that location, as was already considered in the section on "*Thin weak interlayers*" (pp. 66–68). These mechanical interactions between fracturing layers may cause an irregular spacing of the joints or, if leaving the joints regularly spaced, may violate the proportionality rule L/H = const. for mechanically identical layers. This appears to be the case in Fig. 4.4, where the distance between the systematic joints of the thinner bed is much larger than the thickness of the bed. More often, however, it is found that the distance between regularly spaced joints is only a few tenths of the layer thickness. A good example of such closely spaced joints is shown in Fig. 4.23.

But even in cases where one may be reasonably sure that unfractured interlayers have prevented fracture interaction between neighbouring beds, and therefore a regular joint spacing should scale with the layer thickness and thus not be much smaller than the layer thickness, spacings of only a few tenths of the layer thickness, or even less, are not uncommon (see e.g. Fig. 2.7). This puzzling phenomenon is not yet fully understood. Nevertheless, we shall venture into some possible causes of the "abnormally" close joint spacing in single layers.

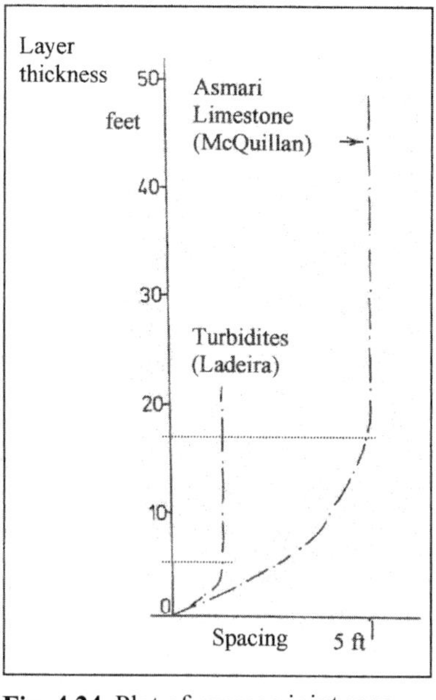

Fig. 4.24. Plot of average joint separation vs. bed thickness for Asmari limestone (McQuillan) and turbidites (Ladeira) (after an internal Shell Research report by N.J. Price (1978))

H. McQuillan (1973) studied the development of closely spaced fractures in thick units of Asmari limestones (SW Iran), and F.L. Ladeira and N.J. Price (1981) reported measurements of fracture separation in thick beds (H > 1.5 m) of Carboniferous turbidites and Jurassic limestones. Based on these data, Price and Ladeira plotted the relationship between joint separation and bed thickness, the essence of which is shown in Fig. 4.24. The dotted horizontal lines in the figure mark the layer thickness at which the relationship between spacing and layer thickness exhibits a drastic change. Up to these lines, the thickness vs. spacing curves may be approximated by straight lines, indicating proportionality between joint spacing and the thickness of fractured layers of the same lithology, whereas above the dotted lines, the spacing is practically independent of the layer thickness. Noting the difference in length scales on the two axes, one notices that L/H is considerably smaller than 1, and varies from less than 0.1 to a few tenths.

Ladeira and Price offer a possible mechanical explanation of why the joint spacing (L) observed in thick layers may no longer depend on the thickness of the layers. The authors argue that, as a tensile fracture develops in a permeable bed the opening of the fracture is temporarily accompanied by a reduction of the fluid pressure inside the fracture which causes a "draw-down" of the pore pressure in an adjacent region, as sketched in Fig. 4.25. The fluid pressure gradient dp/dx is independent of the thickness of the layer, but depends on the permeability, and is therefore also related to the lithology of the rock. Since inside the draw-down region the tensile effective stress σ'_x cannot reach the tensile strength, the fracture nearest to the first one will not develop within a given distance L which is independent of the layer thickness, but depending on the lithology. As noted previously, the draw-down of the pore pressure is a temporary phenomenon and diffusion tends to restore the original fluid pressure. Hence, the proposed spacing mechanism only applies to sets of contemporaneous joints.

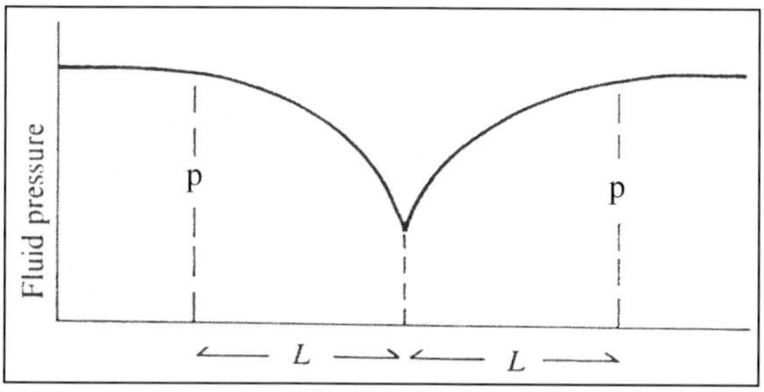

Fig. 4.25. Temporary draw-down of pore pressure by the opening of a tension joint in a permeable rock (from N.J. Price and J.W. Cosgrove (1990) loc.cit., p. 59)

Cleavage (extension) joints. The small L/H ratios reported by Ladeira and Price, and even smaller regular spacings of parallel joints, are not uncommon in sedimentary rocks. Although the hydraulic pressure-draw-down mechanism may be responsible for the small L/H ratios in some cases, in others, one has to search for a different mechanism. In fact, one may even doubt that the densely spaced joints were generated by effective tension, and may instead speculate that the joints were generated by uniaxial compression parallel to the joint planes. Thus, the very closely spaced joints would not be proper tension fractures, but cleavage (extension) fractures. We recall from Chapt. 2 (pp. 18–25), that the growth of cleavage fractures is "driven" by a high effective compressive stress σ'_I parallel to the direction of fracture growth, while the smallest effective stress is zero, or close to zero, possibly even very slightly compressive. Although the magnitude of the driving compressive stress σ'_I is not well known, it is most likely smaller than the uniaxial compressive strength C_o of the rock, but by less than an order of magnitude.

As was argued at the end of Chapt. 2, cleavage fracturing should, in principle, be possible in all tectonic regimes that are prone to compressional tectonic faulting ($\sigma'_I > \sigma'_{II} > \sigma'_{III} \geq 0$, compressive stresses positive). This would seem to concur with the view held by some prominent researchers that the overwhelming majority of macroscopic rock joints are of the cleavage type (e.g., J. Gramberg (1989) *A Non-conventional View on Rock Mechanics and Fracture Mechanics*, Balkema, Rotterdam; P. Bankwitz, D. Bahat and E. Bankwitz (2000) Z. geol. Wiss. *28*, pp. 87–110). However, it should be noted that this opinion seems to be mainly based on numerous observations on "hard rocks", i.e. igneous and metamorphic rocks, where foliations of various types constitute strong mechanical anisotropies with parallel planes of weakness along which the rock may be cleaved. This is different in layered sedimentary rocks with layer-parallel bedding; although these rocks also have some intrinsic mechanical anisotropy, this is commonly of the transversely isotropic type, with the axis of symmetry perpendicular to the bedding plane and mechanical isotropy in all directions parallel to the bedding plane. Since the resistance of sedimentary rocks against splitting is usually lower along the bedding plane, and greatest orthogonally to the bedding plane, the *anisotropy* of layered sedimentary rocks will not favour the growth of vertical fractures in flat-lying beds.

The situation may be somewhat different in flat-lying strata which were built up by the deposition of material on planes that are steeply inclined towards the stratal boundary planes. A typical case is the deposition of "foreset" beds on a prograding delta slope, where the inclined bedding planes are known to affect the dip angle of normal faults. However, it is also commonly held that this strength anisotropy decreases with increasing effective overburden stress and degree of lithification.

Recapitulating what has been stated at the end of Chapt. 2 and in Chapt. 3 on the geological environments of cleavage jointing, vertical cleavage jointing in flat-laying layers is readily envisaged in a compressional regime conducive to strike-slip faulting. In such a regime, the overburden stress (σ_V) is the intermediate principal stress, and the reduction of a compressive, layer-parallel σ'_{III} to approximately zero may be accomplished by lateral extension and/or by overpressuring of the pore fluid. If the reduction of σ'_{III} is achieved by overpressuring alone, without externally applied lateral extension the required overpressures are moderate if the fractured layers can freely accommodate the pressure-induced lateral extension, but "hard" overpressures are needed (p. 29) in layers that are laterally completely confined.

A tectonic stress regime of this kind exists in the forelands of thrust and fold belts, especially if these are slightly convex towards the foreland (N.J. Price and J.W. Cosgrove (1990) *Analysis of Geological Structures*, Cambridge Univ. Press, pp. 312–314). The situation is sketched in Fig. 4.26. Note that, in the present context we ignore the fact that σ'_I decreases in magnitude with distance from the thrust front, and that the σ'_I trajectories are bent slightly downwards from the horizontal because of the frictional resistance at the base of the compressed foreland. Likewise, we ignore that the foreland may be bent down near the thrust front by the load of the thrust belt.

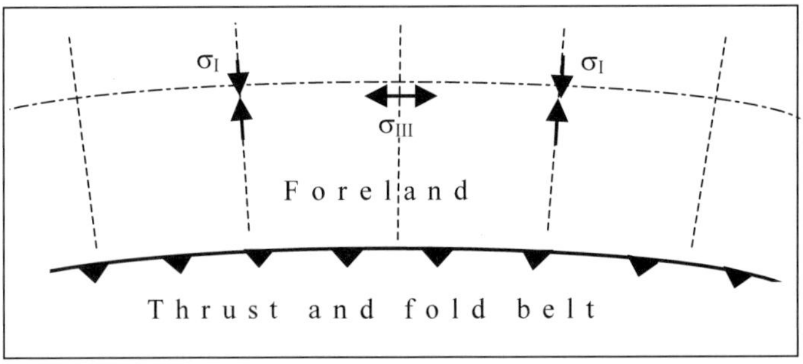

Fig. 4.26. Regional principal stress trajectories in the foreland of an arcuate fold and thrust belt (greatly simplified)

The overpressuring of the pore fluid, caused by the compression of the foreland, may reduce the smallest effective stress σ'_{III} close to zero, or even shift σ'_{III} into the tensile range, and thus generate vertical "internal hydraulic" joints of the cleavage or the tension type. The process is illustrated in the Mohr diagram of Fig. 4.27. for a foreland layer buried under 2 km of overburden. It is assumed that, prior to the compression by the thrust belt, the flat lying sediments were undisturbed and the pore pressure was "normal", i.e. solely due to the weight of the hydrostatic water column. The horizontal effective stresses $\sigma'_h{}^o$ are related to the effective overburden stress by the empirical K_o factor (Eq. 3.7), which we have repeatedly used, and to which we here assign the relatively small value 0.4, which we consider as appropriate for very stiff, undisturbed sedimentary rocks. We further assume, that the rock freely adjusts by horizontal straining to a rise in pore pressure, and thus allows the principal effective stresses to be reduced by the full rise in pore pressure (see Fig. 3.2). (Since we are concerned with the maximal reduction in effective stresses by overpressuring, we disregard a possible component of tectonic extension, which might oppose the build-up of overpressures.)

One then easily finds that the ratio of pore pressure to total vertical stress $\lambda = p/\sigma_v$, must be greater than 0.64 to render a horizontal compressive effective stress tensile. In the figure, $\lambda = 0.65$ allows the horizontal effective stress orthogonal to the thrust direction to reach the tensile strength of –8 MPa. The shaded stress circle in Fig. 4.27 represents the associated critical stress state. On the other hand, the foreland compression not only raises the pore pressure, but also the maximum effective stress σ'_I. In fact, this stress may attain any value short of the value associated with the initiation of wrench faulting. In the figure, the limit value $\sigma'_{I\ limit}$ is defined by the dotted circle through $\sigma'_{III} = 0$, which is the smallest compressional stress circle that touches the Coulomb limit line and thus indicates the onset of faulting in the compression range. For the shear strength parameters assumed in the figure, one finds $\sigma'_I = \sigma'_{I\ limit} = 61$ MPa which probably leaves sufficient scope for the foreland compression to raise σ'_I to the level needed for cleavage jointing.

It may be argued that, in reality, the value of $\sigma'_{I\ limit}$ will even be higher than the value defined by the dotted circle in the figure because in triaxial testing (see Fig. 2.3), a rock sample has to be laterally supported by some confining pressure ($\sigma'_{III} > 0$) since, otherwise, the sample will fail by axial splitting, rather than by faulting (see also Fig. 2.12 and the comments on pp. 22–23).

Thus, we may expect in the overpressured foreland, or at least in its most overpressured parts and at certain distances from the thrust belt, the formation of vertical

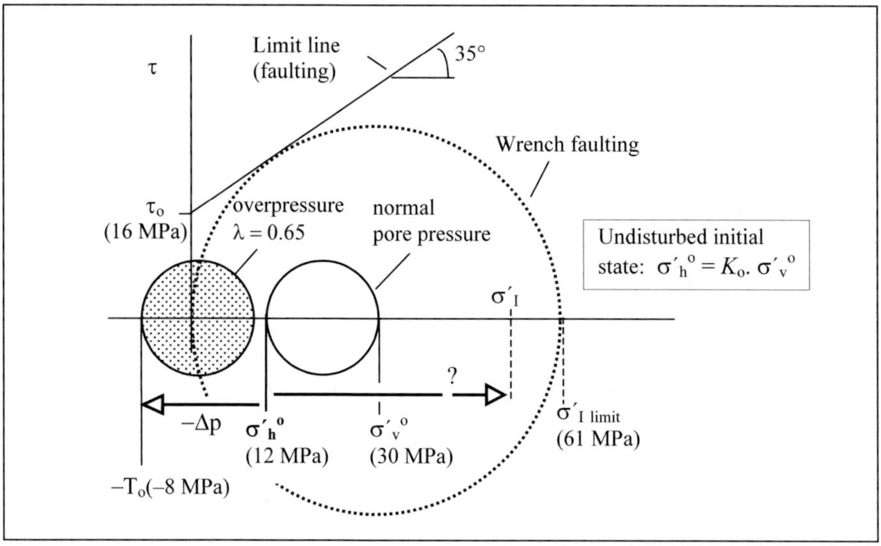

Fig. 4.27. Mohr diagram of overpressuring of the foreland of a fold and thrust belt, possibly causing vertical tension or cleavage joints. (Layer is under 2 km overburden, total overburden stress $\sigma_v° = 50$ MPa, $K_o = 0.4$, pore pressure $p/\sigma_v° = 0.65$, $\tau_o = 2T_o = 16$ MPa)

"internal hydraulic" tension joints and cleavage joints which trend parallel to the σ'_I trajectories. It should not be a surprise to find regions where tension joints, cleavage joints, and conjugate (Andersonian) wrench faults overlap, since the effective stresses in the foreland vary not only in space but also in time.

Vertical cleavage joints can also form in an *extensional* regime, for example, by a gentle passive bending of the layers which may be produced by the emplacement of laccoliths or salt pillows; by the rise of diapirs; or by differential vertical compaction and, more generally, by the flow and local thinning of ductile substrata. But, for several reasons, the conditions of cleavage jointing in an extensional regime are more restricted than in the wrench faulting regime.

Firstly, because the effective overburden stress is now the maximum principal stress, overburdens of at least 2–2.5 km are needed to provide the high compressive stress (σ'_I) required for cleavage fracturing. Secondly, the reduction of a horizontal effective stress to approximately zero is primarily achieved by an externally applied extension, rather than by overpressuring of the pore fluid, since the overpressuring would reduce the overburden stress σ'_v at least as much as layer-parallel normal stresses (see Fig. 3.2). Apart from that, a significant contribution of overpressuring is unlikely, since overburden-induced overpressures are relieved by the extension of the layer. Thirdly, the loading procedure that produces cleavage fractures in nature will seldom follow the experimental procedure of raising σ'_I while maintaining σ'_{III} near zero. The reverse process is more likely, with σ'_{III} being reduced to approximately zero while σ'_I remains more or less constant. The overburden stress $\sigma'_v{}^o$ ($= \sigma'_I$) is then limited by the overburden stress in the smallest limit state of normal faulting, since at a higher overburden stress the reduction of σ'_{III} would lead to normal faulting and, in turn, stop the further reduction of σ'_{III} before it could approach zero. This is illustrated in the Mohr diagram of Fig. 4.28 for a horizontal

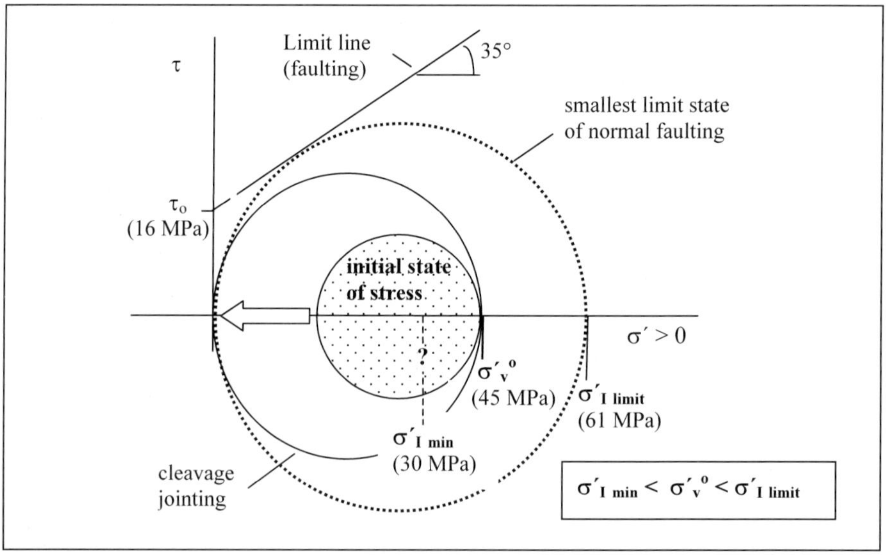

Fig. 4.28. Mohr diagram of extensional unloading promoting vertical cleavage jointing in horizontal layers under constant overburden load ($\sigma'_v{}^o$)(The strength parameters are the same as in Fig. 4.27.)

layer of, say, sandstone or limestone under 3 km of overburden ($\sigma'_v{}^\circ$ = 37.5 MPa); the strength parameters are the same as in Fig. 4.27, and therefore again $\sigma'_{I\ limit}$ = 61 MPa. In other words, the initiation of normal faulting would require an overburden of at least 4 km. Hence, during the extensional reduction of σ'_{III} (indicated by the arrow in the figure), the overburden stress must remain smaller than $\sigma'_{I\ limit}$ to prevent the growing stress circle from touching the limit line.

Thus, in order to induce vertical cleavage fractures in the layer considered in Fig. 4.28, the overburden stress $\sigma'_v{}^\circ$ must be smaller than 61 MPa and greater than the value $\sigma'_{I\ min}$ that would be necessary to start the fracturing process, which was arbitrarily assumed as 30 MPa. Overburden-induced vertical cleavage fracturing under normal pore pressure conditions would therefore be restricted to the depth range of 2 to 4 kilometres.

At this point, we face many unanswered questions; in particular, the fact that we do not know whether any value of $\sigma'_v{}^\circ$ between $\sigma'_{I\ min}$ and $\sigma'_{I\ limit}$ would "drive" the growth of cleavage fractures when σ'_{III} approaches zero. Unfortunately, to our knowledge, uniaxial compression tests on rocks have not yet been carried out in a way that would simulate the loading conditions of the geological process, where the lateral confining pressure is gradually reduced to zero, while the axial compressive load remains more or less constant. It is also regrettable that, despite the innumerable conventional compression tests on rocks, experimental data on the pre-failure growth of axial cleavage fractures seem to be extremely rare. Nevertheless, we propose the hypothesis that vertical cleavage fractures may form at any value of the overburden stress between $\sigma'_{I\ min}$ and $\sigma'_{I\ limit}$, though probably at slightly different σ'_{III}-values close to zero. The theoretical basis for this hypothesis will be discussed a little later.

Spacing of cleavage joints. Next, we ask what may control the spacing of cleavage joints, and in particular, why these joints would be spaced more closely than tension joints. This subject has up to now been little explored. Clearly, the theoretical spacing models for tension joints do not apply to compressive joints, simply because there is no layer-parallel normal stress to be released by the cleavage joints, and therefore no reactive shear stresses are induced to restore the equilibrium of forces. More precisely, if we consider a macroscopic cleavage fracture on the scale that is employed in defining (and measuring) rock stresses, that is, a scale which is large with respect to the grain and pore dimensions of the rock, the fracture appears as a smooth, closed cut. Such a fracture, when cutting perpendicularly across a horizontal layer under horizontal and vertical principal stresses, would not in any way disturb the (macroscopic) stress field. The mechanisms which control the separation of parallel cleavage joints must therefore be sought in the micromechanics of cleavage fracturing.

In Chapt. 2 (Fig. 2.10) we schematically illustrated some mechanisms which, on a pore or grain scale in a rock under compression, give rise to local tensile stresses and tensile microfractures (pp. 20–21). Under uniaxial loading, in the absence of any lateral confining pressure, some micro-fractures are enlarged in a stable way by increasing the axial compression, and will eventually somehow combine into macroscopic cleavage fractures. Although the mechanisms involved are extremely complex, it is generally held that the key role in the formation of cleavage fractures is played by "wing cracks" which nucleate on pre-existing small cracks (Fig. 2.10A). The pre-existing flaws are distributed in all orientations throughout the rock, and the cracks inclined towards the axis of the remote maximum compressive stress are under some shear. When the shear stress on a crack exceeds the frictional resistance of the crack faces, slip can occur and induce small tension regions near the crack tips. Under sufficient local tension, secondary tension fractures ("wing-cracks") are generated and branch off the parent crack (Fig. 2.10A). The axial stress σ'_I required for the initiation of a wing crack is found to be inversely proportional to the square root of the length of the parent crack. The wing cracks grow because they are wedged open at their base by the sliding on the parent cracks. The growth can therefore continue as long as the sliding on the parent cracks is maintained by increasing the axial compressive stress σ'_I. The growth is stable as long as the lateral stress σ'_{III} is zero or compressive, but a tensile σ'_{III} would eventually lead to unstable growth. On the other hand, a compressive σ'_{III}, even if much smaller than σ'_I, severely hinders the growth of wing cracks.

Furthermore, it is important to note, that the individual cracks in a wing crack population, produced by the overburden stress, are rather short, even if the lateral effective confining pressure is zero. Consider, for example, a parent crack of 1 cm length, favourably inclined at 30° towards the vertical, with a friction coefficient $\mu = 0.6$, and assume a typical fracture toughness (see p. 17) $K_{Ic} = 1$ MPa $m^{1/2}$ for the rock. From a fracture mechanical study by M.F. Ashby and S.D. Hallam (1986, Acta metall. *34*, pp. 497–510, Eq. 7), we infer that the formation of wing cracks of 0.25 cm length at $\sigma'_{III} = 0$ would require an effective overburden stress of $\sigma'_I = 67$ MPa (while wing cracks of 0.5 cm would require $\sigma'_I = 88$ MPa). Since the pre-existing cracks are, in general, rather small, we expect the formation of macroscopic cleavage fractures to be achieved by the interaction of a great number of wing cracks and small pre-existing cracks. Furthermore, we know from fracture mechanics, that parallel cracks (statistically aligned with the remote σ'_I direction) never merge, but may terminate on pre-existing cracks. One therefore envisages, on a microscale, that macroscopic cleavage fractures in rocks are produced via the linkage of pre-existing crack-type flaws and wing cracks. The highly schematical sketch in Fig. 4.29 should give a notion of this self-organization of microfractures into a continuous cleavage fracture.

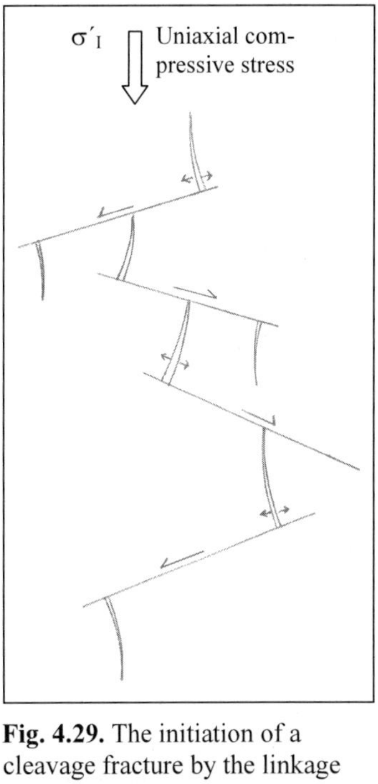

Fig. 4.29. The initiation of a cleavage fracture by the linkage of wing cracks and crack-type flaws (highly schematised)

The important point now is that the wing cracks are open cracks. Therefore, when a uniaxial vertical load initiates a population of uniformly distributed wing cracks which nucleate from uniformly distributed crack-type flaws, the volume of the rock will be increased by the opening of the wing cracks. As the rock layer is laterally confined, the dilatation causes a uniform internal compression that suppresses the further growth of the wing cracks. Hence, the fractures would be "self-confined" (to use a term coined by J. Gramberg) in an embryonic state. One would think that the growth of wing cracks could then only be resumed, and cleavage fractures eventually be formed and propagated, after the internal compressive stress had been relaxed by tectonic extension of the layers and/or internal creep. However, we expect the process, in reality, to be quite different, as we shall explain next.

Instead of considering a uniform dilation of the rock, we envisage that at a very early stage of the wing crack formation the volume dilation turns from a uniform distribution into a non-uniform one, probably even with a certain spatial periodicity. We assume that the formation of isolated patches, or zones, of the type shown in Fig. 4.29 goes hand in hand with the concentration of volume dilation in these proto-cleavage fractures, while the adjacent material is being pushed aside and compressed between neighbouring proto-cleavage fractures. Hence, the volume increase of the vertical proto-cleavage fractures is accommodated by the contraction of the material in between. This is schematically illustrated in Fig. 4.30 for the upper half (above the midplane SS) of a horizontal layer which is loaded by an effective overburden stress σ'_{ov}. The layer contains vertical proto-cleavage fractures (cf) which have not yet reached the layer boundaries. The gradation in shading in the figure is supposed to indicate the transition from the highest dilation (light) and zero horizontal compressive stress $\sigma'_h = 0$ inside the proto-cleavage fractures, to the zones of contraction (dark) and horizontal compressive stress $\sigma'_h > 0$. Clearly, the horizontal normal stresses are now no longer in equilibrium, and the imbalance must be compensated for by horizontal shear stresses (τ), as conjecturally indicated in the figure. Necessarily, the direction of the principal stress σ'_I is deflected by the shear stresses, and σ'_I remains vertical only where $\tau = 0$ inside and midway between neighbouring proto-cleavage fractures. But, inside and above the compression regions, the principal direction is deflected from the vertical to make an acute angle with the shear directions shown in the figure. Therefore, the σ'_I-trajectories (not shown in the figure) will

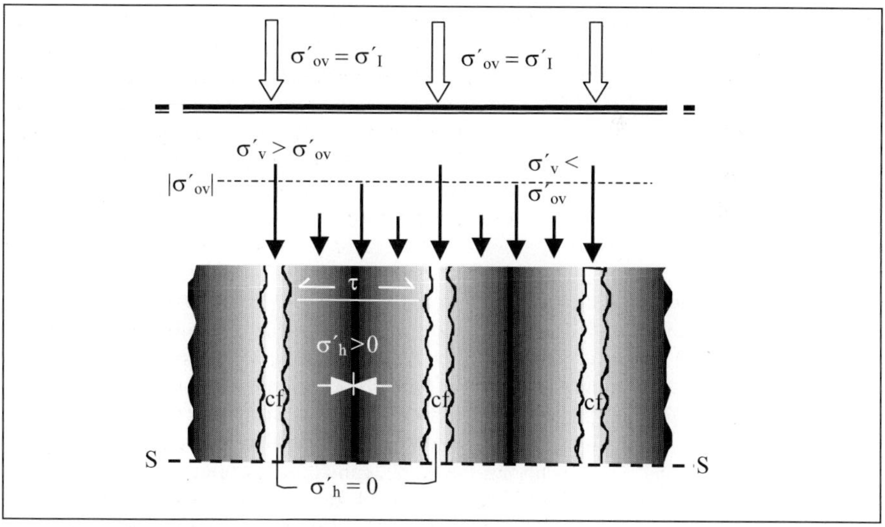

Fig. 4.30. Schematic section of parallel proto-cleavage fractures (cf) inside a layer under high overburden stress σ'_v. (The fracture segments lie above a plane of symmetry SS.) Dilation inside the cleavage fractures is accompanied by contraction (shaded) of the material in between; this produces a horizontal compressive stress ($\sigma'_h > 0$) in the contracted material, while leaving the horizontal stress $\sigma'_h = 0$ in the dilating cleavage fractures. The resulting imbalance of the horizontal normal stresses is compensated for by horizontal shear stresses (τ) which deflect the trajectories of the principal stress σ'_I (not shown in the figure) towards the cleavage fractures, and thereby raise the axial load on the cleavage fractures and reduce the vertical stress on the contracted material, as is schematically indicated in the figure

tend to converge towards the proto-cleavage fractures, and thereby redistribute the overburden load as sketched in the figure. Thus, the vertical stress on the cleavage fractures is raised above the average overburden stress, and is reduced below the average level on the compression regions.

The picture we have drawn of the role of "dilation banding" in the formation of cleavage fractures is purely qualitative and leaves many questions open. Nonetheless, it illustrates how vertical cleavage fracturing in horizontal layers of sedimentary rocks may be achieved by a high overburden stress and a zero confining pressure without resorting to the volume accommodation by a layer extension on a geological time-scale, or by internal creep processes. Instead, the dilational increase in volume is accommodated "instantly" by contraction of the material between neighbouring cleavage fractures.

Thus, the spacing of the parallel cleavage fractures appears to be primarily controlled by a spatial alternation of dilation and contraction. But, since both dilatancy and contractancy in rocks are inelastic phenomena, depending, in particular, on the density and size of the pre-existing micro-cracks, it is difficult to quantify the width of the contracted regions that separate neighbouring cleavage joints. Nevertheless, the model suggests that, depending on the micro-structure of the material, the spacing of the cleavage fractures may become extremely small, perhaps not exceeding the thickness of a cleavage fracture (or rather its dilation band) by more than an order of magnitude. Thus, the model may offer a conceivable, though not necessarily exclusive, explanation of the extremely close regular joint spacings that are sometimes observed in sedimentary rocks.

Summary of joint spacing. In the second part of this chapter we have attempted to explain why parallel tension joints in layers of sedimentary rocks are often regularly spaced, and what controls the distance between parallel joints. The mechanical genesis of the phenomenon is rather complex, as is reflected by the patchwork of quantitative mechanical models, hypotheses and conjectural elements in our presentation. Therefore, a brief review may be useful which focuses on the basic features of the theory, leaving aside various intermediate steps of the foregoing argumentation.

- ***A three-layer unit*** consisting of a stiff uniform layer between uniform lower-modulus beds, is uniformly extended in a lateral direction until tension joints form in the stiff interlayer. Since the layer-parallel (effective) tension is completely released at the fracture walls, the unfractured material of the stiff layer tends to shorten. This tendency is counteracted by the clamping action of reactive shear stresses on the layer boundaries (Fig. 4.12); this confines the reduction of the tensile stresses on either side of a joint to a zone wherein no tension joints can form (assuming a uniform tensile strength). The length (L) of this zone is determined by the resultant force of the interfacial shear stresses (Eq. 4.3), and defines the minimum distance of parallel joints under the actual boundary conditions.

  The crucial factor in determining the spacing of equidistant tension joints is the distribution of the joint-induced interfacial shear stresses which represent the mechanical coupling of the fractured interlayer and the unfractured embedment. Two extreme idealizations of the coupling mechanism have been considered: (*1*) frictional slip and (*2*) firm cohesive bonding of the layers. The first mode of coupling is the basis of N.J. Price's "slip model" of joint spacing (1966), while the second mode is the basic assumption of D.W. Hobbs' "welded layer" model (1967).

- ***Price's slip model*** is simply a static balance of the mechanical forces which act on the material of the stiffer layer (Eq. 4.2). The mechanical coupling mechanism is localized at the contacts between the stiff layer and its embedment, assuming that the interfacial shearing resistance is solely provided by the fully mobilized Coulomb friction (zero interfacial cohesion) which acts on the whole interfacial area between two neighbouring joints. Thus, the shear stresses exert the highest possible clamping action per unit area of the cohesionless interfaces, thereby minimizing the possible distance (L) between neighbouring joints.

  In flat-lying layers, the dimensionless ratio L/H of joint spacing (L) to thickness (H) of the fractured layer turns out to be proportional to the tensile strength (T_o), and inversely proportional to the effective overburden stress (σ'_v). The dimensionless joint spacing therefore decreases with increasing depth of burial at the time of jointing. As long as the effective overburden stress is not drastically raised, the slip model allows for the formation of only one set of regularly spaced tension joints; further extension of the multilayer being accommodated by widening of the existing joints.
- ***Hobbs' "welded" layer model.*** Fracturing in the stiff interlayer disturbs the straining of the three-layer unit that was strictly uniform before the onset of fracturing, and induces layer-parallel shear strains and shear stresses. Since slippage is excluded, the interfacial shear stresses caused by the jointing can only be determined as part of the whole post-jointing stress field of the three-layer unit. In order to obtain an approximate solution to the formidable problem of determining the stress field the following assumptions are made:

  (*1*) the deformational behaviour of the three-layer unit is linear-elastic. (*2*) The layer-parallel normal stresses and displacements in the stiff layer are considered as cross-sectional averages ($\hat{\sigma}'_x, \hat{u}_1$), and tension joints form when the effective average tensile stress $\hat{\sigma}'_x$ reaches the tensile strength T_o. (*3*) The layer-parallel gradient of $\hat{\sigma}'_x(x)$ in the stiff layer is proportional to the fracture-induced local change in the displacement field (Eq. 4.9).

  These assumptions are sufficient to determine the distribution of the dimensionless tensile stress $\hat{\sigma}'_x / \hat{\sigma}'^{max}_x$ between neighbouring joints, with $\hat{\sigma}'^{max}_x$ the maximum tensile stress midway between the joints (Fig. 4.14). But, a further assumption is needed to obtain an explicit relation between the joint spacing (L), the interlayer thickness (H), the elastic stiffness difference between the layers, and the extensional straining (e*) of the three-layer unit. It is assumed (*4*) that the joint-induced layer-parallel shear stress in the ambient layers decreases linearly with the distance from the interfaces, and vanishes at a distance (d) equal to the thickness of the jointed layer (d = H).

  The final result (Eq. 4.20a) then states that the joint spacing to layer thickness ratio (L/H) is proportional to the square root of the stiffness ratio $E_{(1)}$ (stiff)/$E_{(2)}$ (soft), increases with increasing tensile strength T_o of the stiff layer, and decreases with increasing extensional strain of the whole multilayer.

  Thus, both Price's slip model and Hobbs' "welded-layer" model predict a joint spacing that scales with the thickness of the stiff interlayer.
- ***Thin weak interlayers.*** The weak layers in Hobbs' model were supposed to be thicker than the jointed interlayer to make sure that the jointing was not affected by the fracturing in other layers. But Hobbs' theoretical approach may also be applied (even more aptly) to the contrary situation of very thin low-modulus layers (2) separating a set of stiffer layers (1) of identical properties. The weak interlayers are not fractured, but must be that much thinner than the stiff layers, so that joints in one layer (1)

localize the formation of joints in a neighbouring layer (1) in vertically aligned positions (Fig. 4.17). According to the analogue (4.20b) to Hobbs' formula (4.20a), the spacing L/H_1 of the vertically aligned joints in the stiff layers (1) is then proportional to the square root of the thickness ratio H_2/H_1.

Thus, the presence of very thin weak interlayers ($H_2 \ll H_1$) may drastically reduce the spacing of vertically (or nearly so) aligned tension joints.

- ***Infill jointing and "fracture saturation".*** While the slip model allows for only one set of regularly spaced tension joints, the "welded-layer" model allows new joints to be formed midway between existing joints by a continued extensional straining of the intact ambient layers. Since the layers are "welded" together, the external straining is fully transmitted to the joint-bounded sections of the interlayer, where it can raise the midway tensile stress ($\hat{\sigma}_x'^{max}$) until it reaches the tensile strength. Although, in principle, several generations of infill joints can be generated, with each set halving the separation of the former one, the sequence of infill episodes is strongly restricted by the limited strength of the ambient elastic layers.

  More importantly, the fracture condition applied to infill jointing in Hobbs' model refers to cross-sectional averages of the layer-parallel tensile stresses ($\hat{\sigma}_x'^{max} = T_o$), which may strongly vary from the actual stresses σ_x' on a cross-section. Therefore, a more accurate theory of infill jointing should analyze the non-averaged layer-parallel normal stresses σ_x'. Such an analysis (pp. 68–74) reveals that near either interface the tensile stress σ_x' exceeds the cross-sectional average $\hat{\sigma}_x'$, and drops below $\hat{\sigma}_x'$ in a central strip (Fig. 4.19). Tension fractures may therefore nucleate near the interfaces and start growing in the vertical direction before the average stress has reached the tensile strength T_o. Under sufficient tensile straining of the ambient layers, the fractures can propagate across the central region of low or zero tension and thus develop into tension joints proper, i.e. fractures transecting the layer. Interestingly, however, if the already existing tension joints are too closely spaced, σ_x' will become compressive in a central strip (Fig. 4.19), as was demonstrated in a numerical analysis by T. Bai and D.D. Pollard.

  Applying basic fracture mechanical concepts, Bai and Pollard also analyzed the propagation of vertical cracks and found that the fractures cannot penetrate the central compression zone between neighbouring joints when the joint spacing to layer thickness ratio (L/H) is less than 0.55 (or, more realistically, about 0.8). At this critical joint spacing a state of joint saturation is reached where a further rise in external strain cannot produce new infill joints (pp. 70–72).
- ***Models and reality – delamination*** (pp. 74–83). The reactive shear stresses that are induced by jointing in sedimentary rocks attain maxima at the interfaces, which in Hobbs' model are about four times higher than in the slip model (Eq. 4.25). This strongly suggests (Eq. 4.25a), that the joint terminations at the interfaces, where the interfacial shear stresses are highest, are nuclei of interfacial shear rupture (with the possible exception of layers under very low effective overburden).

  Interfacial shear ruptures which start from joint tips, grow into delamination zones if the jointed layers are of sufficient thickness to release the necessary elastic energy for "driving" the decohesion process against the resistance of interfacial friction and the fracture resistance $2\Delta\Gamma$ (Eq. 4.28). As far as inferred from "guesstimated" interfacial strength data, the length of the delamination zone (Eq. 4.41) will easily exceed the distance between neighbouring joints, while the required minimum layer thickness may lie between a few decimeters and a few meters at the most.

  Thus in sedimentary sequences of alternatingly stiff and weak rocks (e.g. limestone beds interspersed with clayey layers), with the stiff rock beds differing in thickness, the thicker beds may be delaminated by jointing, while the thinner beds are left firmly bonded. It would therefore appear that the "welded" layer model applies to the thinner beds, while joint spacing in thicker beds is more aptly described by the slip model, even if prior to jointing the beds were strongly bonded.

- ***Closely spaced joints.*** Whereas the joint spacings predicted by the slip model and the welded-layer model will not be much less than the thickness of the jointed layer, regular joint spacings of a few tenths or less of the layer thickness are not uncommon in sedimentary rocks (Figs. 2.7, 4.23). As mentioned before, vertically aligned joints in neighbouring layers may be closely spaced if the unfractured interlayers are very thin. In some cases (Fig. 4.24) of permeable rocks, narrow spacing may be attributed to the temporary draw-down of the pore pressure around a newly formed tension joint (Fig. 4.25). In other cases, one may suspect that the vertical or subvertical joints are not tension joints in the strictest sense at all, i.e. joints formed under externally applied extension, but *cleavage (extension) joints*, i.e. joints generated parallel to a high compressive effective stress σ_I' and a zero, or nearly zero, effective normal stress σ_{III}' perpendicular to the fractures (Fig. 2.8).
- ***Cleavage (extension) joints.*** While the phenomenon of cleavage fracturing is well established by uniaxial compression tests on rock samples, an inquiry into the when and where of cleavage joints in the brittle crust encounters many uncertainties. The differences between the rock-mechanical objectives and the subjects of geomechanical analyses do not allow a direct transposition of rock-mechanical results to the geological field (see Chapt. 2). In particular, the development of cleavage fractures in flat-lying strata which, when transected by cleavage fractures are still held together by the ambient layers and the resistance of lateral boundaries, is different from the rock engineer's interest in the uniaxial failure load (C_o) at which a laterally unconfined block fails by splitting.

An appraisal of the conditions of cleavage fracturing in the brittle crust requires first of all, an estimate of the maximum compressive stress σ_I' needed for the initiation of cleavage fractures. Although relevant data for rock layers of some lateral size are lacking, it is a reasonable assumption that the critical stress σ_I' is considerably smaller than the uniaxial compressive stress C_o of unconfined samples, albeit probably by less than a whole order of magnitude.

Vertical cleavage joints in flat-lying layers are mainly expected in the compressional tectonic regime that is characteristic of (Andersonian) strike-slip faulting. Such a stress regime, with horizontal principal stresses σ_I' and σ_{III}', exists in the foreland of fold and thrust belts (Fig. 4.26), where σ_{III}' may be reduced to zero by an arcuate shape of the fold belt, or by the push of the thrust belt generating overpressures of the pore fluid (Fig. 4.27). The overpressures required vary from moderate ($\lambda \approx 0.65$) to very high ($\lambda \approx 0.9$), depending on whether the layers are free to extend in a lateral direction or are completely constrained (see also Chapt. 3, pp. 30, 46). Vertical cleavage joints may also form in an extensional regime (with the overburden stress $\sigma_v'^{o} = \sigma_I'$), although only within a rather restricted depth range.

The formation of a cleavage joint is accompanied by volume dilation and the build-up of an internal lateral compressive stress (pp. 92–93). This follows from the micromechanics of cleavage fracturing, which mainly consists of the formation of open wing- cracks that branch off pre-existing cracks (Fig. 4.29). It is envisaged that at an early stage of the wing crack formation, dilation concentrates in isolated patches which are separated by contracted material. This, as yet hypothetical dilation banding (Fig. 4.30) would determine the spacing of parallel cleavage joints, and might explain the extremely narrow spacing sometimes observed in sedimentary rocks.

Appendix to Chapter 4

A) The mobilized Coulomb friction in Price's model after the formation of a first joint

The fully mobilized Coulomb friction at the interfaces of a horizontal layer is $\tau = \sigma'_{\perp} \tan \varphi$ (with the friction angle φ of the interface). Before the formation of a first joint, the normal stress $\sigma'_{\perp}$ on the horizontal bed surface is the effective overburden stress σ_v', but it is no longer constant along the bed surface when a first joint and the balancing interfacial shear stresses are formed. In the following analysis, we will evaluate the average of the Coulomb friction $\tau_{av} = \hat{\sigma}'_{\perp} \tan \varphi$ over the interfacial slip interval. We make use of the superposition principle of linear elasticity, which allows us to establish the state of stress in a body under given boundary conditions by partitioning these conditions and adding the states of stress associated with the partial boundary conditions.

Following this procedure, we will consider (in Fig. 4.31) the state of stress in the competent layer, *after* the formation of a first joint, as being the sum of the uniform state of stress just *before* jointing (indicated in Fig. 4.31a), and the stresses (indicated in Fig. 4.31b) that are caused by applying the compressive stress $\sigma_x' = T_o$ to the walls of the first fracture. Because of the symmetry of the problem with respect to the horizontal mid-plane (see Fig. 4.12), we only need consider the lower half of the fractured layer. The boundary problem in Fig. 4.31b is stated for the rectangular layer segment (shaded) that extends from the joint to the stress-free vertical section at an unknown distance L. While the top surface in Fig. 4.31b is also free of shear stresses, and likewise of effective normal stresses, the stress distributions on the lower interface are complex, and can only be determined as part of an analysis that includes the deformation of the adjacent elastic material. Fortunately, for the present purpose, we do not need the solution to this difficult "contact problem", but may be content with considering the balances of total forces and force moments on the rectangular body in Fig. 4.31b.

In formulating the boundary problem (Fig. 4.31b), we take care to define the boundary conditions in terms of stresses only. This means, that no part of the boundary is rigid or fixed in place, since this would impose displacement boundary conditions. In particular, the transverse extension, which is associated via the Poisson effect with the contraction due to $\sigma'^{\,post}_{x}$, is not hindered by the boundaries. Similarly, the fracture walls do not necessarily remain planar when the material contracts in the x-direction. We notice that the shear stress τ must vary along the basal interface, since it must decrease near the fracture wall which is free of shear stress, and also gradually decrease to zero near $x = L$. Nevertheless, the average value τ_{av} of the interfacial shear stress is determined by the balance of the total forces parallel to the x-axis: $\tau_{av}.L = T_o.L/2$. (As mentioned previously (p. 56), the contributions of the pore pressure on the vertical boundary planes cancel each other out.) Next we consider the balance of the force moments. As a point of reference, we conveniently choose the point Q midway along the basal plane in Fig. 4.31b, since the moments of the basal shear forces with respect to this point are zero. The constant pore pressure on the whole boundary does not exert a resultant moment on the rectangular body either. But, there remains the clockwise moment of the fracture force $T_o.H/2$ with the "moment arm" H/4 to be compensated by an anticlockwise moment of magnitude $T_o.H^2/8$. This counter moment can only be provided by basal effective normal stresses $\sigma'^{\,post}_{\perp}$ that are induced by the "push" force on the fracture walls. Note that, the average $\hat{\sigma}'^{\,post}_{\perp}$ of these stresses must vanish, since there is no counter force to balance the vertical force $\hat{\sigma}'^{\,post}_{\perp}.L$. Therefore, as indicated Fig. 4.31, the distribution of the effective normal stresses on the basal plane consists of both a tensile and a compressive component. In order to provide the required counterclockwise moment around Q, the stresses will be predominantly compressive between Q and x = L, and predominantly tensile between the joint (x = 0) and the reference point Q. Moreover, near x = L, the compressive stress decreases to zero, since the state of stress at x = L is still assumed to be undisturbed by the formation of the first fracture.

The combination of the stress states in Fig. 4.31a and b, will therefore change the effective normal stresses $\sigma'_\perp$ on the interfaces into $\sigma_v' + \sigma'_\perp{}^{post}$, but will leave the mean value $\hat{\sigma}'_\perp$ equal to σ'_v. Since the maximum shear stress, mobilized by the fracture-induced release of tensile stress, cannot exceed the Coulomb limit stress $\tau = \sigma'_\perp \tan \varphi$, the total reactive shear force on an interface cannot exceed $L.\sigma_v'.\tan \varphi$. This value is actually attained if *slip occurs on the whole interfacial segment* from x = 0 to x = L. With this assumption of the "slip" model, τ_{av} in Eq. 4.3 becomes equal to $\sigma'_v \tan\varphi$.

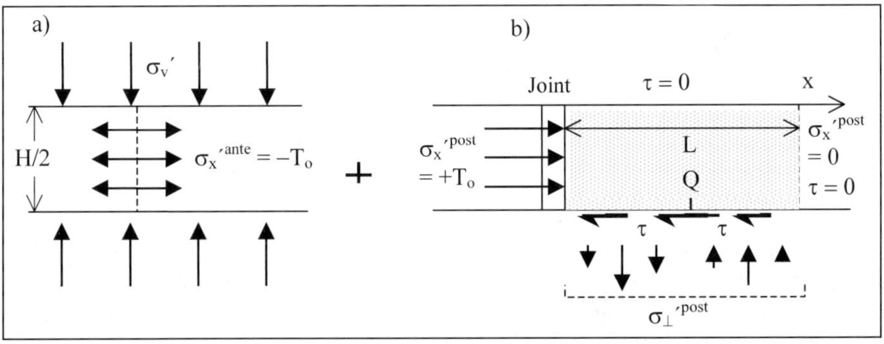

Fig. 4.31a,b. The state of stress after the formation of a first joint, as the result of the superposition of the uniform pre-fracture state (ante) in a) and the disturbance (post) field in b). Note that the resultant basal normal force in b) is zero

B) Partial slippage on the interface

In Eq. 4.37b (p. 80) the unknown average normal stress $\hat{\sigma}'_\perp$ on the sliding part of a horizontal interface was replaced by the overburden stress σ_v' that existed at the onset of jointing. The situation differs from the slip model, where slippage was assumed to occur over the whole distance between neighbouring joints and the basal average stress $\hat{\sigma}'_\perp$ and σ_v' were equal, as shown by the vanishing of the vertical basal force in Fig. 4.31b.

The disturbance field associated with a *partial* slippage on the interface is considered in Fig. 4.32, where, after jointing, normal and tangential stresses act on a cross-section x = ΔL. We consider again the balances of forces and moments of forces that act on the shaded segment of height H/2 in the figure. Choosing conveniently the corner point Q as reference point for the balance of force moments, we notice that only the normal stresses have a moment arm (i.e. the perpendicular distance from the reference point to the line of action of the force). In combination, the horizontal normal stresses on the sections x = 0 and x = ΔL exert a right-turning moment which, like in the former case of Fig. 4.31b, has to be balanced by a suitable distribution of the basal normal stress $\sigma'_\perp{}^{post}$. But in addition, the resultant vertical force $\hat{\sigma}'^{post}_\perp.\Delta L$ must balance the upward directed shear force $\tau_{av}(x = \Delta L).H/2$ at x = ΔL. Hence, the basal disturbance force must be negative (i.e. tensile), and will thus somewhat diminish the effect of the overburden stress σ'_v on the interfacial friction.

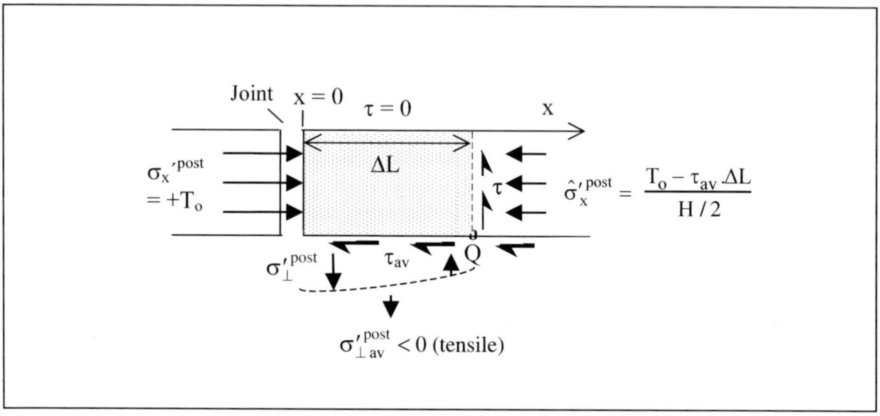

Fig. 4.32. Disturbance stresses (post) after slippage restricted to a part ΔL of the total distance L between neighbouring joints. Note that the basal normal force must be tensile to balance the upward directed shear stress τ on the cross-section $x = \Delta L$

Chapter 5

Multiple Sets of Tension Joints

Systematic and non-systematic joints. As a rule, tension joints in sedimentary rocks are arranged in more than one set of parallel or sub-parallel fractures. The simplest and most frequent arrangement consists of a set of "systematic" joints with straight traces on the bedding, and a set of "non-systematic" joints with less straight traces, roughly orthogonal to the systematic joints. Because of its fairly common occurrence, even in systems of more than two joint sets, this orthogonal or nearly orthogonal pattern is also referred to as the ***"fundamental joint system"*** (germ. *fundamentales Kluftsystem*), a schematization of which was presented in Fig. 4.7 (p. 52). Illustrating cases of the fundamental joint system are given in Fig. 4.4 (p. 50) and Fig. 5.1 below.

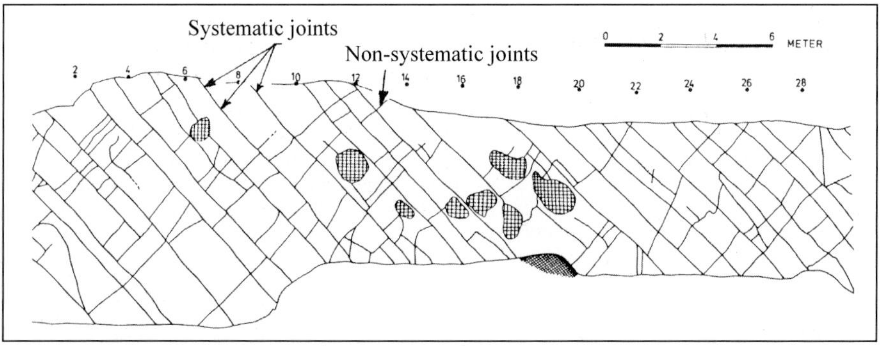

Fig. 5.1. Bedding plane traces of a fundamental joint system of throughgoing systematic (master) joints and non-systematic joints abutting on the systematic joints (Upper Cretaceous, Beckum, Westphalia, Germany; from H. Bock (1989) Z. dt. Geol. Ges. *131*, pp. 627–650)

Systems of systematic and non-systematic cross joints occur on a local scale, in particular, when associated with folds or faults, or on a regional scale where "lineaments" of systematic joint traces may cut across whole sedimentary basins, even in completely undisturbed flat-lying layers (see Fig. 1.4 and Fig. 5.2). Naturally, the systematic joints are formed prior to the non-systematic joints which they have terminated. Since the systematic joints form normal to the direction of greatest tensile stress σ'_{III} (< 0), the formation of the orthogonal set of the non-systematic joints requires a rotation of this principal stress direction through 90°. The switch in the σ'_{III} direction is easily understood when the jointing was produced by a *biaxial stretching* of the layers. Sketched in Fig. 5.3 is a layer that is subject to layer-parallel extensional strains e_x and e_y, with $|e_x| > |e_y|$. Assuming layer-parallel isotropy, σ'_{III} acts parallel to the greatest extensional strain (e_x), and is reduced to zero when the first joint set is formed. From this moment on, the greatest tensile stress acts parallel to the first joints, and can be raised to the tensile strength of the rock by a continued increase in the extensional strain e_y. As discussed in the beginning of Chapt. 4, the new joints will be arrested at the older ones, if those have not yet been healed, but they may intersect joints already healed. Now, the degree of healing also depends on the age of the joints, which differs among a set if the continued biaxial straining results in infill jointing, as was extensively discussed in the preceding chapter. Infill jointing and variations in the degree of joint healing may explain various deviations from the ideal "ladder" pattern of continuous systematic joints and connecting cross joints.

Fig. 5.2. Systematic joint lineaments on bedding planes of flat-lying Upper Cretaceous limestone; Münster basin, Germany (courtesy of Dietmar Meier, Petershagen, Germany)

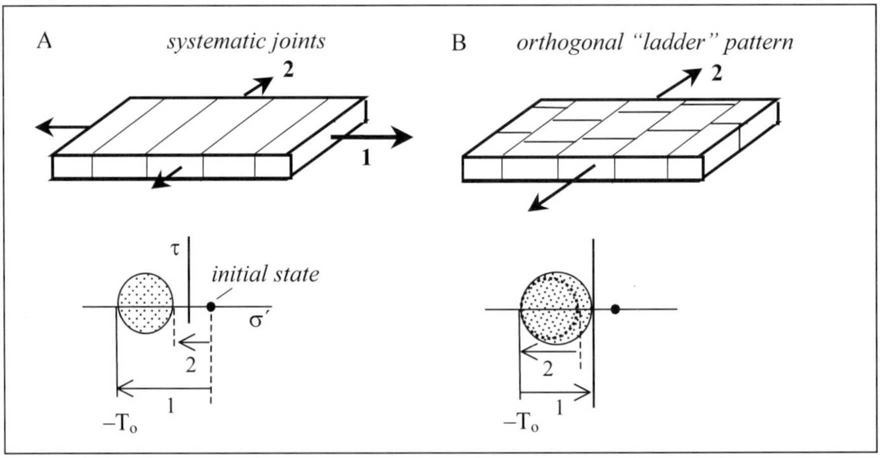

Fig. 5.3. Biaxial extension generating systematic (A) and orthogonal non-systematic (B) tension joints in "ladder"-type position. Mohr diagrams indicate the associated change in the layer-parallel principal stresses

It is even possible that a second set of straight systematic joints forms which cross-cuts the first set without apparent interaction or offsets. Such a case was shown in Fig. 4.8 where, however, the two sets of systematic joints do not intersect orthogonally but at an angle of 70°. This acute angle may cause some doubts as to whether the joints are correctly interpreted as tension fractures, or should rather be considered as conjugate shear joints. This delicate question will be taken up in a separate section on shear joints. Here, it suffices to point out that the tension joint interpretation requires that the two sets were not formed simultaneously, contrary to what would be required for the formation of conjugate shear fractures, and that the older joint set has gained shearing strength by mineralization. (Less likely, though not impossible, a frictional shearing strength could also be imposed by a change of the fracture-normal tension into compression.) Only then can the joints of the younger set cut straight across the healed joints of the older set. This can even be the case, when after the formation of the first set, the pair of layer-parallel principal stress directions of the regional stress field have experienced some rotation. But since the rotation induces shear stresses on the older joints, the rotation angle should remain relatively small to avoid slippage on the healed joints, and thus cause segmentation of the younger joints.

Non-orthogonal sets. If the joints of the first set are still open at the time the second set is formed, and the σ'_{III}-axis makes an acute angle with the strike of the first joints, the joints of the second set will be non-orthogonal and abut the joints of the first set. This is very well illustrated in an experiment by T. Rives and J.-P. Petit (1990) that is schematically shown in Fig. 5.4, taken from the paper by the authors. In the experiment, a PVC plate coated with a brittle varnish was first slightly deformed by torsion producing an oblique fold axis (Fig. 5.4A); then a shortening was applied (Fig. 5.4B) which caused a rotation of the fold axis and tension at the curved plate surface. The first set of tension fractures cut the varnish nearly parallel to the fold axis. As the shortening was continued (Fig. 5.4C), the fold axis was rotated further, while the tension on the curved plate surface became nearly parallel to the shortening direction. The tension eventually produced a second set of joints which strike orthogonally to the axis of shortening and abut the open fractures of the first set.

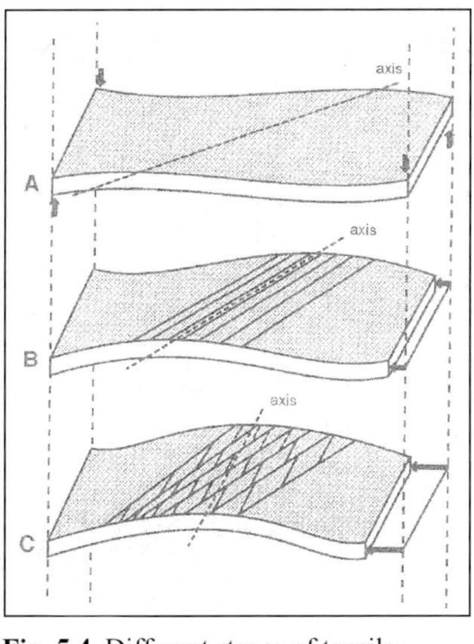

Fig. 5.4. Different stages of tensile fracturing of a PVC plate under a combined torsional and flexural deformation (see text) (T. Rives and J.-P. Petit, (1990); in: Rossmanith (ed.) *Proc. Intern. Conf. Mechanics of Jointed and Faulted Rocks*, Balkema, Rotterdam, pp. 205–211)

The combined bending and torsion in the experiment suggest a tectonic mechanism which, at least on a local scale, produces a non-orthogonal change in the σ'_{III}-direction.

If the systematic first joints are open, they act as free surfaces and are therefore principal stress planes in the rotated regional stress field. Assuming that $\sigma_I > 0$ is normal to the layer, near a joint plane, one of the layer-parallel principal stresses must act parallel to the strike of the joint and the other one, σ_{III}, must act normal to the joint plane, where it vanishes. (Naturally, the principal stress axes are indetermined when all normal stresses vanish near the open joint.) If it were not for the presence of the open first joints, the stress field rotated around the σ_I axis would be uniform with the layer-parallel principal stresses $\sigma_{II} > 0$ or < 0 and $\sigma_{III} < 0$. Although perturbed by the first joint set, the rotated stress field may be considered as undisturbed at a sufficient distance from each joint, and defined by the directions and values of the "remote" principal stresses. R. Dyer (1988, J. Struct. Geology *10*, pp. 685–699) analyzed the perturbed stress field in a jointed layer of a linearly elastic material. For the special case of a stress rotation through 30°, he found that the horizontal principal stress which is parallel to the joint planes ($\sigma_{\parallel}$) is compressive at the joints, if

$$-3 < \sigma_{II}/\sigma_{III} < -\frac{1}{3} \tag{5.1a}$$

and tensile if

$$-\frac{1}{3} < \sigma_{II}/\sigma_{III} < 1 \tag{5.1b}$$

where again σ_{II} and σ_{III} are the undisturbed "remote" principal stresses. In deriving these results, it was necessary to consider not only the "remote" principal stresses but also the normal stresses perpendicular ($\sigma_{\perp}$) and parallel ($\sigma_{\parallel}$) to the joint planes. The analysis then showed that $\sigma_{\perp}$ has to be tensile in the "remote" field in order to allow a smooth transition to $\sigma_{\perp} = 0$ in approaching an open joint. With the exception of the joints themselves, $\sigma_{\perp}$ then becomes tensile everywhere in the perturbed field.

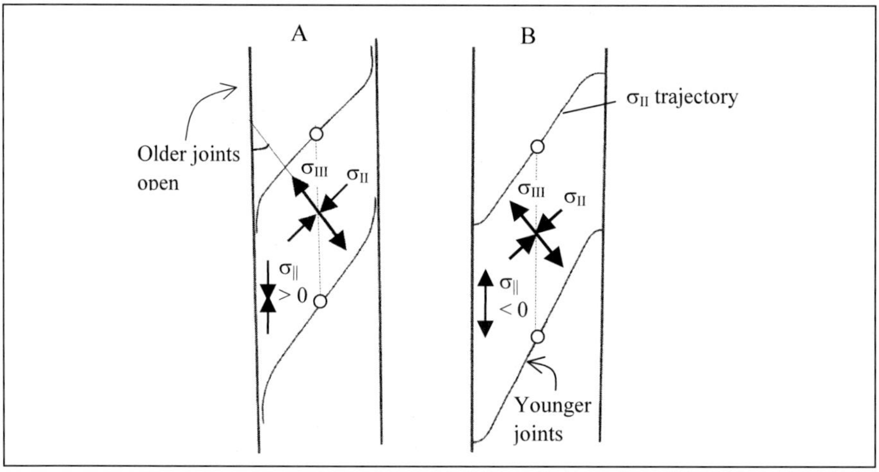

Fig. 5.5. Schematic diagram of bedding plane traces of open bed-normal tension joints (systematic joints) and hypothetical fracture paths of younger (non-systematic) joints which are generated by bed-parallel loading with the direction of the greatest tensile stress σ_{III} (< 0) at an acute angle with the older joints. While the joint-normal tensile stress $\sigma_{\perp}$ vanishes on the open joints of the first set, the joint-parallel principal stress ($\sigma_{\parallel}$) at the joints is either compressive (A) or tensile (B), depending on the far-field values of the principal stress ratios σ_{II}/σ_{III} (see Eqs. 5.1a and 5.1b). Accordingly, the σ_{II} trajectories are curved near the open joints, as sketched in the figure (figure slightly modified after R. Dyer (1988), J. Struct. Geol. *10*, pp. 685–699; see text for further comment)

The growth of a mode I fracture in a non-uniform stress field with curved principal stress trajectories poses a formidable problem in linear elastic fracture mechanics, since the eigen-field of the growing tension fracture, i.e. the stress concentration at the fracture tips which is characterized by the stress intensity factor K_I (Chapt. 2, pp. 15–16), depends on (and also affects) the non-homogeneous "guiding" stress field and the shape of the growing fracture. The present problem can probably only be tackled in a rigorous way by advanced numerical methods (see e.g. A.L. Thomas and D.D. Pollard (1993) J. Struct. Geology *15*, pp. 323–334). But in the light of observations in the field, and in experiments, it seems reasonable to obtain a first approximation to the solution by assuming that a mode I crack will propagate in a field of non-uniform principal stress directions with each crack increment perpendicular to the local σ_{III}-direction.

In our case, this means that a tension fracture grows along the σ_{II} trajectories, as sketched after Dyer's paper in Fig. 5.5. The fractures will start midway between the older joints where the uniform field of the remote stresses is least affected by the joint planes and the tensile stress σ_{III} has its highest value. The state of stress at the mid-plane between two joints is assumed equal to the state of the imposed remote stresses. Near the older joints the σ_{II} trajectories will then be curved, as indicated in Fig. 5.5, either aligning with the open joints (Fig. 5.5A) or terminating on them orthogonally (Fig. 5.5B). Hence, when sufficient remote tension (σ_{III}) is applied, a mode I fracture will follow a σ_{II} trajectory until it terminates either orthogonal or parallel to an open joint. Note, in Fig. 5.5B the abrupt change in fracture orientation when the fracture closely approaches an unhealed older fracture. (According to Dyer's analysis the change occurs at a distance as close as about one tenth of the total height of the older joint.)

It needs to be borne in mind that the analysis ignores interactions of the eigen-field of the growing tension fractures and the non-uniform stress field between the older joints. One may expect this interaction to be most significant in the vicinity of the open joints and possibly to affect the angle at which the two joints meet. This point was given special attention by Thomas and Pollard (1993) in connection with their laboratory and numerical experiments on the interaction of echelon fractures.

Further causes of orthogonal jointing. Some other mechanisms, besides biaxial tectonic extension, have been proposed as possible causes of orthogonal cross-jointing. First, as a set of open systematic joints cuts a layer into narrow bands, a gentle local bending of the bands produced, for example, by local differences in overburden load or by differential compaction of substrata, may be sufficient to induce the necessary tension for the formation of non-systematic cross joints.

Secondly, the elastic *Poisson "cross strain" effect* has been suggested in the literature as a possible cause of non-systematic cross jointing. This effect implies that a bed which is extended in one direction tends to shorten in the orthogonal directions. If the shortening is prevented, the elastic material responds by tensile stresses. Suppression of the Poisson cross strain has been suggested as a possible cause of tension fractures in uplifted sedimentary basins and compressionally folded beds. In the following, we shall examine the two cases in some detail.

Jointing in basins. Let us first consider a flat-lying layer of a sedimentary basin that is *uplifted*, while the basin surface is maintained in position by the erosive removal of overburden. We assume that, the strongly lithified material responds linearly elastic to the reduction of overburden. Uplifted beds, therefore, not only thicken but, due to the Poisson effect, also tend to shorten horizontally. M.R. Gross (1993; J. Struct. Geol. *15*, pp. 737–751) considered how the shortening of weak layers with high Poisson ratios (ν) was restrained by the smaller shortening of stiffer interlayers, and conjectured that the hindrance of shortening would produce sufficient tension in the weak layers to generate vertical tension fractures.

Let us more closely examine the possible role of the Poisson effect in uplifting. Consider a horizontal layer (Fig. 5.6) with layer-parallel x and y axes that is uplifted over Δz kilometers, while an overburden of the thickness Δz is simultaneously removed by erosion. Let us further assume, that the uplifted layer remains laterally completely constrained ($\Delta e_x = \Delta e_y = 0$). This appears a reasonable assumption, at least for a central region of the basin, because in reality most of the deformation of the uplifted sedimentary layers is accommodated by faulting and folding at the basin margin. Note that, the exclusion of any horizontal shortening ensures maximal tensional effect of the suppressed Poisson shortening. We also consider the layer as mechanically isotropic, and the stress reduction is therefore the same in all horizontal directions, $\Delta\sigma_x = \Delta\sigma_y = \Delta\sigma_h < 0$.

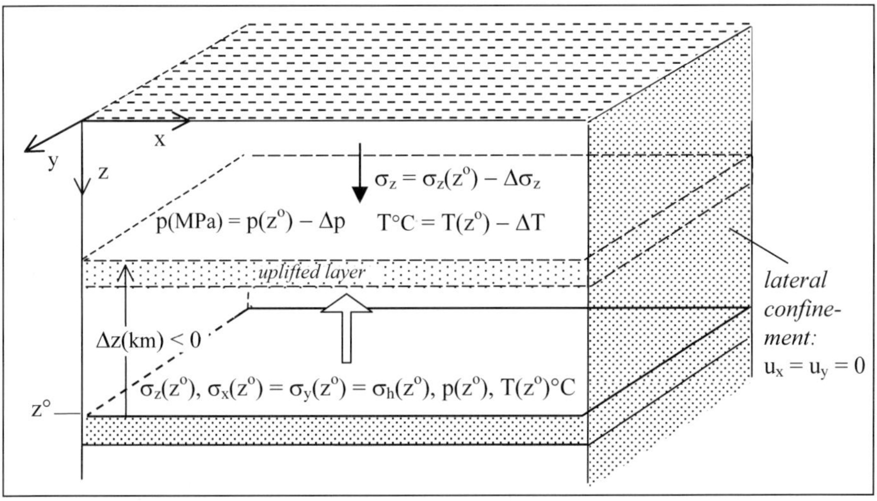

Fig. 5.6. Uplift of a laterally confined horizontal layer and erosive removal of overburden maintaining the surface position at z = 0. Decrease in overburden stress σ_z, pore pressure p and rock temperature T

It is important to notice, that horizontal stress changes during uplift of the laterally constrained layer are not only due to the suppression of the Poisson effect, but also result from the decreases in pore pressure and temperature of the uplifted rock. The elastic reduction in the total horizontal normal stress $\Delta\sigma_h$ is therefore a linear function of the decreases in total overburden stress $\Delta\sigma_z$(MPa), in pore pressure Δp(MPa) and in temperature ΔT(°C). This functional relationship is derived from the linear poro-thermo-elastic equations (see e.g. G. Mandl, FBR, Sect. 5.1):

$$\begin{aligned} \Delta e_x &= \frac{1}{E}\left[\Delta\sigma_x - \nu\left(\Delta\sigma_y + \Delta\sigma_z\right)\right] - \frac{a}{K}.\frac{\Delta p}{3} - \alpha\frac{\Delta T}{3} \\ \Delta e_y &= \frac{1}{E}\left[\Delta\sigma_y - \nu\left(\Delta\sigma_z + \Delta\sigma_x\right)\right] - \frac{a}{K}.\frac{\Delta p}{3} - \alpha\frac{\Delta T}{3} \end{aligned} \tag{5.2}$$

Here, the equations are written in terms of *total* stresses, where $a = 1 - K/K_s$ with K and K_s being the *compression moduli* of the bulk and the solid material of the porous rock, respectively. The coefficient α represents the *thermal volumetric expansion* of the unconstrained porous rock. Typical values of α for rocks are of the order of 10^{-5} per °C.

Inserting as stated before, $\Delta\sigma_x = \Delta\sigma_y = \Delta\sigma_h$ and $\Delta e_x = \Delta e_y = 0$ for the laterally completely constrained layer, Eqs. 5.2 reduce to the relationship for the increment in total horizontal stress:

$$\Delta\sigma_h = \frac{\nu}{1-\nu}\Delta\sigma_z + \frac{E}{1-\nu}\cdot\left[\frac{a}{K}\frac{\Delta p}{3} + \alpha.\frac{\Delta T}{3}\right] \tag{5.3}$$

Note that all terms in this equation are negative, indicating tensile stresses. To quantify the terms we assume an average rock density $\rho = 2.5$ gm/cm^3, and a geothermal gradient $\Delta T/\Delta z = 25°C/km$ which is typical for continental areas. We further assume, that the erosive removal of overburden keeps pace exactly with the amount of uplift Δz(km) which is counted as negative. With these data inserted in Eq. 5.3, uplifting over Δz kilometers under the simultaneous removal of overburden of thickness Δz, reduces the horizontal total stress by

$$\Delta\sigma_h(\text{MPa}) = \frac{\nu}{1-\nu}.25(\text{MPa})\Delta z(\text{km}) + \frac{E}{1-\nu}\cdot\left[\frac{a}{K}\frac{\Delta p}{3} + \alpha\frac{25\Delta z(\text{km})}{3}\right] \tag{5.4}$$

The only unknown parameter left on the right-hand side of the equation is the decline in pore pressure Δp. We consider the pore pressure as approximately equal to the pressure of the overlying hydrostatic fluid column. Assuming that the pore fluid is water, the pore pressure declines by $\Delta p = 10\Delta z$ MPa during an uplift of Δz(km). (Recall that Δz is counted as negative.)

We apply Eq. 5.4 to a *rock of intermediate strength and stiffness* with the typical parameters: $E = 0.2 \times 10^5$ MPa, $\nu = 0.2$, $K = 1.1 \times 10^4$ MPa, $K_s = 3.6 \times 10^4$ MPa, $a = 1 - K/K_s = 0.69$, and $\alpha = 1.5 \times 10^{-5}/°C$. (Note that the values of the Poisson ratio ν and the bulk compression modulus K should be understood as mean values, since these parameters vary somewhat with the changes in differential stress and effective mean stress during the uplift of the rock.) With these rock parameters inserted, Eq. 5.4 yields

$$\Delta\sigma_h = -14.6 \text{ MPa/km} \quad \text{or} \quad \Delta\sigma'_h = -4.6 \text{ MPa/km} \tag{5.5}$$

It is interesting that the suppression of the Poisson effect, represented by the first term on the right of Eq. 5.4, has contributed only 43% to $\Delta\sigma_h$ in Eq. 5.5, while the rest is provided by the suppression of the shrinkage of the bulk volume that would have been caused by the decrease in pore pressure and temperature in the laterally unconfined layer.

But even though we have generalized the Poisson effect by including the shrinkage due to the decline in pore pressure and temperature, the decrease in *effective* horizontal stress $\Delta\sigma'_h = -4.6$ MPa/km for the chosen rock parameters is *not enough* to establish tensile (negative) stresses σ_h'. This we conclude by considering the compressive horizontal stresses $\sigma_h^{o\prime}$ that existed before the uplifting of the layer, and is the result of the increasing overburden load during the subsidence phase of the sedimentary basin. It is a reasonable assumption that, at least in the central region of a subsiding basin, the build-up of the horizontal stresses takes place under complete lateral confinement (zero horizontal extension) of the subsiding layers (Mandl 2000, loc.cit., Sect. 5.2.1). But, as already noted in Chapt. 3, the horizontal stress response to the increasing overburden load is not elastic in nature, and is therefore characterized by an empirical parameter K_o, which in soil mechanics is known as the "coefficient of earth pressure at rest" (see p. 30). The final state of the horizontal effective stress $\sigma_h^{o\prime}$ after subsidence of the laterally confined layer is then proportional to the effective overburden stress $\sigma_z^{o\prime}$:

$$\sigma_h^{o\prime}(z) = K_o . \sigma_z^{o\prime}(z) = K_o .15(\text{MPa/km}).\left| z(\text{km}) \right| \tag{5.6}$$

Together with Eq. 5.5, the relation 5.6 determines σ_h' after an uplift of Δz km:

$$\sigma_h'(z - \Delta z) = K_o .15\left| z \right| - 4.6 \left| \Delta z \right| \tag{5.7}$$

Thus, we notice that the stress $\sigma_h'(z - \Delta z)$ after uplift can be tensile (negative) only if $K_o < 0.3\ |\Delta z|/|z|$. But, since K_o will rarely have a value smaller than 0.4 or 0.45, provided the layer was not extensionally disturbed by tectonic processes prior to uplifting, and because $|\Delta z|/|z| < 1$, the uplift of the laterally confined layer cannot produce tension joints. In fact, it will even produce horizontal effective *compressive* stresses of several MPa at shallow depth (Mandl 2000, loc.cit., p. 181). Only if the uplifting takes place under a much higher geothermal gradient $\Delta T/\Delta z$ than chosen in Eq. 5.4, could tension stresses be produced. For example, a high geothermal gradient of 50°C/km would allow the formation of tensile stresses after the layer (with initial $K_o = 0.4$) was uplifted over 80% of its original burial depth.

A higher gradient in pore pressure $\Delta p/\Delta z$ would have to be taken into account in Eq. 5.3 if the pre-uplift pore pressure was abnormally high and the beds were uplifted under hydraulic insulation. This "undrained" uplift of a sealed, overpressured layer was discussed at length by Mandl (2000, loc.cit., pp. 182–188), however, without finding theoretical indications of the formation of tensile stresses σ'_h.

The rock parameters used above may be considered as fairly typical for the majority of sedimentary rocks. If, however, the uplifted rocks are *extremely stiff*, Eq. 5.3 may determine tensile reductions $\Delta\sigma'_h$ per kilometer uplift, the sum of which may then convert the initially compressive stresses $\sigma_h^{o\prime}$ into a tensile stress σ'_h. Choose, for example, the parameters $\nu = 0.15$, $E = 10^5$ MPa and $a/K = 8.3 \times 10^{-3}$ for a very stiff rock; Eq. 5.4 then determines $\Delta\sigma'_h = -9.4$ MPa/km. With this value replacing the coefficient –4.6 in Eq. 5.7 and K_o being given the small value of 0.4, Eq. 5.7 then shows that any uplift beyond 64% of the original burial depth z^o would produce horizontal tensile effective stresses. But, recall from Chapt. 3 (pp. 32–33) that these internal hydraulic fractures are formed as "hairline" cracks.

Likewise, in *very weak* rocks, vertical hydraulic tension fractures may be generated during the uplifting, as may be seen by inserting, for example, the parameters $\nu = 0.4$, $E = 3 \times 10^3$ MPa, $a/K = 8.3 \times 10^{-3}$ in Eq. 5.4, which then determines $\Delta\sigma'_h = -10.1$ MPa/km. Assuming further that $K_o = 0.6$, which implies $\sigma_h^{o\prime}(3\ \text{km}) = 27$ MPa, we notice that σ'_h turns from compressive into tensile at a depth of 325 m. If the layer is unfractured, σ'_h would become –3.3 MPa at the surface. Hence, near the surface, the tensile effective stress in the very weak (e.g. clayey) rock is likely to cause vertical tension fractures.

Thus, it would appear that only in rocks which are very stiff or very weak, or in rocks under very high geothermal gradients, could uplifting under lateral confinement produce vertical tension fractures. In the majority of sedimentary rocks, the combined reductions of overburden, pore pressure and temperature will *not suffice in producing subvertical tension joints in uplifted layers*, even though Poisson's cross strain is fully suppressed.

So, what other factors might contribute to the build-up of the necessary tensile stress for subvertical tension jointing during uplift? The factor that first comes to mind is a *reduction of the K_o value prior to the uplift*. This could render the right hand side of Eq. 5.7 negative. Recall that K_o is the ratio of the effective layer-parallel normal stress to the effective overburden stress, established during the subsidence phase of the basin in a laterally completely confined horizontal layer. Obviously, any relaxation of the lateral confinement during the accumulation of overburden, will affect the K_o value that enters Eq. 5.7 for the subsequent uplift of the meanwhile lithified sediments. Unfortunately, evaluations of changes in K_o during the downwarp of a basin are obstructed by the complexities and uncertainties of the basin histories in terms of sedimentation and boundary geometry, and by the inelastic build-up of effective stresses under the increasing overburden. However, despite these impediments, we may obtain an impression of the K_o changes during the subsidence phase from an extremely simplified basin model that was analyzed by N.J. Price (1974; Proc. 3rd Int. Conf. Soc. Rock Mech., Denver, 1A, p. 487).

Price drew attention to the fact that in large sedimentary basins the curvature of the Earth's surface may have a considerable effect on the straining of subsiding layers. As a hypothetical example, he considered the elongated basin illustrated in Fig. 5.7. During subsidence a layer initially at the surface experiences shortening along its long axis AB

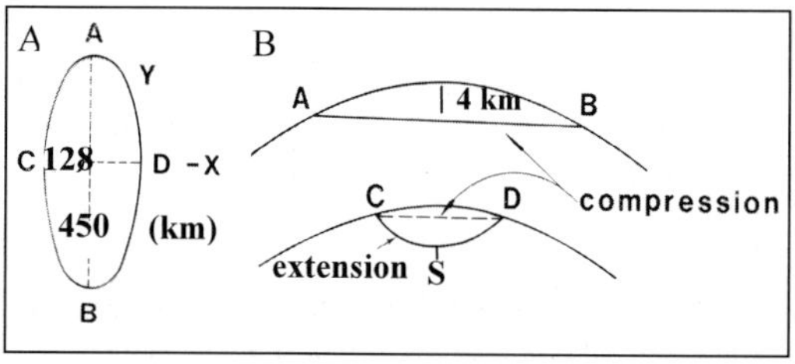

Fig. 5.7. Subsidence of large hypothetical basin; straining effect of the curvature of the Earth's surface. A) Elongated basin in plan; B) sections through the long and short axes (after N.J. Price (1974); see also Price and Cosgrove (1990), loc. cit., pp. 214–217)

(Fig. 5.7B) down to a depth of 4 km, but extension along the short axis CD after shortening during subsidence down to merely 0.3 km. The strains are considered as uniformly distributed along the cross sections CD and AB. With the radius of the earth (6400 km) and the length of the surface arc AB (450 km) in Fig. 5.7A, one simply derives the length of the associated chord in Fig. 5.7B to be about 449.9 km. Hence, during subsidence to a depth of 4 km, the layer experiences the total compressive strain $e_y = 0.1/450 = 0.22 \times 10^{-3}$.

In the cross-section CD the subsiding layer experiences the compressive total straining $e_x = 0.086 \times 10^{-3}$ during the first 0.3 km of subsidence (indicated by the stippled line in Fig. 5.7B). But in order to determine the extension of the layer during further subsiding to the central depth of 4 km, an assumption has to be made on the shape of the subsided layer in the cross-section CD. According to Price's original illustration in Fig. 5.7B, the shape of the subsided layer is a circular arc (greatly exaggerated in the figure) of the circumscribed circle of the triangle CSD. We consider here, the more realistic shape of the section CD shown in Fig. 5.7C. The length of the boundary curve is

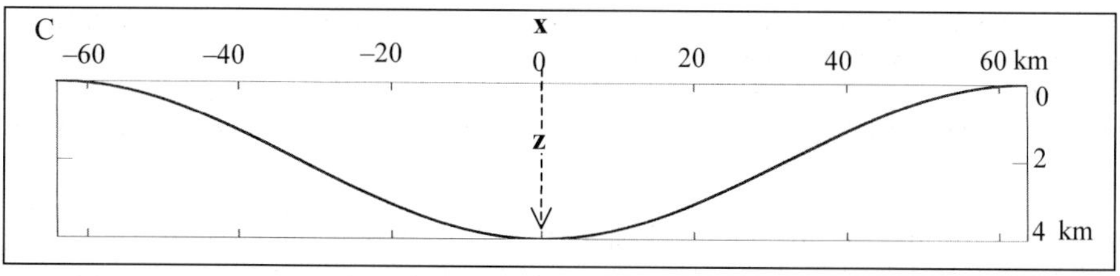

Fig. 5.7C. Sinusoidal basin section CD of Fig. 5.7A: $2z(x) = 1 + \cos(\pi . x/64)$

then 128.31 km, which implies that the surface arc CD has experienced the total extensional strain $e_x = -2.4 \times 10^{-3}$ during subsidence.

In reality, the straining of the inner region of the subsiding sediment body is very likely considerably less than predicted by the simplistic basin model above, since the extensional strain will not be uniformly distributed along the subsiding layers, but will be mainly accommodated along the basin margin (parallel to the AB axis) by flexural stretching and/or the formation of small grabens and normal boundary faults which eventually form inside the marginal flexure zone. Nevertheless, it would appear likely that sufficient tensile strain is left in the undisturbed inner parts of the subsiding basin for a substantial reduction in the value of K_o. Hence, in accord with Eq. 5.7, the subsequent uplifting of the meanwhile lithified sediments may produce a tensile layer-parallel stress σ_{CD}', and generate vertical tension joints which strike parallel to the long basin axis (AB).

Strictly speaking, Eq. 5.7 was derived under the condition $\sigma_{CD}' = \sigma_{AB}'$, and thus applies to circular basins. In such a basin, with the diameter CD used in Fig. 5.7B, the uplift would produce an irregular orthogonal pattern of mutually abutting vertical joints. In elongated basins, the pre-uplift values of K_o will differ along the x- and y-axes, being reduced in the x-direction and increased along the y-axis. Hence, a set of systematic joints will develop which trend parallel to the long y-axis of the basin. Note that the joints form while under strike-parallel compression (σ_y'), – a point we shall refer to later.

But, vertical joints which trend parallel to the long basin axis do not only develop during uplift, but also *prior* to the uplift phase if the pore fluid is *highly overpressured*. The accretion of overburden in the subsidence phase, and possibly the shortening along the long basin axis (AB), are likely to generate high overpressures in thick shale beds, from where the overpressures can invade interbedded layers of porous rocks, such as sandstones or carbonates. The associated changes in effective stress are cursorily outlined by the Mohr diagrams in Fig. 5.8.

Figure 5.8 refers to a horizontal layer under an overburden of 4 km. Under a normal hydrostatic pore pressure, the effective vertical stress $\sigma_v'^{o}$ is 60 MPa, and in the absence of tectonic compression, represents the maximum principal stress. (Note that this value is not affected by a water column on top of the basin, since the total normal stress on the basin surface is the same as the pore pressure in the top layer.) The stippled limit circle in Fig. 5.8A represents the state of stress that would initiate normal faulting. But, we may reasonably assume that new normal faults do not form once the subsidence has ceased. Therefore, without overpressure of the pore fluid, the smallest principal stress (σ_{CD}') must certainly be greater than the minimal principal stress necessary for normal faulting. It defines, together with $\sigma_v'^{o}$, the (solid) circle that represents the stresses on all cross-sectional elements parallel to the long basin axis AB. The stress $\sigma_{AB}'^{o}$, parallel to the long axis, is the intermediate principal stress, whose magnitude we do not know. Since the basin boundaries no longer deform when subsidence has ended, the layer at 4 km depth may be considered as laterally confined. Hence, while a rise in pore pressure Δp reduces $\sigma_v'^{o}$ by Δp, the horizontal stresses $\sigma_{CD}'^{o}$ and $\sigma_{AB}'^{o}$ are reduced by the smaller amount $\beta.\Delta p$. The overpressuring, therefore, not only shifts the original stress circle towards the tension side of the stress plane, but also reduces the diameter of the circle, as was explained in Chapt. 3 (Fig. 3.2B). Since tensile fracturing requires that the new stress circle (shaded in the figure) intersects the negative σ'-axis at the tensile strength point, the required magnitude of $\beta.\Delta p$ depends on the tensile strength T_o and the value of $\sigma_{CD}'^{o}$ under normal hydrostatic pore pressure conditions. Assuming that the rock is sufficiently lithified to elastically respond to the overpressuring, the value of the parameter β is determined by the elastic moduli of the rock (see Eq. 3.6). (Also, note that the β value chosen in the diagram as 0.48

satisfies the condition $K_o < \beta$ of Eq. 3.13 with $K_o = \sigma_{CD}'/\sigma_v' \cong 0.3$.) In the diagram, the overpressure parameter $\lambda = p/\sigma_v$ attains the value 0.85, which is indicative of a "hard" overpressure.

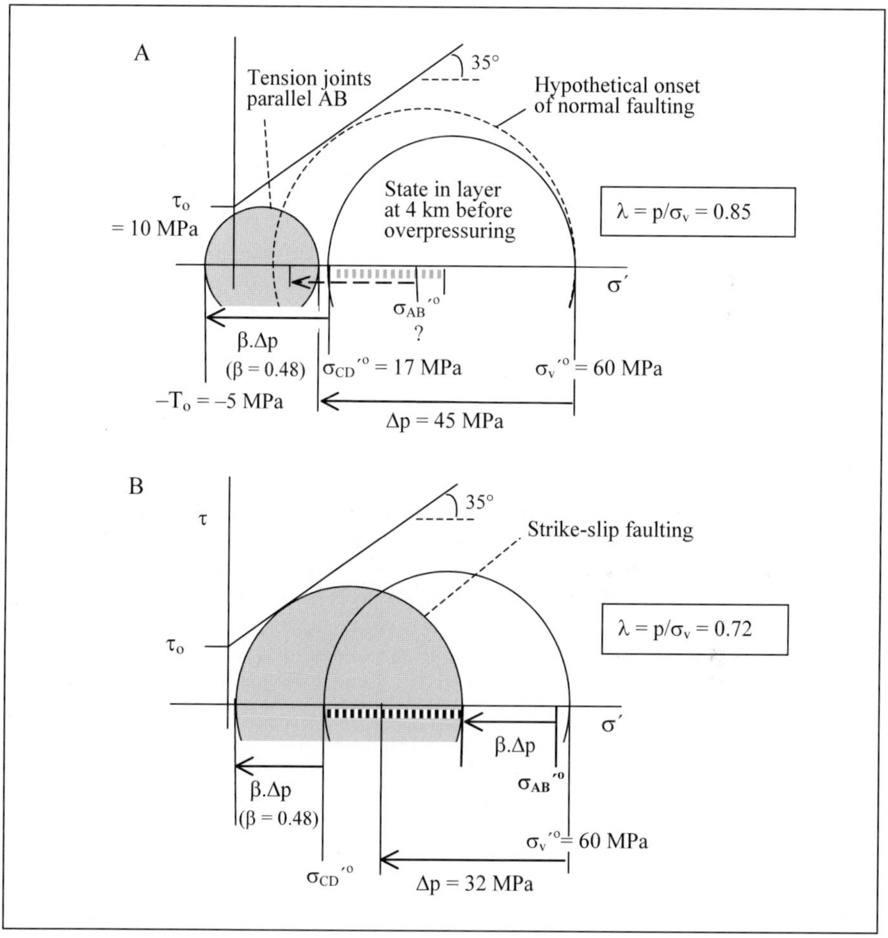

Fig. 5.8. Mohr diagrams of overpressured stress states (hypothetical) in a subsided layer at 4 km below the surface of an elongated basin. (The subscripts CD and AB refer to the short and long basin axes in Fig. 5.7. The basin walls are immobilised at the end of the downwarp phase.)
A) Vertical internal hydrofracturing parallel to the long basin axis AB by "hard" overpressuring ($\lambda = 0.84$). Note that the effective stress σ_{AB}' parallel to the AB axis remains the intermediate principal stress if, without overpressuring, its value ($\sigma_{AB}'^{o}$) lies inside the hatched interval of the σ'-axis. Otherwise, the roles of the principal stress σ_{AB}' and the overburden stress σ_v' would be interchanged by the rise in pore pressure.
B) If, in the absence of overpressures, the intermediate principal stress $\sigma_{AB}'^{o}$ was only a little smaller than the overburden stress $\sigma_v'^{o}$, the overpressuring would produce (minor) strike-slip faults instead of tension fractures

Since the shaded stress circle in Fig. 5.8A represents the effective normal stresses on the cross-sectional elements which are parallel to the AB axis, and the maximal tensile stress acts parallel to the CD axis, the overpressuring will generate vertical internal hydro-fractures that trend parallel to the long basin axis. But, note that a sufficient overpressuring is not the only pre-requisite for these joints to be formed. Recall, that the restrictive condition Eq. 2.3 for tensile fracturing requires that the maximum compressive stress remains smaller than $3T_o$. This implies, in terms of the notations in Fig. 5.8, as was already discussed in Chapt. 3 (p. 30):

$$\sigma_v'^o = 15(\text{MPa/km}).z(\text{km}) < \left(4 + \frac{1-\beta}{\beta}\right).T_o + \frac{\sigma_{CD}'^o}{\beta} \qquad (5.8)$$

We obtain an impression of the depth range of the internal hydro-fracturing by inserting the values of T_o, σ_{CD}'/σ_v', and β from Fig. 5.8A into the above condition. This results in the limiting depth z_{lim} = 4.03 km, which indicates that the layer in Fig. 5.8 has just subsided to the level where high overpressures can still produce vertical tension joints. But recall that the values of β and T_o in Fig. 5.8 were chosen to represent sedimentary rocks of intermediate stiffness and strength. A greater stiffness, and hence a greater value of β (<1) in Eq. 5.8, would result in a smaller limit depth z_{lim}, while a higher tensile strength would increase the limit depth.

So far, we have not taken the intermediate principal stress $\sigma_{AB}'^o$ into consideration. Although it does not enter the condition for tension fracturing, it is of interest. In the Mohr diagram of Fig. 5.8A, $\sigma_{AB}'^o$ on the σ'-axis lies inside an interval marked by the dotted line. It is immediately seen that for any value of $\sigma_{AB}'^o$ inside this interval, the stress σ_{AB}' remains the intermediate principal stress during the process of overpressuring. However, if $\sigma_{AB}'^o$ lies to the right of the marked interval, it is converted by the overpressuring into the maximum principal stress $\sigma_{AB}'^o - \beta\Delta p$, while the overburden stress $\sigma_v'^o$ turns into the intermediate principal stress $\sigma_v'^o - \Delta p$. Therefore, vertical tension joints will then develop parallel to a relatively high horizontal stress. As we shall see later, this is an important factor in stabilizing the fracture path.

The reader can also verify, with reference to Fig. 5.8A, that a smaller pore pressure rise Δp = 35 MPa would reduce $\sigma_{CD}'^o$ to zero, and thus shift the shaded circle only to the origin of the stress plane. A high $\sigma_{AB}'^o$ of, say, 50 MPa, would then be reduced to $50 - \beta.\Delta p$ = 33 MPa, and the effective overburden stress to 25 MPa. It would then seem likely that in some sandstone or limestone layers at $\sigma_{III}' = \sigma_{CD}' = 0$, the maximum principal stress of 33 MPa could generate vertical *cleavage fractures* parallel to the long basin axis. Naturally, this conclusion remains somewhat speculative because of the complete lack of appropriate experimental data and the difficulty (or impossibility) in distinguishing between tension and cleavage fractures by inspecting the fracture planes in the field or in the laboratory.

Finally, if in the absence of overpressuring the intermediate principal stress ($\sigma_{AB}'^o$) differs from the overburden stress ($\sigma_v'^o$) by about as little as indicated in Fig. 5.8B, the overpressuring can no longer produce tension or cleavage fractures, but (minor) strike-slip faults instead.

In summary, the processes of jointing considered above take place during or after the downwarp phase and during subsequent uplift. The idealised processes are supposed to operate in tectonically undeformed or only slightly deformed layers on a regional scale, that is, over large parts of sedimentary basins. In considering the downwarp phase, the basin is assumed as large enough to allow the curvature of the Earth's surface to impose significant strains on the subsiding layers. If the basin is of an elongated shape, a subsiding layer experiences an extension parallel to the short basin axis. The associated reduction of the corresponding horizontal effective stress contributes during a subsequent uplift phase to the combined tensional effects of the suppression of the Poisson strain and the decline in pore pressure and temperature. Together, the four factors may sufficiently reduce the effective stress parallel to the short basin axis in an ascending layer to allow the formation of vertical tension joints parallel to the long basin axis. Likewise trending tension joints, albeit of the internal hydraulic fracture type, can be produced by very high overpressures prior to the uplifting.

While these processes explain, at least partly, the formation of systematic vertical tension joints parallel to the long basin axis, the question remains as to how, on a regional scale, a second set of joints may be formed that trend orthogonally, or nearly so, to the first set of joints. Thus, the joints of *the second set of the fundamental joint system* should be roughly parallel to the short basin axis.

We fail to see how these joints could be produced by overpressuring of the subsided layer in Fig. 5.8. On the other hand, when the subsidence continues beyond the 4 km depth, compression along the AB axis (Fig. 5.7B) would turn into extension. Hence, after the overpressuring has produced the longitudinal hydraulic joints at a depth of, say, between 3 and 4 km, further subsidence by a few more kilometers might eventually produce regional joints parallel to the short basin axis (CD). This is the situation of the biaxial extension sketched in Fig. 5.3, with the first joint set reducing the tensile stress σ_{CD}' to zero and allowing the stress σ_{AB}' to become the minimum principal stress.

But, since uplifts of large sedimentary basins over more than 3–4 km seem to be rare, the observation of orthogonal joint nets in uplifted layers would suggest that the second joint set was formed during the uplift phase. Indeed, the uplift would tend to reverse the downwarp compression along the AB axis (Fig. 5.7B). However, as mentioned before, in reality, the uplifting during basin inversion proceeds along boundary faults, partly by the reactivation of normal boundary faults of the subsidence phase (for an illustrated discussion of the inversion of graben-type basins see: Mandl (2000), loc.cit., Sect. 7.4). Thus, most of the extension of the uplifted layers will be accommodated near the basin margins, an observation on which we based the assumption of complete lateral constraint (Fig. 5.6) of the uplifted layers in a circular basin. Here, we allow for some minor extension along the AB axis in the tectonically undeformed interior of the basin. The extension is supposed to be just sufficient to trigger the release of "locked-in" stresses which would then perform most of the fracturing work in generating the second joint set of a fundamental joint system. The concept of "locked-in" stresses needs some further comment.

"Locked-in" or "residual" stresses (germ. *Eigenspannungen*). Let us first revert to the uplifting of sedimentary rocks as was indicated in Fig. 5.6, and consider two horizontal layers, firmly "welded" together by cohesive bonds, and strongly contrasting in elastic stiffness. One layer is of intermediate stiffness and strength, the other layer is very weak. To characterize the elastic response of the rocks, we use the material parameters listed in Fig. 5.9A, which are the same as used before (pp. 101–102). In Fig. 5.9A, we consider the stress states in the two layers after subsidence to a depth of 3 km. The parameters and stresses of the stiff and weak layer are indicated by the superscript "s" and "w", respectively. The horizontal stresses $\sigma_h^{s\prime}$ and $\sigma_h^{w\prime}$, for simplicity assumed as isotropic, are determined by the respective K_o values of 0.4 and 0.6 in the laterally completely confined layers. Figure 5.9B shows the uplifted layers near the surface (z = 0), still under complete lateral confinement. The horizontal effective stresses recorded in the figure are calculated by Eq. 5.4, thereby taking the decrease in pore water pressure as 10 MPa per km uplift. After uplifting, a horizontal compressive stress of 4.2 MPa is left in the stiff layer, and a tensile horizontal stress of –3 MPa in the weak layer, if the layer withstood the tensile stress without being fractured. It is, however, very possible that the weak layer was jointed at a few hundred meters below the surface. This would cause the effective horizontal stress to drop to zero, and to remain zero during the final ascent, when the fractures are not healed. This alternative result is indicated in brackets in Fig. 5.9B. Also note that the difference between total and effective stress vanishes at the surface.

The uplifted multilayer in Fig. 5.9B is still under complete lateral confinement. In this state, *in-situ* stress measurements should give horizontal stress values as indicated in the figure. Since these stresses are the remainder of the stresses that existed prior to the

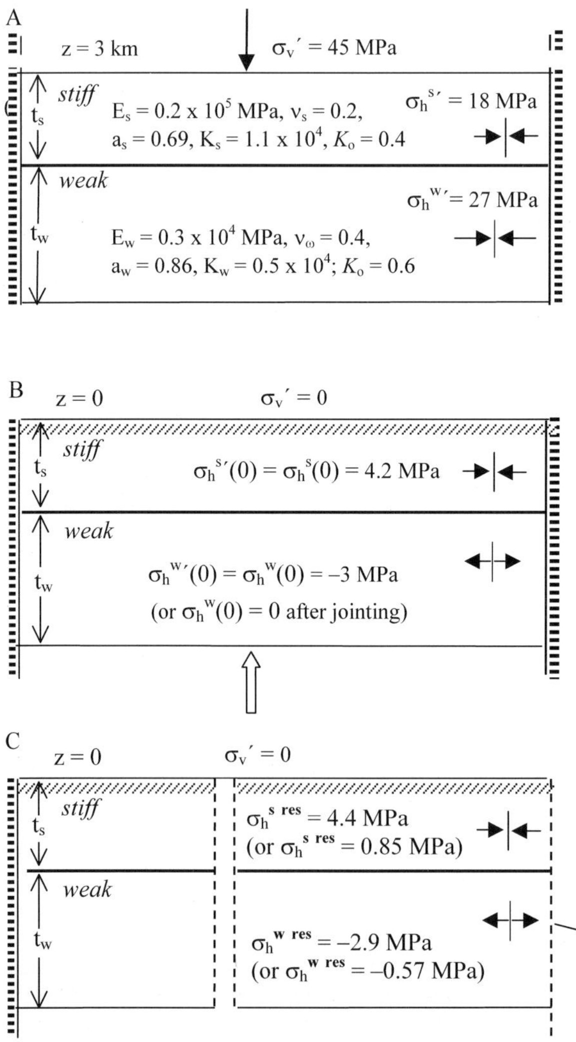

Fig. 5.9. Residual stresses in "welded" multilayer of strong (s) and weak (w) rocks: A) in constrained state after subsidence, B) uplifted to surface while still constrained, C) after lateral unloading

$$\sigma_h^{s\ res} = \frac{\sigma_h^s - R\sigma_h^w}{1 + R.T}$$

$$\sigma_h^{w\ res} = \frac{R.\sigma_h^w - \sigma_h^s}{R + T^{-1}}$$

$$R = \frac{E_S}{E_W} \cdot \frac{1 - \nu_W^2}{1 - \nu_S^2}, \quad T = \frac{t_S}{t_W}$$

G.R. Holzhausen and A.M. Johnson (1979) Tectonophysics *58*, pp. 237–267

$$\sigma_h^{s\ res}.t_s + \sigma_h^{w\ res}.t_w = 0$$

uplift, they are referred to as *remnant (or remanent) stresses*. But what happens to the stress field when the lateral constraints of the near-surface layers are removed by natural or artificial processes, as schematically indicated in Fig. 5.9C. The unloading causes the layers to extend. But since the layers are firmly welded together, they are forced to extend as a unit. Therefore, the unloading reduces the compressive horizontal stress in the stiff layer and adds compressive stress to the tensile stress in the weak layer. Thereby, the horizontal force across a vertical cross-section in the stiff layer must be balanced by the force across a vertical cross-section in the weak layer, to meet the requirement of zero force at the vertical free boundaries (see the box next to Fig. 5.9C). Based on this condition, and on Hooke's law, Holzhausen and Johnson (1979) determined the stresses which remain "locked" in the welded layers; the corresponding formulae are presented in the box next to Fig. 5.9B (for further details also see Mandl (2000) loc.cit., Sect. 5.2).

Thus, the horizontal stresses in the welded layers are self-equilibrating in the sense that the forces are in balance inside any adequate subdomain of the rock body, and remain in balance when the external loads of the rock body are completely removed. By definition, these locked-in stresses are referred to as "*residual stresses*" (germ. *Eigenspannungen*). It is these stresses that would be released when the cohesive bonding and interlayer friction of the layers were removed.

In the unloaded layer unit of Fig. 5.9C the residual stresses are moderate. Much higher residual rock stresses may be locked in "pockets" of a few grains in cemented or fused contacts. Probably the first experiments which demonstrated the existence of this type of residual stresses in rocks were conducted by N.J. Price on unweathered samples of nodular, muddy limestone. After several cycles of large uniaxial compression, the samples began to expand against the axial load (N.J. Price (1966) *Fault and Joint Development in Brittle and Semi-brittle Rocks*, Pergamon Press, Oxford, pp. 46–51). Price concluded that the distributed residual stresses in the rock samples, which gave rise to the expansion, were of the order of 60 MPa. Very careful studies on residual strains and stresses in rocks have been undertaken by M. Friedman (1972; Tectonophysics *15*, pp. 297–330) and by Friedman and J.M. Logan (1970; J. Geoph. Res. *75*, No. 2, pp. 387–405), who applied X-ray diffractometry to measure distortions of the crystal lattice (spacing of lattice planes) in quartz grains at polished surfaces of quartzose sandstones. (The interested reader may find an extensive review on residual and remnant stresses in T. Engelder (1993) *Stress regimes in the lithosphere*, Chapt. 10, Princeton University Press.)

Friedman demonstrated that the residual strains in individual quartz grains are relaxed when the grains are freed from the constraints by its nearest neighbours. This implies, that the residual strains are indeed elastic distortions which were locked in individual grains by the cementation or "welding" of quartz grain boundaries while the rock was still under load. This immediately raises the question: will the propagation of a macrofracture release locked-in strains and stresses, and how far from the fracture walls will the relaxation occur? To answer these questions, the X-ray technique was applied to a quartzite and a sandstone before and after propagating a macrofracture through the rocks. It was found that the residual elastic strains (and stresses) were relaxed within a narrow zone of not more than 5 mm on either side of the macrofracture.

One may further suspect the orientation of macrofractures to be affected by an anisotropy of residual strains and stresses in a similar way as it is affected by the material anisotropies because of bedding or grain orientation. In studying possible fracture control by residual strain anisotropy, Friedman and Logan (1970) induced fractures in disks of sandstone and quartzite by applying compressive axial point loads or diametrical loads. When tensile macrofractures and shear fractures were produced by applying the load normal to the bedding, the strike of the fractures turned out as normal to the orientation of the most tensile residual stress in the bedding plane that was determined by X-ray diffractometry before the loading of the specimens.

After this interlude on residual stresses, we return to the intriguing question of how a set of tension joints may be generated parallel to the short axis of an uplifted layer. We recall that during the downwarp of the large basin, the layer was compressed along the long basin axis (AB in Fig. 5.7B), but most of the shortening occurred by folding and faulting near the basin boundaries, while the main part of the subsiding layer remained nearly undeformed. Let us now envisage how, in this part of a layer, residual stresses can develop during a subsidence-uplift cycle.

First, note that a compressive load is not uniformly transmitted through a rock with a granular structure, but is carried by a framework of grain bridges, as was clearly demonstrated by photo-elastic experiments and numerical simulations (see e.g., G. Mandl, FBR, pp. 92–93). In Fig. 5.10, we consider, greatly simplified, a load carrying grain bridge which is roughly aligned with the direction of a high layer-parallel compression, and is laterally supported by some compressive stress, such as applied by the overburden. The grains which in Fig. 5.10A are still in cohesionless contact, become cemented in Fig. 5.10B. Assuming that, after hardening, the cementing material is weaker than the grain material, the grains and grain contacts will still carry most of the load. When the indurated rock is unloaded by a reduction or removal of the horizontal compression in the direction indicated in Fig. 5.10C, the granular framework tends to expand in that direction. But, as long as grains and cement remain firmly bonded, the extension of the granular framework is hindered by the cementing material which is now put under tension. On a macroscale, this tensile pre-stress constitutes a layer-parallel strength anisotropy, with the effective tensile strength now being smallest across planes at right angles to the direction of the horizontal unloading. In the uplifted layers, this should facilitate the formation of vertical tension joints normal to the long basin axis.

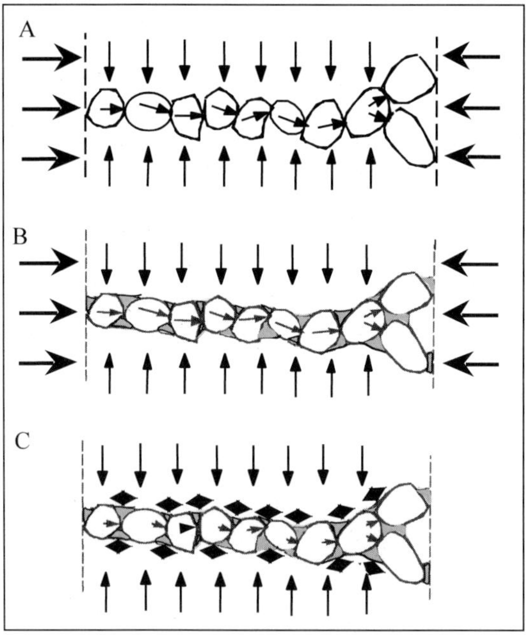

Fig. 5.10. Locking-in of intergranular stresses in a loaded grain bridge: A) before cementation; B) after cementation; C) stress transfer after unloading (double arrows indicating tension in the cementing material)

On the other hand, the removal of overburden during uplift implies a vertical unloading which puts the cementing material under vertical tension (not indicated in Fig. 5.10C). Pockets of vertical tensile residual stress may initiate exfoliation fractures parallel and close to the topographic surface (see Price and Cosgrove (1990) loc.cit., pp. 109–110).

Although, during uplifting, the decrease in the compressive stress σ_{AB}' (parallel to the long basin axis) initiates residual tension which, when relieved may facilitate the formation of vertical tension joints parallel to the shorter basin axis, the question remains whether σ_{AB}' will be sufficiently reduced to relieve the tensile pre-stress. Let us therefore first recall what was stated before (p. 108), with respect to the uplifting under complete lateral confinement: in the very stiff and the very weak rocks the horizontal effective stress becomes tensile at some distance from the surface. Comparing this with the uplift situation in an elongated basin, one should expect the onset of tension to be suppressed along the short axis by shortening and advanced along the long axis by extension. Hence, in a very stiff or very weak layer, the uplift should produce vertical tension joints parallel to the short basin axis long before the layer has approached the surface. Thus, the joints produced by extension along the long basin axis and the associated relief of residual tensile stresses, would be orthogonal to the systematic joints which strike parallel to the long basin axis.

However, the formation of these cross-joints poses a more difficult problem if the uplifted layers are of intermediate stiffness. In this case, the uplift under complete lateral confinement leaves the horizontal effective stresses compressive, even near the free surface, as we have seen before (cf. Fig. 5.9C). Also, recall that this result was obtained even though the horizontal Poisson tension caused by the vertical unloading was fully accounted for. Using the material parameters of the stiff layer listed in Fig. 5.9A, the uplift formula (Eq. 5.7) for a laterally completely constrained layer would predict a horizontal compressive effective stress $\sigma_{AB}' = 9.2$ MPa after an uplift from a depth of 3 km to 1 km. Thus, ignoring mechanisms which operate at the surface, such as weathering and buckling of exfoliation sheets, we have to invoke, again, extension along the long basin axis. To reduce the effective compressive stress of 9.2 MPa to zero by uniaxial extension along the long basin axis would require an extensional strain of 0.44×10^{-3}. This is about four times the uniformly distributed extensional strain that would be gained by a complete reversal of the vertical subsidence from 1 to 3 km in the hypothetical basin of Fig. 5.7. Thus, it seems very likely that some extra extension has to be imposed, for example, by slightly upbending the ascending layers, to allow the relief of locked-in tensile stresses and the formation of cross-joints.

An interesting question in relation to residual tensile stresses has not yet received much attention: how does the relief of these locked-in stresses affect the distance between neighbouring parallel joints? As already mentioned, Friedman has demonstrated that the propagation of a tensile macrofracture in a quartzite or a sandstone relaxes the residual strains within a distance of not more than 5 mm from the fracture walls. Would this not suggest, that the release of residual tensile stresses in the intergranular cement may control the distance between neighbouring joints, and perhaps be a reason for the very narrow spacings of tension joints (or cleavage fractures) that are sometimes observed, but not accounted for by the models in Chapt. 4?

Orthogonal joint sets in compressional folding. A further geological setting where fundamental systems of orthogonal or nearly orthogonal joints are commonly observed is compressive folding. Folds of this probably most common type are initiated and amplified by compressive forces, which are roughly parallel to the layering in its undeformed state. The sketch in Fig. 5.11A serves to distinguish compressional folding from the bending mechanism in Fig. 5.11B, where a stack of layers is being bent in response to the weight of the rock, and to the local uplifting by igneous intrusions, rising salt or clay domes and ridges, or by the differential, subvertical movements of basement blocks that force the overlying layers into draping the basement steps.

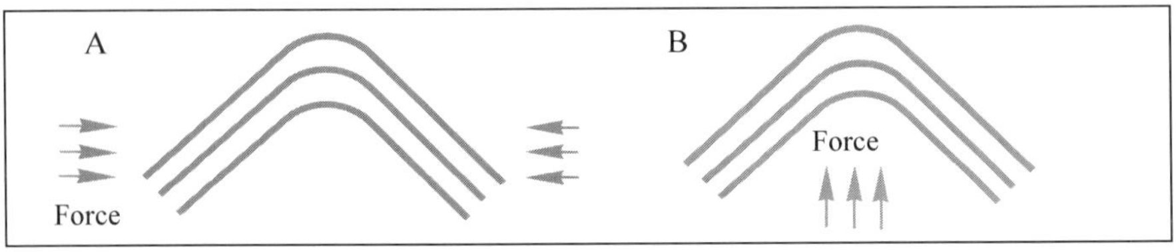

Fig. 5.11. Compressive folding (A) and drape folding (bending) (B) of layered rocks

In dealing with jointing and faulting in compressional folds, N.J. Price and J. Cosgrove (1990, loc.cit., pp. 252–253) refer to one of the simpler results of the elastic plate theory which concerns the bending of a plate whose upper and lower surfaces are free. If the plate is bent parallel to the yz plane in Fig. 5.12A, the material on the surface above the neutral surface (middle surface of zero extension) is extended along an arc in the yz plane, and a set of tension joints will eventually develop parallel to the hinge line. At the same time, the Poisson effect tends to shorten the material in a crosswise direction, i.e. parallel to the hinge line in the xz plane. Hence, the two principal curvatures (i.e., the curvatures of the orthogonal intersections of the plate surface with the xz and yz planes) have opposite signs ("anticlastic"). However, when the plate is very long in the x and y directions, the plate will be bent to a cylindrical surface as indicated in Fig. 5.12B, and the Poisson shortening by anticlastic bending is suppressed. For the present purpose, it suffices to note that the suppression of the hinge line curvature created a tensile stress $\sigma_x = \nu.\sigma_y$.

The cylindrically bent plate in Fig. 5.12B may be compared to a thick competent bed inside a buckling multilayer, whose buckling resistance is actually controlled by the competent bed. Although the fold is of finite length, the anticlastic bending is definitely

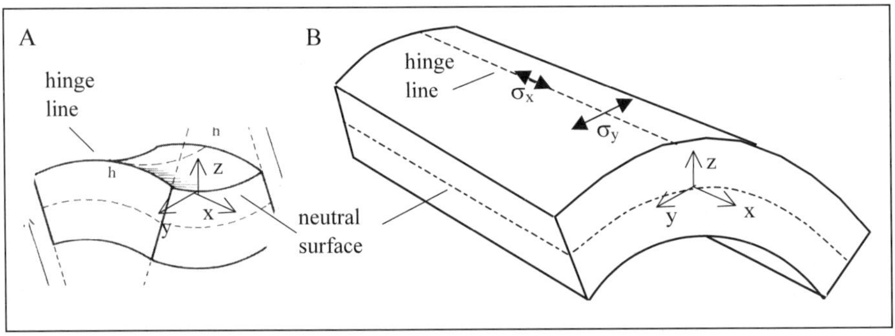

Fig. 5.12. Poisson effect in the buckling of an elastic plate:
A) Anticlastic bending of a short plate with free upper and lower surfaces implies orthogonal stretching of the outer arc and shortening parallel to the hinge line above the neutral surface. (The principal curvatures have opposite signs – "anticlastic" surface.)
B) If an extensive plate is bent to a cylindrical surface, shortening along the hinge line suppressed, which results in the tensile stress $\sigma_x = \nu.\sigma_y$

suppressed by the confinement of the competent bed between other beds which continue outside the fold. Despite the mechanical interaction of the thick layer with its less competent neighbours (of higher Poisson ratio and lower E-modulus), we may safely conclude that the tensile stresses (σ_x) induced along the hinge line of the competent layer by the Poisson effect will not be greater than ν-times the tensile stress (σ_y) across the fold hinge.

Thus, as in the case of basin uplift, other mechanisms are needed in addition to the suppression of the Poisson effect, in order to raise the hinge-parallel tensile stress to overcome the tensile strength T_o. A first mechanism of this kind is associated with the upwards convex curvature of the hinge line in buckle folds which have the well-known periclinal ("whale back") shape (e.g., Fig. 1.6). As the buckle fold grows, the length of the hinge line increases and its curvature varies. The hinge line curvature is greatest in the axial end regions of the fold, which migrate outwards as folding progresses. In these end regions, the tension induced by the curvature and the Poisson effect together may be sufficient to produce joints at right angles to the hinge line ("cross joints").

However, in the central region of the fold the hinge line curvature may be practically zero, and one cannot expect the Poisson effect on its own to raise the hinge-parallel tension to $-T_o$. On the other hand, cross joints are commonly observed over the whole length of the fold axis. This then suggests that *prior* to folding, tensile pre-stresses existed in the flat lying layers, which acted parallel to the layering and normal to the direction of the greatest horizontal compression. Pre-stresses of this kind are indeed encountered in the forelands of thrust and fold belts, as was discussed in the preceding chapter (pp. 88–89, Figs. 4.26 and 4.27). These tensile pre-stresses may have generated a regional system of tension joints parallel to the direction of maximum compression, and thus more or less at right angles to the later fold axes.

On the other hand, pre-folding pre-stresses in flat lying layers may also be of the residual type. Consider, for example, the process sketched in Fig. 5.10 in terms of cementation of compressed foreland sediments. When the cemented layers are folded, the layer-parallel compressive stress normal to the fold axis is reduced, and the stress transfer sketched in Fig. 5.10C puts the cementing material under tension. On a macro-scale, the locked-in tensile forces represent a maximum tensile stress along the trajectories of former maximum compression. The relief of this tension may, therefore, produce longitudinal jointing, i.e., jointing parallel to the hinge line. This may offer an explanation of the existence of longitudinal joints outside the hinge zone of a fold, in particular, in fairly straight fold limbs.

On the other hand, hinge-parallel joints in straight fold limbs are often clearly associated with the inter-layer shearing in flexural-slip folding. An example of this was presented in Fig. 4.3. Also, the jointing by flexural slippage, to which we shall revert in Chapt. 7, may be greatly facilitated by the relief of residual tensile stresses.

Undoubtedly, there are still other, as yet unexplored factors involved in the formation of orthogonal joint systems. A vexing problem, for example, concerns the reason why systematic joints are often surprisingly straight and continuous over great distances (see Figs. 1.4 and 5.2). We shall look into this next.

The straightness of systematic joints. If a Mode I fracture propagates in a uniform stress field perpendicularly to the tensile σ'_{III} axis, local heterogeneities of the material or neighbouring fractures are likely to deviate the propagating fracture from a straight path. What are the stress conditions that push a kink-type or smoothly curved deviation back to a straight path. In other words, under what conditions is the straight path of a Mode I fracture stable or unstable? B. Cotterell and J.R. Rice (1980; Int. J. Fract. *16*, pp. 155–169) analysed the problem by determining the stress intensity factors at the tip of a small kink-type or smoothly curved crack path extension. The result of the rather intricate analysis is simple: the straight propagation path under pure Mode I loading is *stable,* if the difference between the remote normal stress $\sigma'_{\parallel}{}^{\infty}$ parallel to the fracture and the remote normal stress $\sigma'_{\perp}{}^{\infty}$ perpendicular to the fracture is positive:

$$\sigma'^{\infty}_{\parallel} - \sigma'^{\infty}_{\perp} > 0 \tag{5.9}$$

Note that, compressive stresses are considered positive (contrary to the usage of Coterell and Rice). If the differential stress is negative, the straight fracture path is unstable.

This result may be understood in a heuristic and qualitative way. First, note that in a uniform field of principal stresses any (geometrical) cross-section parallel to the intermediate principal stress, and inclined with respect to the directions of σ_I and σ_{III}, experiences the shear stress

$$\tau = \frac{\sigma_I - \sigma_{III}}{2} \sin 2\delta \tag{5.10}$$

where δ is the angle between the trace of the cross-sectional element in the σ_I, σ_{III} plane and the σ_I axis, as illustrated in Fig. 5.13. (A simple derivation of the formula is given in the Appendix on Mohr's stress circle.)

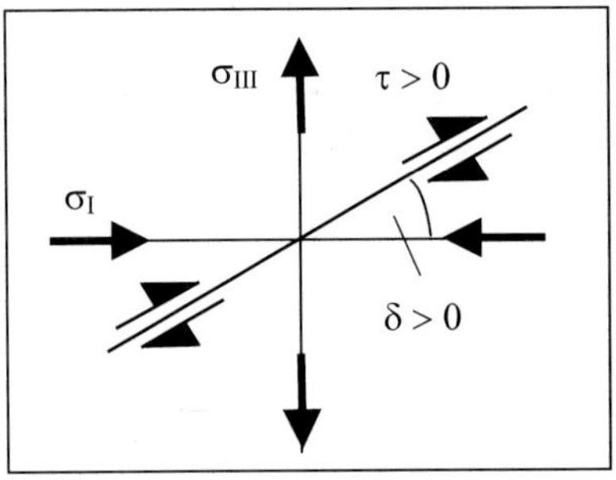

Fig. 5.13. Principal stresses and shear stress

Next, consider in Fig. 5.14A the straight mode I fracture and the remote principal tensile stresses $\sigma'_{III}{}^{\infty} = \sigma'_{\perp}{}^{\infty} < 0$, $\sigma'_{I}{}^{\infty} = 0$. The mode I fracture has propagated by a small kinked element, which is no longer under pure mode I loading but, according to Eq. 5.10, is also affected by the remote shear stress

$$\tau^{\infty} = -(\sigma'^{\infty}_{III}/2)\sin 2\delta = -(\sigma'^{\infty}_{\perp}/2)\sin 2\delta > 0 \tag{5.10a}$$

which induces dextral shearing along the kinked element, as is schematically indicated in Fig. 5.14A. In contrast, a remote tensile principal stress $\sigma'_{I}{}^{\infty} = \sigma'_{\parallel}{}^{\infty} < 0$ parallel to the mode I fracture in Fig. 5.14B induces sinistral shearing on the kinked element, while the remote compressive principal stress $\sigma'_{\parallel}{}^{\infty} > 0$ in Fig. 5.14C would induce dextral shearing. If both normal and parallel principal stresses are acting, the sign of the remote principal stress difference in Eq. 5.9 decides the sense of the shear on the kinked fracture element. The shearing sense is dextral (positive) when $\sigma'_{\parallel}{}^{\infty} > 0$ and $\sigma'_{\perp}{}^{\infty} < 0$, and thus condition 5.9 is satisfied.

While the stress field near the tip of a mode I crack is symmetrical with respect to the straight extension of the crack, this is no longer so near the tip of the kinked extension, since the material is compressed on one side by the shear couple, and extended on the opposite side (Fig. 5.14D). Therefore, the principal stress trajectories are asymmetrically distorted in the near-tip region. This is schematically indicated, for the trajectories of the maximum tensile stress σ'_{III}, in Fig. 5.14E. No matter what the exact shape of the trajectories may be, in principle, the trajectories have to swing around the tip of the fracture increment as indicated. Figure 5.14F shows the mirror image situation with the negative deviation angle δ; in accordance with Eq. 5.10, the shearing sense on the kinked fracture increment is now sinsistral (negative).

Although, because of the shear couple, the fracture increment is of a mixed mode I–II, the propogation of the element will be predominantly controlled by the maximum tensile stress, and the fracture path will intersect the tensile stress trajectories at right angles, or nearly so, as indicated by the dotted curves in Fig. 5.14E and F. In both cases, the propagating fracture is forced to resume the straight propagation path normal to the remote tensile stress. In other words, the fracture path is stabilised in accordance with condition 5.9, by the shear couple which is induced by a *compressive* stress $\sigma'^{\infty}_{\parallel}$.

Examples of geological settings where this stabilizing condition is fulfilled were encountered in the previous discussion on systematic jointing in basins and compressional folding. Tension joints are formed parallel to the long axis of large elongated basins during the uplift phase, or by high overpressures during or after the subsidence phase. These longitudinal joints form under strike-parallel compression (pp. 110–112). Similarly, in folding, vertical cross-joints (i.e., joints orthogonal to the fold axis) may develop prior to folding while following the direction of greatest compression.

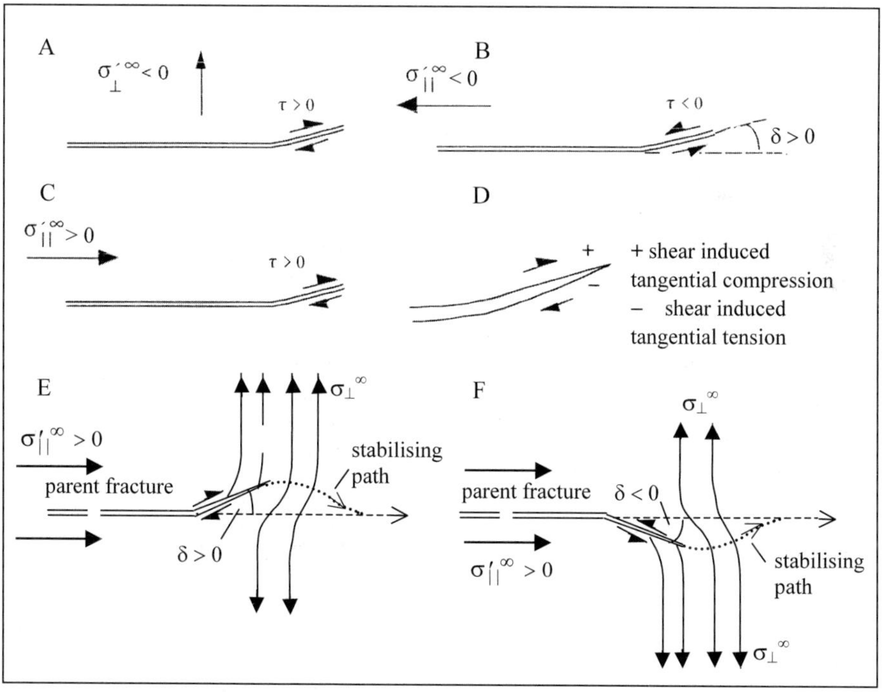

Fig.5.14. Mode I fracture extended by a kinked element in a field of remote principal stresses normal and parallel to the parent fracture. The shear stresses induced on the kinked fracture element and near its tip (A, B, C, D) cause asymmetric distortions of the trajectories of the maximum tensile stress (E, F). If the normal stress parallel to the parent fracture is *compressive*, the distortion of the trajectories forces the propagation path back to the straight extension of the mode I fracture

Certainly, one may envisage yet other geological settings where tension joints form under stress conditions in accordance with the Cotterell-Rice condition (Eq. 5.9). Nevertheless, there remain some doubts as to whether the Cotterell-Rice condition does adequately explain the straightness of very long joints. Note that, the condition was derived for fractures that have developed independently of other fractures, and not by the coalescence of fractures that initiated from non-coplanar sources. In fact, one would rather expect major joints to have formed by the coalescence of shorter fractures that have grown from nuclei off the plane of the main fracture. Thus, the straightness of major joint traces still poses some puzzles.

Another factor in promoting the straight propagation of tension joints is most likely the effective strength anisotropy that may result from residual tensile stresses. We have argued before, that the uniaxial unloading of a cemented granular material (see Fig. 5.10) produces tensile stresses in the intergranular cement; the mean orientation of the residual intergranular tensions being parallel to the direction of unloading. Therefore, when relieved, these residual stresses constitute a reduction of the tensile strength (or even represent a tensile stress component) across the macroscopic planes perpendicular to the unloading direction. Thus, when propagating along such a plane of macroscopic fracture anisotropy, the tension joint is likely to relieve the residual tensile stresses just ahead of the fracture tip, and so produce a swarm of intergranular microfractures which are preferentially aligned parallel to the joint plane. It seems very probable that the microfractures are also most densely arranged along the geometrical continuation of the joint plane, since the tensile macro-stress is greatest there. Thus, this crack concentration may act as a "fracture guide" which keeps the path of the growing tension joint within a straight strip of only a few grains in width.

Summary of multiple joint sets. Joints generally occur in several, differently oriented sets of parallel or sub-parallel fractures. Although the systems of sets may be quite complex, they commonly allow a *fundamental system* of two orthogonal or nearly orthogonal sets to be discerned, with a set of "systematic" joints striking straight and continuously over larger distances, and a second set of commonly shorter, "non-systematic" joints whose somewhat curved traces trend roughly orthogonal to the systematic joints (Fig. 5.1).

- ***The origin of systematic and non-systematic tension joints.*** The formation of orthogonal joint sets requires a rotation of the σ_{III} direction through 90° after the first (systematic) joint set has formed. This switch in principal stresses is most readily achieved by biaxial stretching of the layers (Fig. 5.3). If the joints of the first set are still open (or unhealed) when the joints of the second set are generated, these (non-systematic) joints will terminate at the systematic joints. If the principal stress axes were rotated through less than 90°, the non-systematic joints will trend at an acute angle with respect to the systematic joints which they intersect or abut on, depending on whether the older joints are healed or still open. In the latter case, the non-systematic joints terminate by curving to become parallel or orthogonal to the systematic joints (Fig. 5.5).
- ***Jointing during basin subsidence and uplift.*** The formation of joint sets in sedimentary basins can only be understood in terms of the stress history. However, a rigorous (numerical) analysis of the stress history is severely impeded by the uncertainties about the sedimentation history, the development of the basin geometry and the boundary stresses, and also by the material laws which govern the inelastic deformation behaviour (compaction, faulting and folding) of the accumulating sediments. Despite these impediments, insight into the mechanics of large-scale jointing in basins may be gained by invoking some reasonable, though fairly drastic, simplifications.

The analysis considers a "reference model" consisting of a horizontal, mechanically isotropic layer which subsided during the downwarp of the basin and ascended to the surface during basin inversion. During the whole process, the overburden surface remains horizontal and fixed in position by sedimentation and erosion during subsidence and uplifting of the basin respectively. Noting that during downwarp and uplift, in reality, most of the deformation is accommodated by folding and faulting at the basin margin, it is assumed that an inner part of the subsiding or uplifted layer remains laterally completely constrained, allowing neither layer-parallel extension nor shortening. During subsidence the material responds inelastically (compaction) to the increase in effective overburden; the horizontal effective stress ($\sigma_h{}^{o\prime}$) in the subsided layer is therefore related (Eq. 5.6) to the effective overburden stress $\sigma_v{}^{o\prime}$ (= $\sigma_z{}^{o\prime}$) by the empirical K_o factor (< 1). Any superimposed horizontal extension or compression would reduce or increase the value of this factor respectively. In contrast to subsidence, the uplifted layer is sufficiently lithified to respond (linear) elastically to the reductions in overburden weight, pore pressure and geothermal temperature (Eq. 5.3).

For this reference model it was then found: (*1*) that during subsidence vertical tension jointing of the laterally completely confined inner parts of the layers is only possible if the pore fluid is highly overpressured (pp. 110–111, Fig. 5.8A); and (*2*) that uplifting cannot give rise to vertical tension joints in layers of intermediate stiffness and tensile strength (p. 107). The suppression of the shortening tendencies of the Poisson effect and the decline in pore pressure and temperature in the uplifted layer does not sufficiently reduce the layer-parallel effective compression stress ($\sigma_h{}^{o\prime} = K_o.\sigma_v{}^{o\prime}$). Possible exceptions, are very stiff and very weak rocks (p. 108), or very high geothermal gradients.

Hence, according to the reference model, vertical tension jointing during uplifting would require a substantial reduction of the layer-parallel compressive pre-stress ($\sigma_h{}^{o\prime}$) before uplift. This may be accomplished, even without additional tectonic complications, in large basins, where the curvature of the Earth's surface causes subsiding layers to be significantly strained. Considering Price's model of a large elongated basin after a central downwarp of 4 km (Fig. 5.7 and the hypothetical basin section in Fig. 5.7C), it appears likely (p. 109) that the undisturbed inner region of a subsided layer accumulates a tensile strain along the short basin axis (CD), which effectively reduces the K_o factor in that direction, while along the long basin axis K_o is increased by compression. The associated reduction of the horizontal pre-uplift stress ($\sigma_{CD}{}^{o\prime}$) parallel to the short basin axis may then allow the combined tensile effects of the suppressed Poisson strain and the decline in pore pressure and temperature to produce systematic tension joints parallel to the long axis of the basin.

The formation of vertical tension joints parallel to the short basin axis, and thus, orthogonal to the systematic joints ("cross-joints"), is more difficult to understand. While the layers in a large elongated basin experience shortening along the long basin axis during subsidence (Fig. 5.7B), the uplift may reverse the process and impose some extension parallel to the long axis. Recall, that the reference model predicts layer-parallel effective tension in very stiff or very weak layers when the uplifted layers approach the surface. Hence, the additional elastic extension imposed along the long basin axis by the curvature of the Earth's surface may suffice to trigger the release of residual tensile stresses (see below) parallel to the long basin axis. This should allow cross-joints to be formed long before the layers have approached the surface. Most of the fracturing work may thereby be performed by the released residual tension.

In contrast to very stiff and very weak layers, uplifted layers of intermediate stiffness and strength in the reference model remain under lateral compressive effective stresses all the way up to the surface. Thus, a set of cross-joints can only form when an additional uniaxial extension is imposed, for instance, by regional upbending of the layers.

- ***Residual (locked-in) stresses.*** Residual stresses in rocks are stresses inside a rock body, or in parts of it, which remain in equilibrium after all boundary loads have been removed. A typical example of residual stresses on a meso-scale is a sedimentary multilayer consisting of alternating weak and stiff layers which are firmly "welded" together (Fig. 5.9). When, after uplifting, multilayered blocks are cut out of the sequence and are thus completely unloaded (Fig. 5.9C), layer-parallel normal stresses of an opposite sign remain locked in adjacent layers by the interlayer bonding.

  Higher residual stresses may exist on a microstructural scale. In sedimentary rocks, they may typically originate from the lithification processes. In a granular material, boundary loads are transmitted through a framework of load carrying grain bridges (Fig. 5.10). If, after cementation of the grain contacts and hardening of the cement, the indurated rock is unloaded, the extension of the granular framework is hindered by the cementing material, which is thereby put under tension (Fig. 5.10C). These locked-in intergranular tensile stresses constitute a reduction in the effective tensile strength of the rock.

  Applied to subsidence and uplifting of an elongated basin, this means that the subsiding sediments are indurated by cementation while being compressed parallel to the long basin axis. As the compression is reversed into extension during a subsequent uplift, tensile stresses are initiated in the cementing material, which facilitate the formation of layer-normal joints parallel to the short basin axis.
- ***Fundamental joint systems in compressive folds*** consist of bed-normal tension joints which are (sub-) parallel and (sub-)normal to the fold hinge (line of maximum curvature). *Hinge-parallel* tension joints (longitudinal joints) are caused by the extension along hinge-normal arcs (Fig. 5.12B) and/or by the relief of residual stresses in the cementing material of compressed foreland sediments. As in the aforementioned basin uplift, the residual tensile stresses are initiated in the cementing material during the transition from compression to extension (p. 119), and are therefore parallel to the direction of the pre-folding compression.

  Hinge-normal (cross-fold) tension joints seem to be the result of several factors: Firstly, suppression of the elastic anticlastic bending (Fig. 5.12) which implies a tensile Poisson stress parallel to the hinge line. But this stress alone is too small to produce cross-fold joints. So it may be, secondly, supplemented by tension due to the hinge line curvature of periclinal ("whale-back" shaped) folds. And thirdly, regional sets of tension joints which strike roughly at right angles to the fold axes may already have been formed in the flat-lying foreland sediments of arcuate fold and thrust belts (p. 88, Fig. 4.26), by horizontal tensile stresses normal to the direction of the greatest horizontal compression.
- ***The straightness of systematic joints.*** Straightness of tension fractures has been ascribed to a stabilizing shear couple induced by a compressive stress in the direction of fracture propagation (Fig. 5.14). Very likely, the fracture path may also be stabilized by an effective strength anisotropy caused by residual tensile stresses. Eventually, some straight joints of an alleged tension character might, on closer examination, turn out to be shear joints a quite different joint type which is the topic of the next chapter.

Chapter 6

Shear Joints

In the introductory Chapt. 1, a joint was defined as an approximately planar fracture that was formed without visible movement parallel to its plane; in contrast, a fault is a fracture or fracture zone that was formed by shear displacements which are still clearly visible. This purely phenomenological distinction hinges on the term "visible" which, unfortunately depends on the scale of observation. Thus, a joint may be a tension or a cleavage (extension) fracture, as discussed in Chapt. 2, where the relative displacement of the fracture walls is normal to the fracture plane, or it could be a shear fracture with a shear displacement of the fracture walls that is "invisible" at the scale of observation. Such shear fractures are commonly referred to as "*shear joints*". They are to be distinguished from tension fractures that have experienced shearing at a later stage.

"Shear joints" vs. tension joints and faults. In the field, it may be extremely difficult to clearly identify fractures as shear joints. For example, Fig. 6.1 shows two non-orthogonal, continuous sets of discontinuities that cut perpendicularly across the bedding of a straight fold limb. Are these fractures shear joints or tension joints? (The same question might be raised with respect to the two non-orthogonal sets of systematic joints in Fig. 4.8.) Why is this question important? Simply, because from a mechanical point of view, the two kinds of features are completely different, since they are generated by different stress systems, and thus, possibly associated with different tectonic episodes. The alternative stress systems are illustrated in Fig. 6.2.

Although a systematic shear displacement on the discontinuities of Fig. 6.1 is not obvious we can speculate that the discontinuities originated as conjugate shear planes, as

Fig. 6.1. Beds of Namurian fold limbs with two non-orthogonal, continuous fracture sets ("Fisherstreet" Bay, Irish West coast); tension joints or conjugate shear planes?

sketched in Fig. 6.2A. In favour of this interpretation we hold that the two sets are straight, equally well developed, and have evidently been formed simultaneously. It further seems, that the sediments were only weakly consolidated at the time of folding, which makes tensile fracturing less likely. But most importantly, the acute angle between the two sets of planar discontinuities is bisected by a line which could very well coincide with the direction of the maximum principal stress σ_I during compressive folding. The principal effective stresses would then be compressive (positive) as indicated in Mohr diagram B).

In principle, σ_{III}' could also be slightly tensile; but this would imply hybrid "extension-shear" fractures (see Fig. 1.8C, and G. Mandl, FBR, pp. 94–95) which, however, would likely have a rougher and less continuous appearance than seen at the location of Fig. 6.1. Further notice, that in Fig. 6.2A σ_I acts parallel to the bedding plane, indicating zero friction on the bedding plane. Although bedding plane friction in flexural slip folding may be very low, commonly due to high pore pressures, it will generally cause the direction of σ_I to be inclined towards the bedding plane at a certain angle; thus, both the σ_I- and σ_{II}-axes are rotated through this angle around the bedding-parallel σ_{III}-axis. Therefore, the σ_{II}-axis is no longer normal to the bedding; and since the intersection line of two conjugate shear planes is parallel to the direction of the intermediate principal stress (see e.g. G. Mandl, FBR, Fig. 6.67), the intersections of the conjugate shear planes on the fold limb should deviate from the bedding normal. The deviation should be in the co-rotational sense with the bedding plane slip. Although, regrettably, this feature was not inspected at the location of Fig. 6.1, it should prove to be an important indicator of conjugate shear joints.

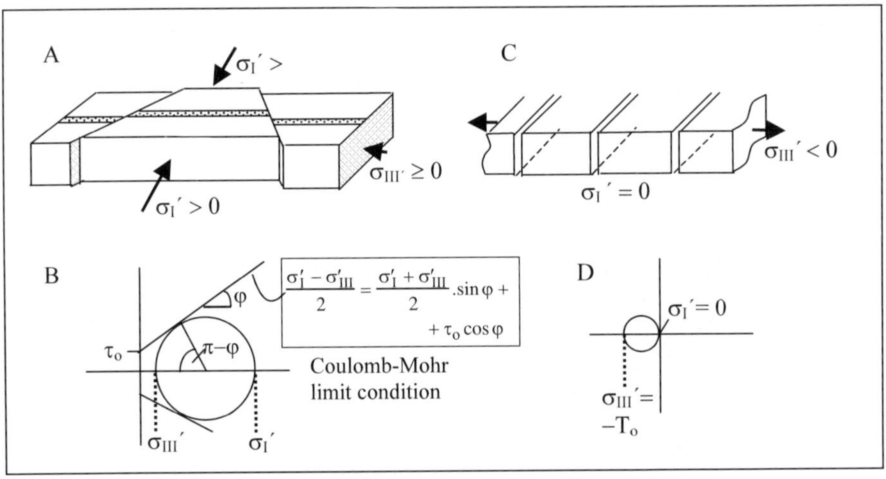

Fig. 6.2. Alternative mechanical interpretations of the fracture sets in Fig. 6.1.
A) Block diagram of wrench-type conjugate shear planes with σ_I- and σ_{III}-directions in bed-ding plane; σ_I-direction bisecting the acute angle between the shear planes; B) Mohr diagram of bedding-parallel effective stresses ($\sigma_I' > \sigma_{III}' > 0$, compressive) associated with A); C) Block diagram of a single set of tension joints with bed-parallel principal tensile stress σ_{III}' normal to fracture planes; D) Mohr diagram of bed-parallel stresses associated with C)

Nevertheless, we cannot be sure that the fractures in Fig. 6.1 are shear joints, although more elaborate constructions of biaxial or two-phase straining would be required to explain the two non-orthogonal sets of discontinuities as ("systematic") tension joints. Assuming that the sediments were sufficiently consolidated to attain any tensile strength at all, the reader should be reminded of several prerequisites for the formation of two non-orthogonal sets of systematic tension joints: (1) contrary to conjugate shear joints, the two sets of tension joints can only be formed in succession; (2) the first set must have acquired sufficient shearing strength to allow the straight intersection by the younger set (see the summary on p. 54); (3) while orthogonal sets of tension joints may be generated by biaxial extension with an interchange of the roles of σ_I' and σ_{III}', without rotation of the principal axes (see p. 102, Fig. 5.3), non-orthogonality would require a rotation of the principal axes around the σ_{II}' axis, either in addition to the switch between σ_I' and σ_{III}', or as rotation of the principal axes through the whole angle between the two joint sets.

In compressional folding, the last condition could be fulfilled in at least two ways: first, by a combined flexural and torsional deformation of the layers (p. 103, Fig. 5.4) when the first set is healed before the second set is formed by the progressing deformation. Secondly, the first set has already been formed (and healed) in the flat-lying foreland sediments of an arcuate fold and thrust belt (p. 88, Fig. 4.26) parallel to the compression direction of later folding, and the second set is formed parallel to the fold axis during later folding. As the directions of the regional compression and the fold axes may deviate from strict orthogonality, the two joint sets may be non-orthogonal (see also the anticlinal joint sets in Fig. 1.6, p. 3). In any case, the tension joint interpretation of two sets of discontinuities as shown in Fig. 6.1, requires that at least one set of discontinuities exhibits clear evidence of being healed in the past.

We may be more certain about the mechanical character of the fractures in Fig. 6.3. The family of sigmoidal fractures form a duplex structure connecting two parallel gliding horizons of a right lateral overthrust. Although shear displacement is hardly visible, closer inspection shows slickensiding on the fracture surfaces. The structure is typical of the upstepping of bedding-plane thrusts along secondary shears. (Various mechanical aspects

Fig. 6.3. Family of sigmoidal shear surfaces connecting two parallel gliding horizons of dextral overthrusting in turbidites. Slip surfaces quartz-coated and slickensided. (Widemouth Bay, North Cornwall, England)

of the process were discussed in G. Mandl, 1988 (1993), *Mechanics of Tectonic Faulting*, Elsevier; henceforth referred to as MTF, see Sect. I.5.3.) Observations in the field and the mechanical analysis strongly suggest that the fractures were generated as shear fractures rather than tension fractures which were sheared later.

Although, as indicated by the last example, in some field cases fractures may be clearly identified as shear fractures, much more often it will be impossible to decide between tension or extension (cleavage) joints, and shear joints, and to draw conclusions regarding the nature of the causative stress system.

The concept of shear joints was criticized as "sheer nonsense" in an excellent review article by David Pollard and Atilla Aydin (1988; Geol. Soc. Am. Bull. *100*, pp. 1181–1204), because by combining "shear" and "joint" the concept mixes fracture types of quite different mechanical origin. The authors argue, that a fracture generated in a shearing mode should be called a "fault", and the term "joint" should be restricted to fractures with field evidence for dominantly opening displacements. If the history of relative displacements and the fracture modes involved are uncertain, the observed discontinuities should be simple referred to as "fractures".

Despite the logic in this argument, we shall adhere to the term "shear joint" for three reasons: first, a tectonic fault is a narrow shear band, generally produced by shear concentration in a damage zone, rather than developed by the growth of a single shear fracture (see G. Mandl, FBR, Sect. 3.4). Further, an important attribute of faulting is the reduction in shear strength of the fault material by the shearing process. Hence, we use the term "*shear joint*" to distinguish a single *shear fracture* from a *proper fault*. Secondly, following N.J. Price (N.J. Price and J.W. Cosgrove (1990) loc.cit.) we also apply the term to *shear bands* whose *aspect ratio*, i.e. the ratio of the maximum shear displacement to the total length of the shear zone, is a hundred or thousand times smaller than the aspect ratio commonly estimated for tectonic faults of a wide range in length. Thirdly, we consider shear bands as shear joints, if they formed at a "*pre-peak*" state of stress of the sheared frictional material, as indicated in Fig. 6.4, and further shearing stopped before the maximum shear stress could pass the peak value. Therefore, shear zones of this kind lack the characteristic softening of true faults. Although shear bands of still hardening material are rather unconventional in geology, we consider them of great practical importance, and shall elaborate on their mechanical genesis at the end of this chapter.

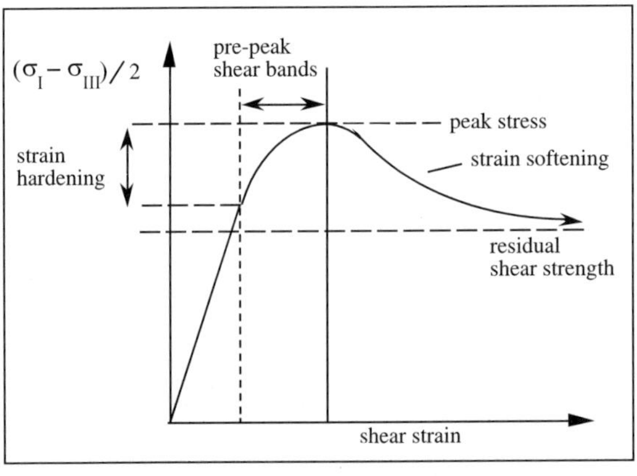

Fig. 6.4. Shear strain response of material elements under constant effective mean stress. Pre-peak shear banding under plane strain conditions in the hardening range (see further discussion at the end of this chapter)

Origins of shear joints. In any case, shear fractures or shear bands are considered as shear joints, regardless of their origin, if they did not develop into faults *sensu-stricto*. Then, our main problem is to explain why the shearing processes failed to develop shear discontinuities into proper faults. Probably the most important reason for the early arrest of shear displacement is the *stress drop* associated with the formation of a brittle fracture or shear band. In a cohesive brittle rock, only a very small shear displacement is needed to break the cohesive bonds between fracture walls or inside an embryonic shear band, thereby reducing the shear resistance of the fracture or the shear band. Similarly, the shear resistance of shear bands in loose material is reduced by dilatancy, grain breakage, or the reorientation of grains, etc. This causes the differential stress in the surrounding area of the shear fracture or shear band to drop below the critical level necessary for the fracture or shear band to develop further. Hence, the differential stress would have to be built up again to allow shear fractures and shear bands to grow in length and shear displacement. The situation is similar to seismic faulting, where after every slip event the differential stress is built up again by the ongoing tectonic deformation. If after a first slip event the unloading is not undone, no further slip will occur.

This is, in particular, the case when *compressional residual stresses are unloaded* by a single slip event. Such residual stresses may have quite different origins, as was discussed in the section on *"Locked-in" or "residual" stresses* in the preceding chapter (pp. 113–117, 124). For instance, the uplifting of sedimentary basins leaves horizontal compressive stresses locked-in in near-surface layers.

In other cases, a switch of principal stresses may cause shear jointing. Such a tectonic situation is sketched in Fig. 6.5A, showing, in plan view, a horizontal layer that is compressed by the horizontal stress σ'_{h1} (= σ'_I). The layer is laterally confined and assumed to lie on a lubricating substratum. Because of the lateral constraints, a transverse horizontal compressive stress σ'_{h2} is induced by σ'_{h1}, which may be considerably greater than the overburden stress σ'_v, as stated in the Mohr diagram of Fig. 6.5A′. When the horizontal compression by σ'_{h1} is relaxed, σ'_{h2} remains "locked-in" (Fig. 6.5B), and a new state of stress is established, as indicated by the shaded stress circle in Fig. 6.5B′. The new state of transverse compression could be released by shear joints of the strike-slip type, as sketched in Fig. 6.5B.

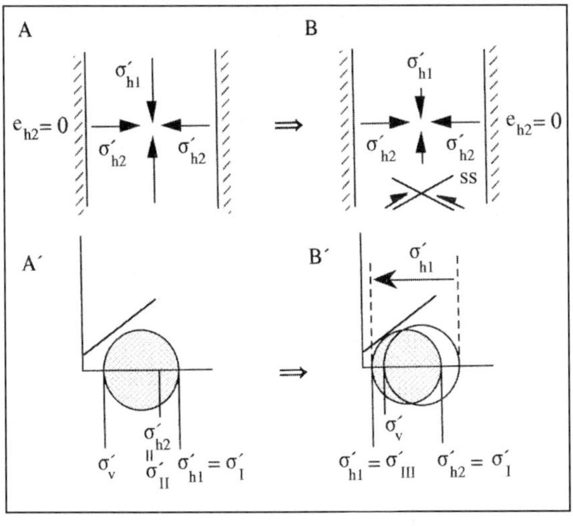

Fig. 6.5. Strike-slip shears induced in laterally confined horizontal layers when the original thrust force (A) is relaxed and the roles of principal stresses are changed (B)

Another mechanism that may produce shear joints rather than faults is associated with the *"bookshelf" mechanism* (see G. Mandl, FBR, Sect. 6.4). We consider the "domino"-style rotation of antithetic Coulomb shears in a layer under direct shear, as sketched in Fig. 6.6A. The flattening of the shear fractures is accompanied by thinning and extension of the sheared layer, as indicated in Fig. 6.6B. In Fig. 6.6C the slip Δs on the rotating fractures is determined and related to the fracture length s, which leads to an extremely simple formula for the aspect ratio Δs/s as a function of the initial inclination angle α the rotation angle β, and the ratio of the thickness of the rock slices t and the layer thickness H.

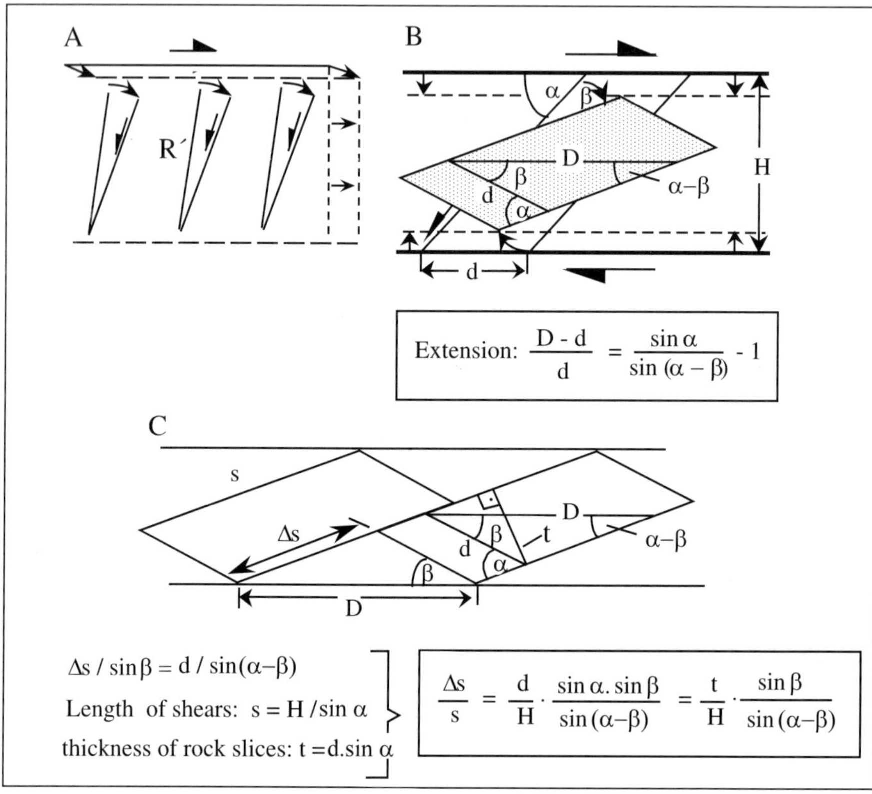

Fig. 6.6. Domino-style rotation of en-échelon shears:
A) Rotation of parallel antithetics by external dextral shear; B) state of a rock slice after rigid-body rotation through angle β, and extension of the fractured layer; C) rotational slip Δs and aspect ratio Δs/s

The domino-style formation and collective rotation of parallel shear fractures in a layer under direct shear can only operate as long as the layer-parallel normal stress $\sigma'_{\parallel}$is smaller than the normal stress $\sigma'_{\perp}$ on the layer boundaries (again see G. Mandl, FBR, Sect. 6.4). If, however, the extension of the layer during the collective rotation of the rock slices is impeded by lateral boundaries, the layer-parallel normal stress $\sigma'_{\parallel}$ will rise rapidly and stop the rotational process, even though the driving shear stress is maintained, or even increased. As mentioned before, an early arrest of the rotation may result in the formation of shear joints rather than faults. This is expressed, in more quantitative terms, by means of the formulae in Fig. 6.6. Let us assume, for example, that the parallel fractures are initially inclined at $\alpha = 60°$, and the confinement of the brittle layer permits an extension of not more than, say, 0.5%. The extension formula in Fig. 6.6B then predicts a maximum rotation $\beta = 0.5°$, with a maximum aspect ration $\Delta s/s = 0.01.t/H$. Since the length of the fractures transecting the layer is $s = 1.15H$, the rotational slip Δs on the fractures is about 1% of the thickness of the joint-bounded slices (fracture spacing) t. This illustrates the difficulty one may face in distinguishing shear joints from tension or extension joints in en-échelon arrays of relatively densely packed fractures.

Échelon arrays of strike-slip joints are not only produced by the shearing displacement of stronger and stiffer boundary rocks, such as the wall rocks of faults, overriding thrust sheets or lateral graben boundaries, but may also be generated in a stationary field of horizontal principal stresses σ_I and σ_{III}. Consider, in Fig. 6.7, a large sheet of strong and brittle rock loaded by an overburden and resting on a weaker substratum. The layers are uniformly shortened in one horizontal direction and extended in the direction perpendicular to it, allowing horizontal principal stresses σ_I and σ_{III} to develop in the brittle layer. As the uniform deformation continues, the principal stresses approach a critical level in the brittle sheet, and shear fractures start forming at randomly distributed flaws or other material heterogeneities. Since the principal directions are constant throughout the brittle layer, at least on a scale that is large relative to the dimensions of material heterogeneities, the fractures develop into vertical strike-slip planes of conjugate orientations associated with the σ_I direction. Such shear planes may be arranged into en-échelon arrays by the interaction of several mechanisms which mainly result from the interference of the local stress fields of neighbouring fractures (for an exploratory account see G. Mandl, FBR, pp. 403–404).

A point of special interest, is the sense of the échelon arrangement of shear fractures, i.e. the sense of the step-over from fracture to fracture. In frictional materials, the near-tip stress field of shear fractures is asymmetric with respect to the fracture plane and first attains the critical state for shear failure at the receding sides (–) of the fractures. Therefore, in en échelon arranged (Andersonian) strike-slip fractures, the direction of the step-over should depend on the sense of slip on the individual fractures: right lateral (or left-lateral) shear joints should step to the right (or left) side, as shown in Fig. 6.7.

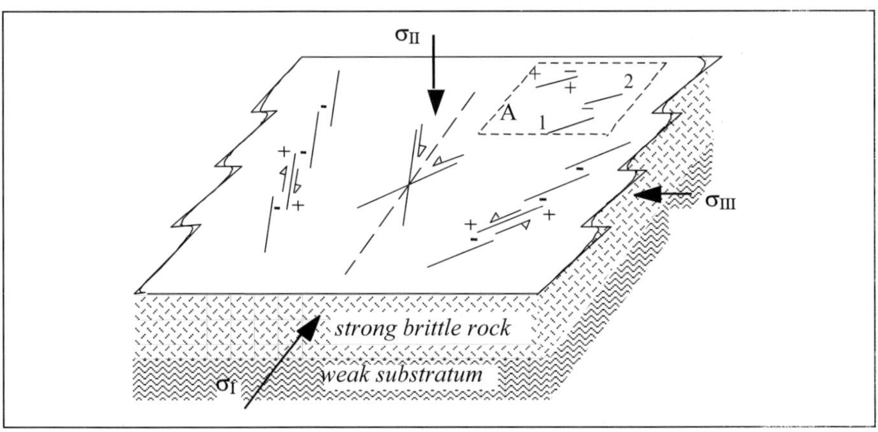

Fig. 6.7. Conceptual échelon arrays of strike-slip shear joints (or faults) in a large horizontal sheet of brittle rock under uniform horizontal shortening and extension. Step-over of shear planes to the receding side (–) are indicated by the insert A

Yet another interesting mechanism was proposed by N. J. Price (see N.J. Price and J.W. Cosgrove (1990) loc.cit., pp. 227–238) to explain the occurrence of large regional joint sets in tectonically undisturbed sediments. Consider, in Fig. 6.8 an undisturbed horizontal sediment sequence which is gently *tilted*. In the figure, it is assumed that the tilting is caused by a set of normal basement faults, while more ductile interlayers prevent the basement faults from printing through into the overburden. One can also envisage the tilting of the sediment layers by the gradual thinning of ductile substrata in unidirectional flow (substratal squeeze flow). In any case, the tilting imposes a vertical simple shear, as indicated in Fig. 6.8.

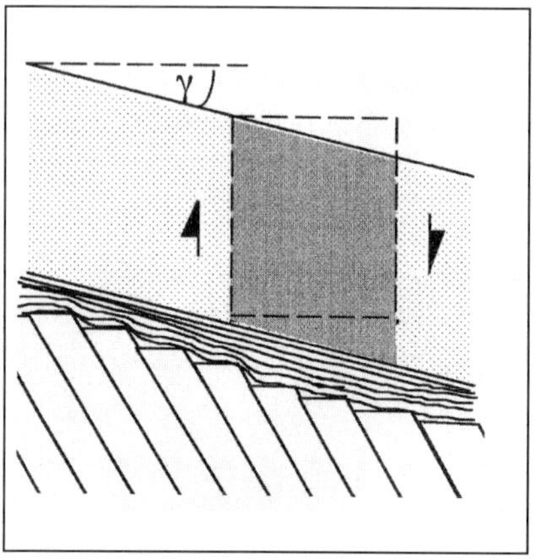

Fig. 6.8. Regional tilting imposing vertical simple shear on an undisturbed layer

We assume that prior to tilting the overburden stress is the maximum principal stress $\sigma_I'^{\circ}$ as indicated in Fig. 6.9A for the vertical section through the axis of $\sigma_{III}'^{\circ}$. The value of $\sigma_I'^{\circ}$ in the Mohr diagram of Fig. 6.9A′ is chosen for a layer with normal fluid pressure under 2 km overburden, and the tectonically undisturbed effective stress $\sigma_{III}'^{\circ} = K_o.\sigma_I'^{\circ} = 0.4.\sigma_I'^{\circ}$. Following N.J. Price, we assume that the layer-normal and layer-parallel normal stresses are not changed when a vertical simple shear τ (Fig. 6.9B) is superimposed. Thus, the stress components on a layer-parallel section are $\sigma_I'^{\circ}$ and τ as shown in the Mohr diagram of Fig. 6.9B′. As τ increases the stress circle expands concentrically until it touches the limit lines. The orientation of the new maximum principal stress σ_I' and the directions of the potential shear planes are determined by the pole method (see the Appendix on the Mohr circle). Notice that the trace of one of the shear planes in Fig. 6.9B′ is almost vertical. The stresses on this shear plane are the coordinates $\sigma'_\perp{}^*$, $-\tau^*$ of the tangent point on the lower limit line, which therefore satisfy the condition $|\tau^*| - \tau_o = \sigma'_\perp{}^*.\tan\varphi$.

It is important to note that the normal effective stress $\sigma'_\perp{}^*$ on the subvertical shear planes at 2 km depth is about 14 MPa. Hence, the shear planes are firmly closed, and opening of the fractures would require a rise in pore pressure of at least 20 MPa. In contrast, when interpreted as tension joints, the fractures would have acted, or may still do so, as preferential conduits for fluid migration. This illustrates the practical importance of distinguishing between joints and shear joints.

It is interesting, and perhaps surprising, that the required tilt angle is very small. This can be demonstrated by a simple estimation. Because $\sigma'_\perp{}^*$ and τ^* differ relatively little from σ'°_{III} and τ, respectively, we approximate the limit condition of the near-vertical shear plane by

$$\left|\tau\right| - \tau_o = \sigma_{III}'^{o}.\tan\varphi = K_o.\sigma_I'^{o}.\tan\varphi \tag{6.1}$$

We further assume a predominantly elastic behaviour of the sedimentary rocks during tilting – an assumption that appears justified if the sediments are well consolidated and the shear angle remains of the order of tenths of a degree. The imposed shear stress τ (Fig. 6.9B) is then related to the shear angle γ by $\gamma = \tau/G$, where G is the elastic shear modulus. From Eq. 6.1, we then obtain the following estimate of the shear angle

$$\gamma(\text{rad}) \approx \frac{\tau_o}{G} + \frac{K_o}{G}\sigma_I'^{o}.\tan\varphi \qquad (6.2)$$

Let us assume that the sedimentary rock has the shear modulus $G = 10^4$ MPa. Then, with $K_o = 0.4$, $\varphi = 30°$, $\tau_o = 5$ MPa, the above expression predicts the extremely small tilt angle $\gamma \approx 1.2 \times 10^{-3}$ rad, i.e. 0.07°. Accommodation of this tilt angle by parallel vertical shear planes with separation d would require an average slip $\Delta s \approx 10^{-3}.d$ on each plane. In reality, the displacements on the shear planes may be even smaller, such as when a proportion of the vertical shear is accommodated by continuous deformation.

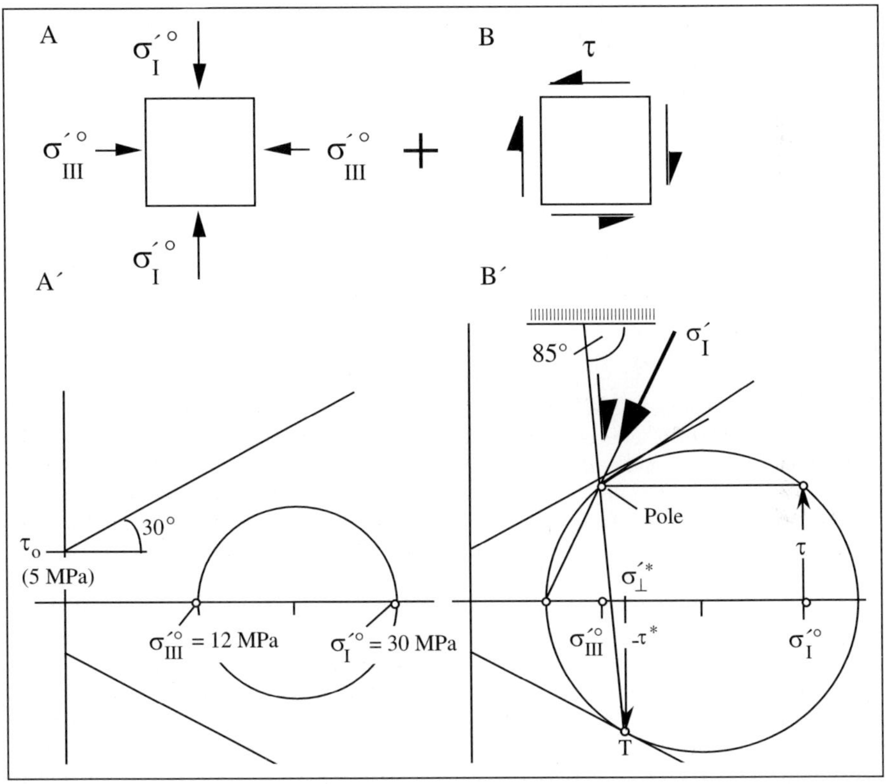

Fig. 6.9. Vertical simple shear imposed on a layer initially under vertical maximum compressive stress ($\sigma_I'^{o}$). A) Vertical cross section of the unit element in initial state, with corresponding Mohr diagram A´. B) Superimposed vertical simple shear τ, and Mohr diagram B´ of the resulting state of stress with pole construction of potential shear planes

Regarding the extreme smallness of the tilt angle, N.J. Price considered it very likely that most sediments experience several periods of tilting in different directions. If the tilt axes differ sufficiently in direction, several sets of differently striking subvertical shear joints could develop.

Furthermore, it is obvious from the construction in Fig. 6.9B´, that changing the material parameters φ, τ_o, G, K_o affects the dip angle of the steep shear plane, and the value of the necessary vertical shear stress τ. The reader may easily verify, for instance, that increasing φ from 30° to 35°, while leaving all other parameters and the initial state of stress unchanged, would imply that τ has to be raised from 11.8 MPa to 13.4 MPa, and the steep shear planes would become vertical. Hence, to generate these shear planes, the tilt angle has to be increased slightly. The influence of the effective overburden load on the value of τ is more important: if, for instance, the layer considered in Fig. 6.9 is buried under 4 km of overburden (σ'°_{I} = 60 MPa) instead of 2 km, τ would have to attain 18 MPa, and the steep shear plane would dip at 81.5°.

The increase in shear stress τ with depth does not necessarily require a proportional increase in tilt or shear angle γ (= τ/G), since one may expect the elastic shear modulus G to increase as the material gains in stiffness under higher effective overburden stress. Nevertheless, one should expect a tendency in the process of sub-vertical shear jointing to progress from near-surface layers to greater depth as the tilt angle is increased.

As we have seen, the dip angle of the steep joints is not only affected by the effective overburden, but also by the material parameters φ, τ_o, G, K_o. Although the variation in the dip angle will hardly exceed a few degrees, it may cause kinks or "*facets*" in the sub-vertical joint plane, which, as suggested by N.J. Price, may be diagnostic features of vertical-shear joints.

Pre-peak shear bands. So far we have considered shear joints which, although initiated by the same stress conditions as faults, got stuck in an embryonic state of development, since the differential stresses were not restored after the drop caused by the softening of the shear band material. From these "embryonic" faults, we now turn to the completely different type of shear joints that are formed while the sheared material is still *in the hardening state*; that is, where the shear resistance of the intrinsic material still rises with further shearing (Fig. 6.4). Such a "pre-peak" shear localization is an uncommon concept in geology, where shear concentration in a uniform material is usually attributed to local strain softening of the material, and/or to inhomogeneous boundary conditions promoting inhomogeneous deformations. It is indeed quite plausible that shearing in the deforming rock begins to concentrate at a place where a slight inhomogeneity of the material, or of the stress field, promotes an early decrease in local shear resistance, and allows shearing to progress with ever greater facility, while the surrounding material may still be hardening. But it is not plausible at all, that shear bands can form while the frictional material is still hardening *everywhere*, and we therefore have to take a critical look at the available evidence of "pre-peak" shear localization in frictional materials.

Since it is practically impossible to decide whether a shear band in the field has formed before or after the loading peak, primarily one will look for *experimental* evidence. It is indeed not difficult to find cases reported in the literature on biaxial or triaxial rock testing where shear bands or shear fractures have appeared *before* the differential boundary stress had attained the limit which the rock specimen could withstand. However, for several reasons, it is difficult to draw firm conclusions from these observations. First of all, the development of a shear band causes a macroscopic inhomogeneity in the mechanical properties of an initially homogeneous and uniformly stressed specimen, accompanied by a redistribution of the stresses. The material then no longer deforms as a homogeneous continuum, and the measured boundary stress-strain response need no longer represent the stress-strain behaviour of the intrinsic material. In other words, the stress-strain curve of the shear zone *material* will separate from that of the specimen as a whole, which is called the *system*. We have to add, however, that the point of separation on the loading path cannot be very accurately determined, since the shear bands develop rather gradually. Moreover, experiments also indicate that the shear concentration first occurs in separate patches which subsequently connect into a continuous band.

It is therefore clear, that only under very special testing conditions can one decide whether the material of a shear band, that forms prior to the peak of the differential boundary stress on the rock specimen, is in a softening state or still hardening. A suitable test, indicated in Fig. 6.10A, is the plane strain uniaxial compression of a perfectly homogeneous rectangular sample, which is confined between lubricated front and back walls, and loaded axially via well-lubricated end platens, and laterally loaded by a constant fluid pressure p. Consider then, in Fig. 6.10B, a shear band that asymmetrically intersects the sample, without terminating at the loading platens. If the development of this band requires an increase in the vertical boundary stress, the shear-band material is obviously still in a hardening state, since softening of the asymmetric shear band (i.e. post-peak state on the material stress-strain curve) would cause a drop of the load-bearing capacity of the system. By contrast, such a conclusion could not be drawn if shear bands developed in a symmetric fashion, as sketched in Fig. 6.10C. It would seem possible that, even if the shear bands are softening, the internal stress field, and, in particular, the σ_I directions, may be redistributed so as to support a further raise of the external load.

Experiments of the type shown in Fig. 6.10B were carried out with sand, by Tatsuaoka, Nakamura, Huang and Tani (1990; Soils and Foundations *30*, No. 1), and very carefully analyzed. It was found that a continuous shear band developed asymmetrically across the sample before the uniaxial load reached its peak. Regrettably, similar plane strain experiments on hard rocks have not yet been performed. Therefore, instead of relying on more direct evidence for the existence of hardening shear joints, we have to content ourselves with circumstantial evidence.

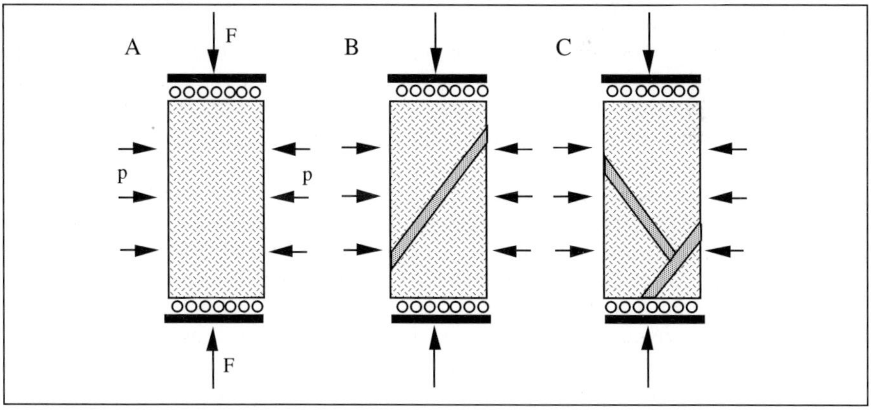

Fig. 6.10. Rectangular rock sample in uniaxial plane strain compression: A) before shear localisation; B), C) incipience of shear banding

An observation in support of the formation of hardening shear bands was made in a model experiment with dry sand, instigated by the author at Shell Research (1984–1985), to study processes of faulting by X-ray computer imagery. Figure 6.11 shows two X-ray tomography scans of a vertical section through a sandpack on a horizontal base which consists of two wooden plates covered by a rubber sheet. The sheet was glued to the plates, except for a central strip that was left free, and could be uniformly stretched parallel to the plane of the section in the figure, by pulling the base plates apart. The sand pack in Fig. 6.11A shows an extension of the rubber strip of 5%, at which the adjacent grains

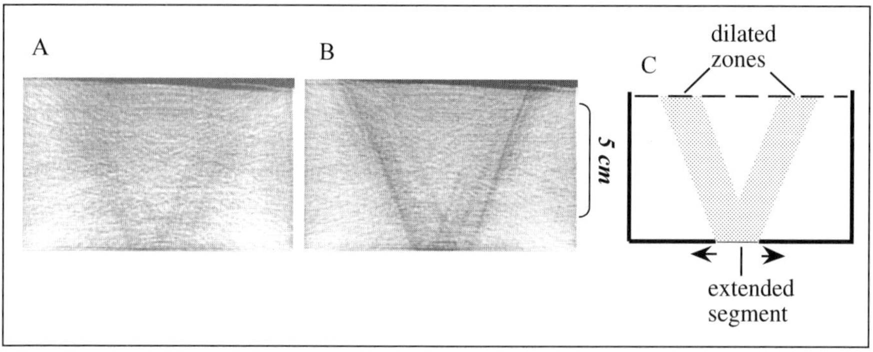

Fig. 6.11. Incipience of shear bands in a broad zone of dilated material: X-ray tomography image of sand pack section after 5% (A) and 10% (B) horizontal extension of a central strip of the rubber base, as schematically illustrated in (C)

experienced a relative displacement of only about 5% of the average diameter of the grains used in the experiment. Although higher up in the sand pack the grain displacement was even less, the dilatancy concentration was already clearly visible. In Fig. 6.11B, the extension of the basal strip is 10%, and the photograph shows two darker zones, indicating increased dilatancy in these zones, since the X-rays are absorbed less where the pore space is increased. The photographs give an indication of the extreme sensitivity of the method in detecting early dilatancy concentration.

Inside each dilatancy zone in Fig. 6.11B, dilatation has already concentrated into several closely spaced, parallel shear bands. The important point now, is that the inelastic volume strain $\varepsilon^p = \delta V_p/V$ which accompanies the shearing of a frictional material begins long before the differential stress $(\sigma'_I - \sigma'_{III})/2$ reaches its peak (cf. Fig. 6.4). In granular materials under moderate effective load, the inelastic shear deformation proceeds by loosening the interlocking of densely packed grains, and thereby increasing the intergranular pore space, by intergranular slippage, and reorientation of load-carrying grain contacts (see e.g., Fig. 5.10, p. 116). Up to a certain state of dilation of a volume element, the operation of these micro-processes requires an increase in load on the volume element, which implies that the material is hardening while dilating. Only when the loosening of the granular aggregate has reached a point where the collapses of load-carrying grain bridges outnumber the new formations, does the material start to soften while continuing to dilate, though at a decreasing rate ($d\varepsilon^p/d\gamma$, relative to the shear strain γ. We may therefore safely conclude, that the dilatations in Fig. 6.11, and the early dilatation observed in similar CT-experiments, took place *prior to peak stress* of the granular material. Consequently, the shear bands that formed during the early dilatation period are also very likely pre-peak phenomena. We further draw attention to the fact, that not all shear bands in Fig. 6.11 have a similar density, but that in one pair of conjugate shear bands at least, the dilatation has proceeded further. These bands might be seen as precursors of a selective development of pre-peak shear bands into true faults.

Shear dilatancy is a common feature of shear zones throughout the brittle crust of the Earth. But, of course, the dilational processes in strong brittle rocks are quite different from those in loose granular materials, and mainly involve the formation and growth of micro-fractures. These differences in the deformation mechanisms at the microscale raise the crucial question: Do shear bands of the type in Fig. 6.10B form while the rock is still hardening throughout?

In answering this question, we anticipate a result of the theory of shear banding, to which we shall turn a little later, which states that pre-peak shear-banding requires *plane or nearly-plane strain* conditions. Hence, on theoretical grounds, pre-peak shear-banding (of the type in Fig. 6.10B) is not to be expected under the axially-symmetrical loading conditions that are commonly applied in rock-mechanical laboratories. Thus, the experimental verification of the theoretical result would require plane strain experiments on hard rocks, which would visualize, or otherwise record, the appearance of asymmetric shear bands. (Actually, it would be sufficient to load cuboidal specimens in a true triaxial apparatus, and to inspect the shear concentration in the mid-plane of the sample parallel to the σ_I and σ_{III}-axes, which for symmetry reasons is under plane strain.) Since, regrettably, such data are not yet available, we have to rely on theoretical analyses, which indeed provide the strongest arguments for a pre-peak formation of shear bands.

Various authors have undertaken computer simulations of shear-banding in frictional (Coulomb-type) materials. Despite the differences in numerical techniques, and in the refinement of approximating the material behaviour, all simulations are based, in one way or another, on the theory of elastic/frictional plastic materials. Here is not the place to indulge in the theory (for a detailed presentation see G. Mandl, FBR, Sect. 9.2; or MTF, II.7), but we should at least reiterate the three basic constituents of the theory. These are (*1*) the stress equations of static or dynamic equilibrium (i.e. including inertial forces); (*2*) the yield condition which must be satisfied by the stresses during plastic deformation (i.e. a Coulomb-Mohr-type limit condition with varying strength parameters φ and τ_o); and (*3*) the elastic/plastic flow rules that connect increments of stresses and strains. Plastic hardening and softening is accounted for in the yield condition by strain dependent changes in the strength parameters φ and τ_o, which include the hardening and softening effects of shear dilatancy. The volumetric changes by shear dilatancy are accounted for in the flow rules by introducing a "dilatancy factor" defined as the increment of volumetric plastic strain $\delta\varepsilon^p$ (> 0) divided by the corresponding increment of the maximum plastic shear strain $\delta\gamma^p{}_{max}$.

In a series of numerical experiments, P. Cundall (1990; in: H.P. Rossmanith (ed.) *Mechanics of Jointed and Faulted Rocks*, Balkema, pp. 11–18) simulated graben faults that were observed in sandbox experiments by W.T. Horsfield (in G. Mandl, MTF, Fig. I.2-23) where plane strain conditions were fairly well approximated. Figure 6.12A,B shows an intermediate state in the development of a typical crestal collapse graben produced in the sandbox by the post-sedimentary tilting of a "basement" block. The numerical simulation was carried out with the large-strain, finite difference program FLAC 2.3, developed by Cundall himself. Figure 6.12C,D gives two representative results in the form of contour plots of the maximum shear strain-rate. Each contour bundle defines a shear band.

The computational results are remarkable for several reasons: first, the elastic/frictional plastic model of the cohesionless material was rigorously simplified by assuming zero dilatancy ($\delta\varepsilon^p = 0$) and by letting the plastic deformation ("yield") in the simulation of Fig. 6.12C start and proceed at a constant friction angle $\varphi = 40°$. Thus, shear softening was completely excluded, and the yield condition was identical to the Coulomb-Mohr limit condition. Surprisingly, shear bands did form, despite the perfect-plastic behaviour (i.e. without hardening and softening). However, because of the lack of softening, the shear bands do not represent faults, but shear joints. In contrast to this, the shear bands in Fig. 6.12D formed during weak strain-softening of the yielding material, whereby φdecreased linearly from 40° to 35° over an accumulated plastic strain of 10%. Since the shear bands have a slightly smaller shear strength than the adjacent material, they represent genuine faults, which are fewer in number than the non-softened shear bands, and similar to the faults observed in the sandbox.

One may further note that the shear joints in the non-softening material are more or less straight, while curvature is associated with strain-softening. According to Cundall, the shear bands start near the free surface, where, due to the greater extension, yielding begins. The initial fault is "locked in" by shear-softening and rotated with the bending layers, while the fault grows downwards towards the hinge in the basement.

The initial stress field in the sediment cover of Fig. 6.12 is non-uniformly changed by the tilting of the basement block, and the perturbation is concentrated in a region above the basement hinge, wherein plastic deformation begins and leads to the formation of shear bands. Although, undoubtedly, stress concentrations can trigger the shear concentration into shear bands, it is important to note that shear bands have also been generated numerically in a completely homogeneous set up. In an interesting paper, A.N.B. Poliakov, H.J. Herrmann, Y.Y. Podladchikov and St. Roux (1994; Fractals *2*, No. 4, pp. 567–581, especially Fig. 3) report shear-band simulations by Cundall's FLAC code, whereby pure shear was applied to a square sample of the same homogeneous non-softening material as

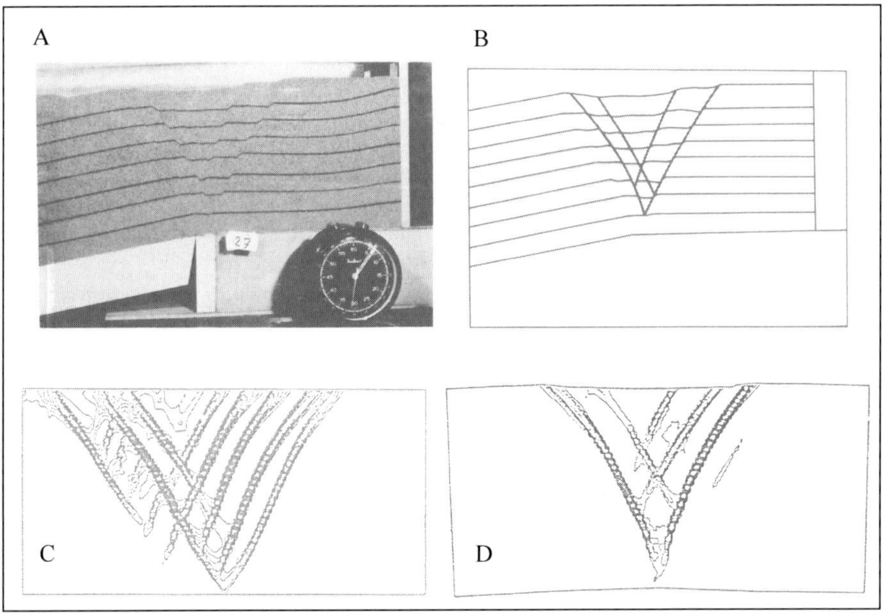

Fig. 6.12. Graben formation by post-sedimentary tilting of a "basement" block, simulated by sandbox experiment and numerical analyses:
A) Photograph of an intermediate stage of graben formation in a sandbox with the left-hand "basement" block tilted towards left.
B) Sketch based on stereo viewing of photos showing active faults (W.T. Horsfield, Shell Research, 1978);
C and D) Numerical large-strain analyses of elastic/frictional plastic deformation under sandbox boundary kinematics by P.A. Cundall (1990, *Numerical modelling of jointed and faulted rock*; in: H.P. Rossmanith (ed.) *Mechanics of Jointed and Faulted Rocks*, Balkema, pp. 11–18).
The bundling of contours of maximum shear strain increments during a deformation step is clear evidence of shear band formation (see text)

used by Cundall in Fig. 6.12C. The calculated shear bands, shown in Fig. 6.13, form a network of conjugate bands, reminiscent of real geological structures. In this simulation study, the shear bands were not triggered by the boundary geometry, material inhomogeneities, or stress concentrations, but by computational round-off errors.

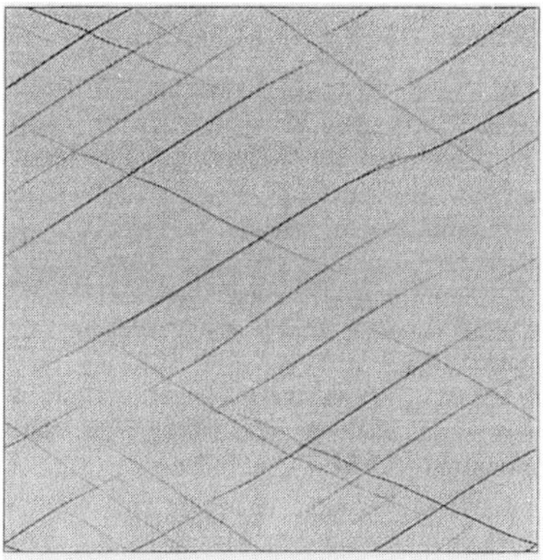

Fig. 6.13. Network of shear bands produced by computer simulation (FLAC-type technique) by A.N.B. Poliakov, H.J. Herrmann, Y.Y. Podladchikov and St. Roux (1994; Fractals *2*, No. 4, pp. 567–581). Simulation of pure shearing of a two-dimensional pressure-dependent elasto-plastic medium under confining pressure; the material was *non-dilating*; and the shear bands in the figure accommodate the accumulated plastic strain after 10^4 time steps

So far, we have mentioned shear-band simulations in non-softening materials which also did not harden. Next, we will refer to the finite element analysis of shear-band formation in a hardening, non-softening frictional material by Y. Leroy and M. Ortiz (1990; Int. J. Numerical and Analytical Methods, Geomechanics *14*, pp. 93–124, see Fig. 10). A rectangular sample of the non-softening and non-dilating ($\delta\varepsilon^p = 0$) material was subject to plane strain uniaxial compression (Fig. 6.10A). The friction angle φ was assumed to increase monotonically (hardening parameter) from 10° to 20° as a function of the plastic strain. A softer element (φ = 17°) at the center of the specimen served as a material imperfection, from which two conjugate shear bands propagated as the axial boundary displacement was raised sufficiently.

It is gratifying that this numerical result is in accordance with a condition for the earliest incipience of shear banding that was derived analytically in the classical paper by J.W. Rudnicki and J.R. Rice (1975; J. Mech. Phys. Solids *23*, pp. 371–394). The elastic/frictional plastic material is unconstrained, homogeneous, dilating, strain hardening and softening, and initially homogeneously deformed and uniformly stressed. It is subjected to a quasi-static continuation of the deformation that *could* give rise to a homogeneous field of strain rates. That is, a homogeneous field of strain rates would be compatible with the equilibrium equations, yield condition, and the elastic-plastic flow rules. However, the analysis reveals that, at a critical state of plastic straining, a highly inhomogeneous field of localized shear rates also becomes possible. In other words, the solution to the continuum mechanical problem *bifurcates* from a homogeneous deformation into a highly concentrated shear band mode. The critical state of the first possible shear band formation corresponds with a critical value of the plastic hardening modulus, which in simple shear would play the same role as the shear modulus G in elastic simple shear. This is illustrated in Fig. 6.14, which shows a schematic shear stress vs. shear strain curve at constant hydrostatic stress, and the relationship between the tangent modulus h_{tan}, the plastic hardening modulus h_{pl}, and the elastic shear modulus G. If the material is still in a hardening state (i.e. pre-peak), the plastic hardening modulus is positive; it vanishes at peak, and is negative in the softening range.

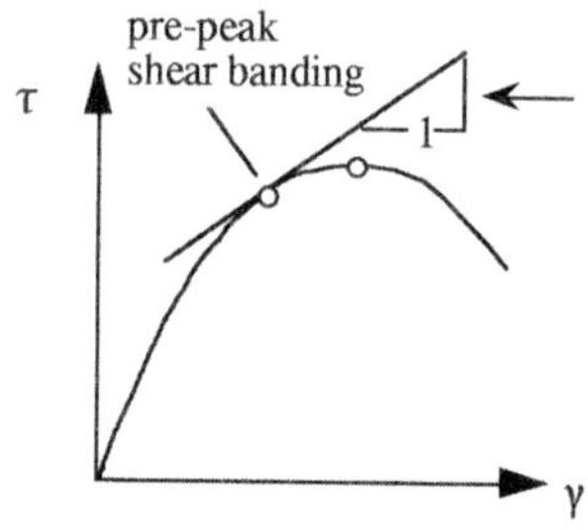

$$h_{tan} = d\tau/d\gamma \quad (1)$$

$$h_{pl} = d\tau/d\gamma_{pl} \quad (2)$$

$$d\gamma = d\gamma_{el} + d\gamma_{pl} = (d\tau/G) + d\gamma_{pl} \quad (3)$$

$$(1), (2), (3) \rightarrow h_{tan} = \frac{d\tau}{(d\tau/G) + d\gamma_{pl}} = \frac{h_{pl}}{1 + h_{pl}/G} \approx h_{pl} \quad (4)$$

Fig. 6.14. Material shear stress vs. shear strain curve at constant confining stress. The tangent modulus h_{tan} (Eq. 1) is related to the plastic hardening modulus h_{pl} (Eq. 2) and the elastic shear modulus G by Eq. 4 in the figure

The formula for the critical plastic hardening modulus $h_{pl\ cr}$ derived by Rudnicki and Rice (their Eq. 20) reduces for *plane strain* to

$$h_{pl\ cr} = \frac{G}{18} \cdot \frac{(1+\nu)^2}{1-\nu} \cdot (\mu - \beta)^2 \tag{6.3}$$

where G is the elastic shear modulus, ν the elastic Poisson ratio, and μ and β are generalized friction- and dilatancy moduli, respectively, which belong to the hardening state. Hence, whatever the exact values of the frictional and dilatancy moduli are, the critical hardening parameter is non-negative.

Therefore, if we exclude the unrealistic cases with $\mu = \beta$, in any plane strain deformation, a range of positive hardening moduli $h_{pl} \leq h_{pl\ cr}$ exists where shear banding becomes possible *before* the peak stress is attained. Moreover, according to the unabridged formula in the Rudnicki and Rice paper, pre-peak shear banding is even possible in deformations that are not strictly planar, but sufficiently close to plane strain. In contrast to this, axially symmetric compression permits shear band formation only in the post-peak state, i.e. when the material is softening. This result was corroborated by other authors who derived a structurally similar formula for $h_{pl\ cr}$ by using the Coulomb-Mohr type formulation of frictional plasticity (see Eq. 8.7 in P.A. Vermeer and R. de Borst (1984), Heron *29*, pp. 1–64):

$$h_{pl\,cr} = \frac{G}{8} \cdot \frac{(\sin\varphi_h - \sin\psi_h)^2}{1-\nu} \tag{6.4}$$

In this formula, the angle φ_h is not a constant, but the "mobilized friction angle" which monotonically increases with strain in the inelastic hardening range, until it reaches its maximum at the peak of the stress-strain curve (Fig. 6.14), where it becomes identical to the friction angle of the Coulomb-Mohr theory. Similarly, ψ_h is a "mobilized dilatancy angle", which attains its maximum ψ at peak stress. The dilatancy factor sin ψ_h is defined as the ratio of the increment of volumetric plastic strain $\delta\varepsilon^p$ to the corresponding increment of the maximum plastic shear strain $\delta\gamma_{pmax}$.

$$\sin \psi_h = \frac{d\varepsilon^p}{d\gamma^p_{max}} = \frac{-dV^p / V}{d\gamma^p_{max}} \tag{6.5}$$

where the negative sign is inserted, since an increase in volume is counted as negative in this book.

Note, that the dilatancy angle ψ_h is defined here by means of the sine function, in analogy with the sin φ term in the Coulomb-Mohr limit condition (Fig. 6.2B). But a slightly different dilatancy angle may be defined by replacing the left hand side of Eq. 6.5 by tan $\underline{\psi}_h$. To avoid confusion, the difference between the two dilatancy angles is pointed out in the diagram of incremental strains in Fig. 6.15.

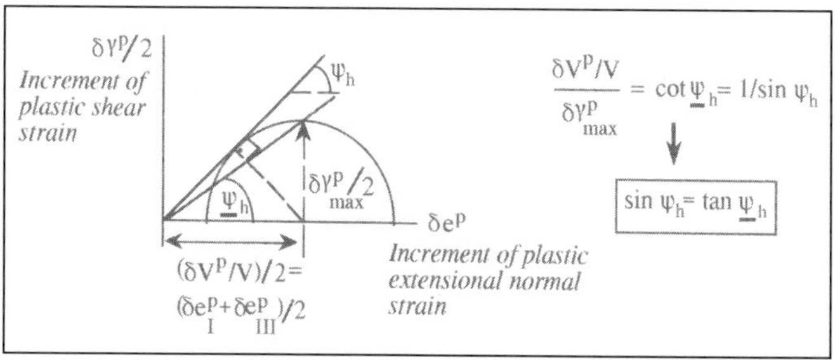

Fig. 6.15. Definition of the slightly different dilatancy angles ψ_h and $\underline{\psi}_h$ in the Mohr circle of extensional strain increments

While there is still a lack of experimental work on pre-peak shear localization in rocks and soils, there exists an impressive amount of theoretical research papers on the subject. But it is difficult to evaluate the various quantitative results regarding the onset, orientation, and stability of shear bands, because of the differences in the applied plasticity models, and of the various simplifications and suppositions introduced to bridge the gaps in our knowledge of the natural process, and to facilitate mathematical treatment. Nevertheless, with some caution, certain trends can be determined which could be of geological relevance. In particular, concerning the *orientation* of shear bands, there are strong theoretical indications that pre-peak shear bands in plane strain form at an inclination angle Θ to the σ_I axis, which is closer to $\Theta = \pm[45° - (\varphi + \psi)/4]$ than to the Coulomb orientation $\Theta = \pm(45° - \varphi/2)$. In contrast, shear bands in the softening range, as is common in faulting, prefer the Coulomb orientation (P.A. Vermeer (1990) Géotechnique *40*, No. 2, pp. 223–236).

In conclusion, the numerical simulations and the mechanical theory of shear localization, and some experimental work, leave no doubt that shear bands in frictional materials under plane strain can spontaneously form while the material is still getting stronger. These shear bands thus form a special class of shear joints. Nevertheless, the mathematical proof of pre-peak shear localization does not really satisfy our wish to understand the physics behind the mathematical formalism. In other words, we wish to understand *why* shear bands can form in the hardening range. Before concluding this chapter, we therefore explore a model that may explain, at least in part, why the formation of a shear band in a frictional (pressure-sensitive) material does not necessarily require material softening.

Mechanism of pre-peak banding. Let us consider in Fig. 6.16A, a potential shear band that is unrestrained in length and, for some reason, experiences a decrease in the band-parallel normal stress across the boundary from a constant value $\sigma_{||}$ outside the band to a smaller value $\sigma_{||} - \Delta\sigma_{||}$ inside the band, but where there are no changes in the stresses

$\sigma_{||}$, $\Delta\sigma_{||}$, $\sigma_\perp$ and τ in the band-parallel x_2-direction. Naturally, the change in $\sigma_{||}$ must be continuous, since a stepped change across the boundary, though compatible with static equilibrium, would invoke an infinite shear stress τ at the band boundary, on either side of which the material is the *same*, but the continuous transition of $\sigma_{||}$ is restricted to a narrow band margin. With reference to Fig. 6.16A, static equilibrium of stresses then requires that $\partial\tau/\partial x_3 = \partial\sigma_\perp/\partial x_3 = 0$ (see e.g., Eq. 4.21 with the specific weight of the material being neglected). Thus, as indicated in Fig. 6.16A, the stresses τ and $\sigma_\perp$ remain constant in- and outside the potential shear band. The postulated reduction in $\sigma_{||}$, which does not affect the shear strength of the material, has been termed "*stress softening*" (P. Cundall, 1990).

The important question now is, how can the reduction in $\sigma_{||}$ promote shearing inside the potential shear band? An answer is provided by the "*bookshelf*" mechanism of direct shear, which was discussed extensively by the author (Mandl, FBR, Sect. 6.4), and was already invoked in Fig. 6.6 as a mechanism of post-peak formation of shear joints. Here, we use the "bookshelf" mechanism, with rotating parallel slip elements, to explain pre-peak shearing of a band of frictional material. The band, loaded by a constant $\sigma_\perp$, can accommodate "simple-shear" movements of its boundaries by the repeated formation of transverse, antithetic shear planes and their collective rotation (Fig. 6.16B).

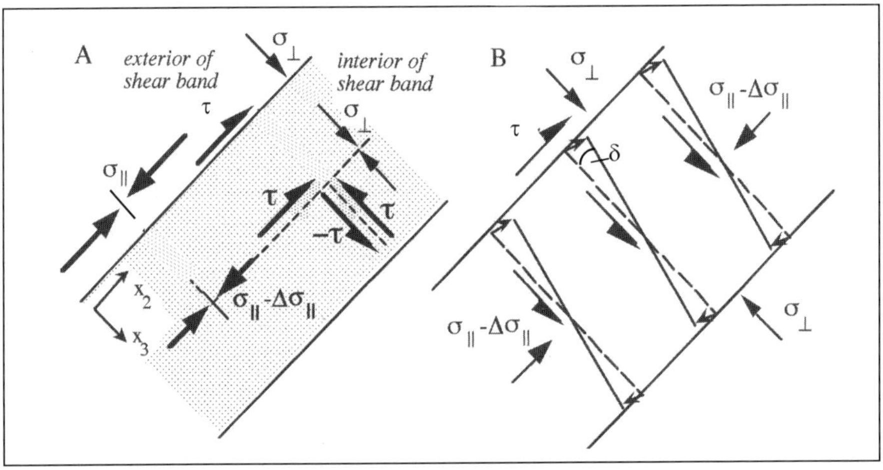

Fig. 6.16. Stress-state "softening" of an incipient shear band:
A) Reduction of band-parallel normal stress $\sigma_{||}$ inside the band.
B) "Domino" style book-shelf mechanism of parallel slip planes (see text)

Note that, by letting the distance between the en-échelon planes and the rotation angle shrink to zero, the kinematics of the process become continuous. Further note, that simple shearing *sensu stricto* would require that, as indicated in Fig. 6.16B, the transverse slip elements are generated orthogonally to the band boundaries. The repetitive process of formation of slip elements and rotation through an infinitesimal angle, accompanied by infinitesimal sliding, would then leave the distance between the boundary planes unchanged, and thus produce a band-parallel shear without dilatancy or contractancy of the sheared material. This was assumed in Cundall's numerical simulation of the graben structures of Fig. 6.12C. In passing, we also note that this simple shearing mechanism was first proposed by the soil mechanician G. de Josselin de Jong (1959; Dr. Thesis, Techn.

University, Delft) as an alternative to simple shearing by sliding along boundary-parallel slip planes ("card-deck" mode).

Now we still face the problem of how the simple-shearing mechanism of Fig. 6.16B can be established by solely reducing the band-parallel normal stress $\sigma_{||}{}'^{o}$ inside a potential shear band, while leaving the exterior of the band in a subcritical (pre-peak) state of stress. This may be shown in Mohr's stress plane. (Readers who are less conversant with the graphical method of stress circles and stress poles are referred to the Appendix at the end of the book.) To start with, we consider a homogeneous, isotropic rock which is horizontally confined by a constant compressive stress $\sigma_{III}'^{o}$, represented by point O in the diagram of Fig. 6.17A. The vertical load is applied by raising a uniform compressive principal stress σ_I'. The shearing strength of the material is represented by the pair of straight Coulomb limit lines in Fig. 6.17A. The straining is planar and parallel to the σ_I^o, σ_{III}^o-plane; and the σ- and τ-axes of the stress plane are identified with the vertical and horizontal directions in real space, respectively. This superposition of space- and stress axes allows us to insert, in the σ_I^o, σ_{III}^o-plane, the trace Γ of a potential band of shear localization at its orientation in real space. We draw the Γ-line at the inclination angle γ towards the vertical direction of σ_I^o through the point O, which serves as pole of the stresses outside the band Γ. The intersection S of Γ with the stress circle C^{ex} then represents the tangential and normal stress components acting on the shear band.

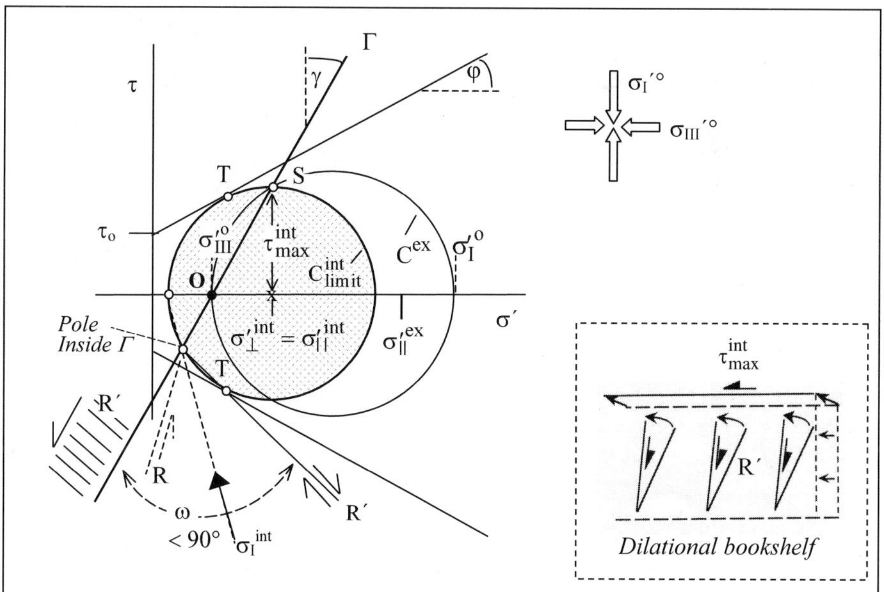

Fig. 6.17A. Mohr Diagram of pre-peak formation of a simple-shear band (Γ) in biaxial loading under vertical stress $\sigma_I'^{o}$ and constant horizontal confining stress $\sigma_{III}'^{o}$. The band-parallel shearing is produced by the "bookshelf"-type collective rotation and slippage of a train of elements R′, as sketched in the insert. The shaded circle C^{int}_{limit} represents the limit state of stress *inside* the shear band, while the circle C^{ex} through O and S represents the stresses *outside* the shear band (see text for further comment)

The onset of shear displacement along Γ requires that: 1)the band is in the frictional plastic limit state, and (2) the maximum shear stress τ_{max}^{int} in the band acts parallel to the band boundary (the material being assumed as isotropic). These two conditions define the stress circle C_{limit}^{int} of the activated shear band Γ in Fig. 6.17A; the circle does not only have to touch the limit lines (in T) but the vertex S of the circle must lie on the Γ-line. At the same point S, the circle C^{ex} of the external stresses must intersect the limit circle C_{limit}^{int} since the interior and exterior of the shear band have the tangential stress τ and the normal effective stress $\sigma_\perp'$ in common (the latter assuming continuity of the pore pressure). We notice that the circle C^{ex} need not touch the limit lines, thus allowing the external stress state to remain subcritical.

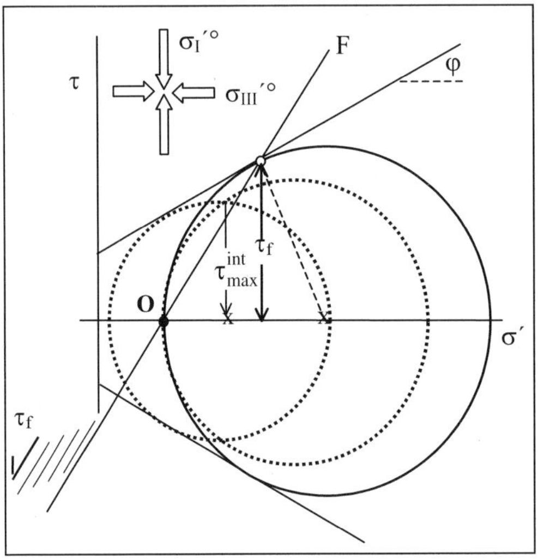

Fig. 6.17B. Comparison of "driving" shear stresses τ_{max}^{int} on the shear band Γ and τ_f on the fault plane F

Further, note that, as shown in Fig. 6.17B, the shear stress τ_{max}^{int} on Γ is smaller than the shear stress τ_f on a fault plane F that would be initiated by the *uniform* limit state at the same lateral confining stress $\sigma_{III}'^{o}$.

Associated with the limit state of stress inside the shear band Γ (Fig. 6.17A) are two sets of conjugate slip elements R and R′, which include the acute angle $\alpha = \pm(45° - \varphi/2)$ with the direction of the new σ_I^{int}-axis. The directions of R, R′ and σ_I are determined by the pole method in Fig. 6.17A. The point where the Γ-line intersects the lower part of the limit circle C_{limit}^{int} is a "stress pole", and the chord drawn to this point from any point $(\underline{\sigma_\perp}', \underline{\tau})$ on the circle is the real-space trace of the planar element (with the normal in the σ_I, σ_{III} -plane) that is acted upon by the stress components $\underline{\sigma_\perp}'$ and $\underline{\tau}$. Thus, in particular, the chords connecting the points of tangency (T) with the pole are the traces of the conjugate slip elements in real space. Likewise, the chord drawn from the circle point $\sigma_{III}'^{int}$, 0 to the pole is the trace of a material element acted upon by the principal stress $\sigma_{III}'^{int}$, and is thus parallel to the σ_I^{int} direction.

An en-échelon array of parallel slip elements of R′-type can generate a simple (or quasi-simple) shearing of the band by repeatedly being formed and collectively rotated, as sketched at the lower end of the Γ-line and in the insert of Fig. 6.17A. It is important to note that the angle ω between the slip elements R′ and the band trace Γ in the figure is slightly smaller than 90°. Therefore, the rotation of the en-échelon elements is accompanied by some thickening of the band ("dilational" mode of the "bookshelf" mechanism) while the band tends to shorten. In contrast, $\omega > 90°$ would imply that the band is sheared in "domino" style (Fig. 6.6) which involves a transverse contraction and a lengthening of the band.

In passing, we note that shearing imposed by the dilational mechanism is not simple shearing *sensu stricto*, but a mode of quasi-simple (or "direct") shearing. Further, quasi-simple shearing by the collective rotations of en-échelon slip elements can only be performed by R′-type elements, and not by the conjugate R-type elements (Fig. 6.17A). This is easily seen by comparing the sense of the rotation-induced slip with the slip due to σ_I^{int} (see also G. Mandl, FBR, pp. 281–282).

Whether the angle ω in Fig. 6.17A is greater or smaller than 90° is of crucial importance. Note again, that the formation of a shear band by subcritical external loading can only be achieved when the band-parallel normal stress inside the band is drastically reduced (from $\sigma_{||}'^{ex}$ to $\sigma_{||}'^{int}$ in Fig. 6.17A). As long as the potential shear band is completely contained inside the rock, the shortening or lengthening of the band by the shearing mechanism will be suppressed by the resistance of the ambient rocks. Thus, the suppression of the shortening tendency of the dilational shearing mechanism will result in a reduction of the band-parallel compression, as is necessary for the formation of a shear band under subcritical biaxial load. In contrast, the suppression of the extensional tendency of a domino-style shearing mechanism would increase the band-parallel compression, and thus impede the pre-peak formation of a shear band. Hence, we expect pre-peak shear bands to form only if $\omega < 90°$.

At first glance, one might expect the key parameter ω to depend on the strength parameters φ and τ_o, and on the inclination γ and position O of the potential shear band Γ. Surprisingly, ω is completely defined in a rather simple way by the angle of internal friction φ, as was pointed out to the author by Norbert Tschierske. Consider in Fig. 6.18 the heavier lined arc ST of the internal limit circle C_{limit}^{int} and the angles extended over the arc from P and the center C. Since, according to an elementary geometrical theorem, the angle extended from any point on the periphery of a circle over a given circle arc is half the angle extended over the same arc from the center of the circle, one immediately reads off the figure:

$$\omega = \frac{\pi}{2} - \frac{\varphi}{2} \tag{6.6}$$

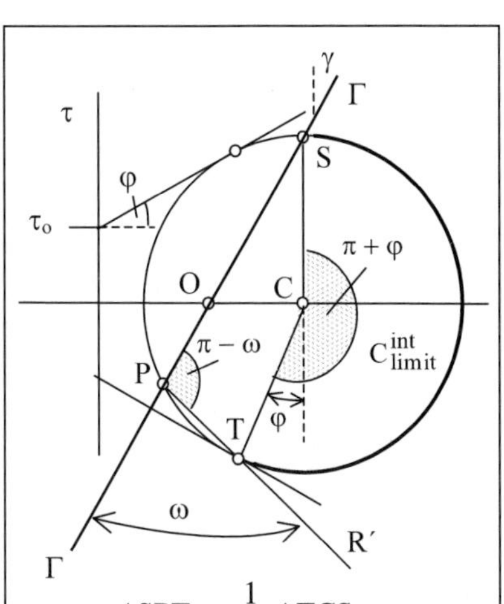

Fig. 6.18. Graphical determination of the angle ω between the shear band Γ and the antithetic slip elements R′

Thus, we know that pre-peak shear banding in a brittle rock with a given friction angle φ is achieved by the same dilational bookshelf mechanism, irrespective of cohesive strength, confining pressure, and in particular, of the inclination γ of the shear band. But, although the "bookshelf" kinematics of the shear banding is independent of the band inclination, the mechanism can only be activated within a certain interval of inclination angles. The reason for this is easily seen: as soon as, with increasing external load, the circle C^{ex} in Fig. 6.17A touches the limit lines (Fig. 6.17B), the mode of shear localization changes into fault-type shearing, i.e. slippage parallel to the shear band ("card deck" mode).

This limits the inclination angle γ of the shear bands towards the external σ_I-direction to the interval determined by the condition

$$\sin 2\gamma + \sin\varphi . \cos 2\gamma \underset{(=)}{>} 1 \tag{6.7}$$

The derivation of this inequality by N. Tschierske is given in the Appendix to this chapter. According to Eq. 6.7, the inclination angle γ lies in the interval

$$\gamma_{min} < \gamma < 45° \tag{6.7a}$$

where γ_{min} is defined by Eq. 6.7 when the equality sign is used. For instance: $\gamma_{min} = 15.2°$ if $\varphi = 35°$.

Having obtained upper and lower bounds for γ we finally ask which value the inclination angle will actually attain. The answer is simple: γ depends on the state of the (uniform) external stress, i.e. the stress circle C^{ex}, at the onset of shear localization. Consider in Fig. 6.19, the development of internal and external stresses in a rock with the typical friction angle $\varphi = 35°$. The limit circles $C\underline{lim}^{int}$ and $C\overline{lim}^{int}$, and the associated Γ lines, mark the interval of possible pre-peak stress states and γ-values. Let us further assume, that shear localization commences when the vertical loading stress σ_I^{ex} attains

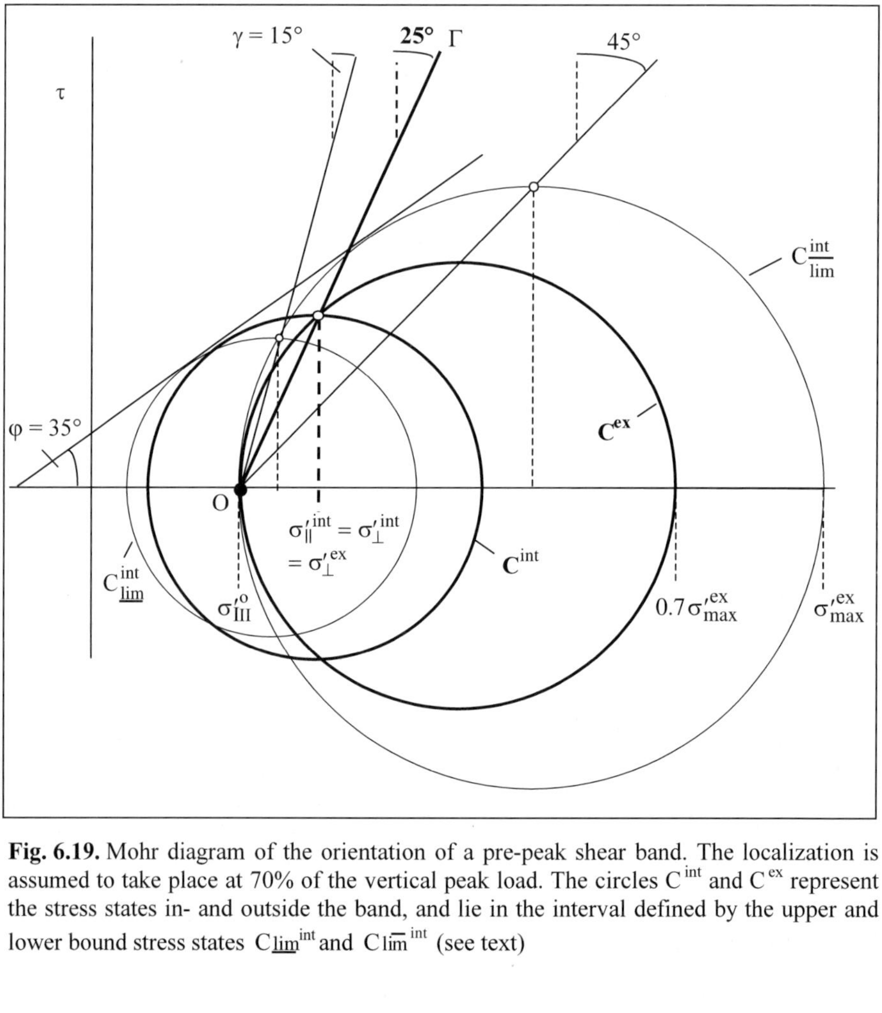

Fig. 6.19. Mohr diagram of the orientation of a pre-peak shear band. The localization is assumed to take place at 70% of the vertical peak load. The circles C^{int} and C^{ex} represent the stress states in- and outside the band, and lie in the interval defined by the upper and lower bound stress states $C\underline{lim}^{int}$ and $C\overline{lim}^{int}$ (see text)

70% of the peak load σ_{max}^{ex}. Then the inclination angle γ of the shear band has the uniquely determined value of 25°.

At this point it should be noted that the above derivations, although referring to an angle γ that was counted clockwise from the vertical, also apply to an anticlockwise γ. In other words, the corresponding constructions in the Mohr plane are each others mirror images with respect to the horizontal σ-axis; hence, the above results also apply to pairs of *conjugate shear bands*.

Is there experimental evidence of the bookshelf mechanism operating in pre-peak shear banding? We think that such evidence does exist, and refer to the narrow bands of en échelon fractures reported by G. Sobolev, H. Spetzler and B. Salov (1978; J. Geophys. Res. *83*, No. B4, pp. 1775–1784). The bands, less than 1 mm in width, formed in biaxially loaded samples of a pyrophyllite lava before the formation of any macrocracks, and at a differential load far below peak level.

But it should also be recalled, that the bookshelf-type shearing mechanism of discrete Coulomb slips becomes a continuous mode of frictional-plastic deformation if, at the limit, the Coulomb slip elements are infinitesimally close, repetitively formed, and rotated through an infinitesimal angle. In fact, the resulting continuous shearing deformation would be predicted by a theory of ideal frictional plasticity that is based on the concept of conjugate Coulomb slips ("double gliding" model) and applies to both *continuous* and *discontinuous* deformations. (The interested reader is referred to the brief account of the theory of frictional-plastic materials presented by the author in FBR, pp. 416–419, or, in more detail, in MTF, Sect. II.7.5). The fact, in itself, that a continuum theory of ideal frictional plasticity is derived from the Coulomb concept of discrete conjugate slips, implies that the actual shearing of the pre-peak band need not be accomplished by an internal array of discernible slip elements, but may also be accommodated by continuous flow of the band material in the limit state.

Regarding this theory and its bearing on pre-peak shear banding, it should be emphasized that the theory deals with an *ideal* frictional-plastic behaviour of the material *without* inelastic hardening or softening.

In concluding this section on pre-peak shear banding, we turn to the intriguing question of what happens to a pre-peak band when the external loading stress σ'_I is raised *beyond* the level at which the shear band was activated. To this aim, we consider, step by step the somewhat crowded diagram of Mohr circles in Fig. 6.20. The solid circles C^{int} and C^{ex} again represent the stress states in- and outside the shear band of Fig. 6.19. Now, the vertical loading stress σ'_I is raised to $\sigma'_I{}^*$, changing the circle C^{ex} into the stippled circle $C^{ex}*$ that intersects the Γ-line at S'. If the shear band Γ is to remain active, the stippled (and shaded) limit circle, with vertex at S', would have to represent the internal stresses of the band. Thus, the vertex S of the former stress circle would have to shift up the Γ-line to S'. But this shift implies an *increase* $\delta\sigma_{||}'^{int}$ of the band-parallel compressive stress, which is quite incompatible with the activity of the dilational bookshelf mechanism.

Thus, the dilational mode of bookshelf-type shearing, although necessary for the formation of a shear band inside a rock under subcritical biaxial load, will block the internal shearing mechanism when the external load is raised beyond the level at shear band formation. In other words, the pre-peak shears "*freeze*" when the external principal stresses are changed. Interestingly, the gradient of $\sigma_{||}'$ across the boundaries of the formerly active band ($\sigma_{||}'^{ex} \rightarrow \sigma_{||}'^{int}$ in Fig. 6.17A) will persist in the new stress field. Thus, the stress field has acquired a "*memory*" of the pre-peak shear band, as was already concluded by P.A. Cundall (1990, loc.cit.) from his numerical analyses.

On the other hand, while immobilizing a shear band, the increase in the loading stress σ'_I may generate new pre-peak bands at greater inclination angles γ. This is indicated in Fig. 6.20 by the dotted limit circle C^{int} * whose vertex S* lies on the new band trace Γ*.

These results lead us to expect that in nature, even minor changes in magnitude and/or direction of the principal stresses immobilize pre-peak shears and generate new ones at altered orientations. The traces of the pre-peak shears, possibly of hairline appearance, would then crisscross each other in a complex network. On the other hand, when confronted with the vexing appearance of such complex crisscross patterns in nature, one may well be advised to consider a possible pre-peak origin of the fractures.

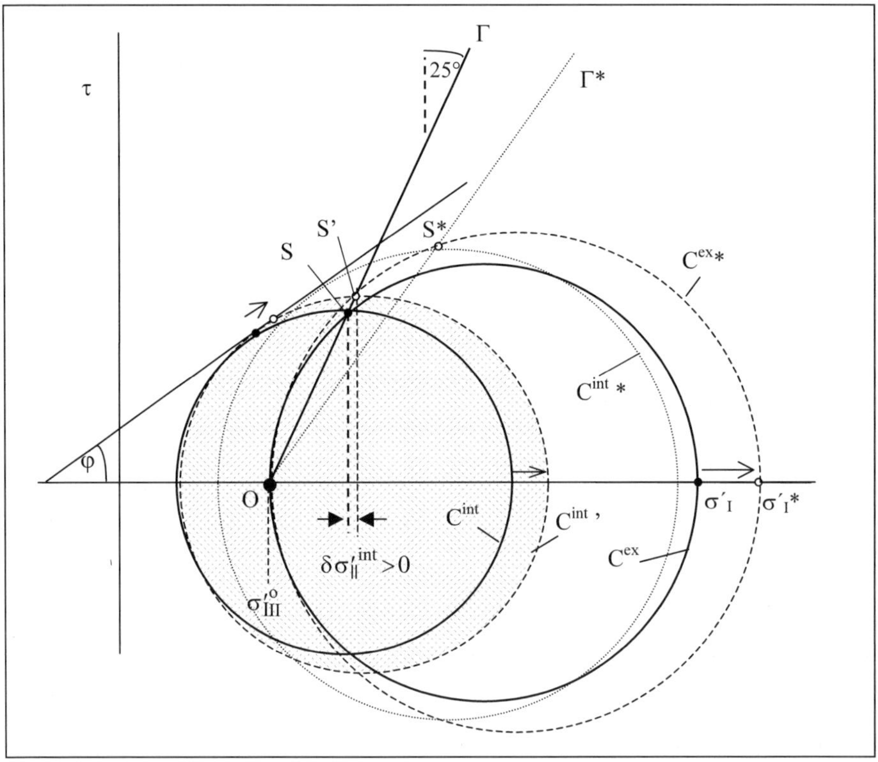

Fig. 6.20. Mohr Diagram explaining how pre-peak shear banding is inactivated when the external load is raised beyond the level at which the band (Γ) was first activated (see text)

Spacing of shear joints. It seems that, in the field, not only are sets of parallel tension joints commonly quite regularly spaced, but so are also joints that are justifiably interpreted as shear joints. Although, in the case of shear joints, the mechanics of regular spacing is much less well understood than in the case of tension joints, we shall briefly comment on the problem. First, recall that, from the mechanical point of view, shear joints may be subdivided into "embryonic" faults and pre-peak shear bands, depending on whether or not the formation of the shear joints is accompanied by shear softening, i.e. a reduction of the shearing strength of the joint material.

As in fully developed faults, the minimal distance between parallel shear joints of the embryonic fault type is controlled by two main factors: the unloading of the driving shear stress by reduction of the shearing resistance of the joint material, and the reactive stresses set up in the ambient rocks. It is the competition between the unloading by shear softening and the stress recovery by boundary reactions which determines the unloading "halo" (or stress "shadow") of a shear joint, inside which no other shear joint can form. The mechanics of the process in faulting have been discussed in some detail in FBR (Sect. 8.1, pp. 387–394). Here, we content ourselves with the summarizing statement that the stress shadow of a shear joint increases with the extent of the strength reduction and the rate at which the reduction takes place in relation to the applied shear. Further, the spacing also increases with the length of the shear joints and the thickness of the jointed layer.

Since in shear jointing the shear displacement may be less than required for the shear strength to drop to the residual level (Fig. 6.4), parallel shear joints may be more closely spaced than fully developed faults. This may be the case, for instance, when parallel strike-slip faults are interspersed between parallel shear joints. Such an arrangement would seem likely in a strike-slip regime, where longer shear joints grow faster into faults than shorter ones, and thereby develop larger unloading halos which stop the growth of shorter neighbours.

In contrast to post-peak shear jointing, virtually nothing is known about the spacing of pre-peak-shears, mainly because diagnostic tools for the identification of these joints are missing.

Summary and comments. Let us recapitulate the essential results of this lengthy chapter.

- ***The recognition of shear joints.*** From a purely phenomenological point of view shear joints are indistinguishable from tension joints, but from a genetic-mechanical point of view the two joint types are essentially different: although both joints develop parallel to the σ_{II}-axis, in contrast to tension joints, shear joints form at an acute angle to the σ_I-axis of the stress field. Shear joints are notoriously difficult to identify; nevertheless, there are certain features which, particularly in combination, may indicate or suggest a shearing origin of the joints: (*1*) the joints occur as two non-orthogonal sets, with both sets equally well developed, and intersecting each other with mutual displacement which indicates a simultaneous development in the same stress field; (*2*) joints are sub-parallel to fully developed faults; (*3*) the fracture walls of presumptive shear joints are striated or slickensided; (*4*) there is evidence of an extremely small aspect ratio of slippage (e.g. 10^{-5}). (*5*) Quite generally, presumptive shear joints should easily fit into a paleo-stress field whose trends are known from contemporaneous tectonic structures.
- ***Origins and types of shear joints.*** Shear fractures or shear bands and proper faults have in common the fact that they are generated by shear stresses on their boundaries. During the shearing process, the resistance against shearing first increases ("hardening" range) until reaching a peak value, and thereafter decreases ("softening" range) to a "residual" shear strength (Fig. 6.4). We distinguish between "pre-peak" shear bands which are generated by a shear stress which did not increase up to the peak value, and "post-peak" shear bands which experienced softening, possibly even down to the residual level of shear strength. In the latter case, the total shear strain is supposed to be of the order of magnitude of the shear strain associated with the whole softening range, while in proper faults the total shear strain exceeds the softening strain by several orders of magnitude. The shear joints occur in sets, because they are generated in fields of pervasive straining.

Post-peak shear joints are initiated by the same stress conditions as faults, but got stuck in an embryonic state of development ("embryonic" faults) for two reasons: (*1*) the "driving" shear stress (and the differential stress) dropped below the level necessary for further activity of the joint, and was not restored afterwards; (*2*) the frictional resistance of the joint was increased as, for example, in the frictional locking of en-échelon shears in "domino"-style rotation (Fig. 6.6). The drop in driving shear stress is mainly caused by the elastic unloading of the surroundings of an active joint in response to the shear-softening of the joint material. Short-lived "driving" shear stresses may be associated, for example, with the switch of principal stresses in a compressional regime (Fig. 6.5), or with the formation of subvertical shear joints in layers above a basement that is gently tilted on a regional scale (Figs. 6.8, 6.9). "Residual" stresses as those, for instance, locked in uplifted layers of sedimentary basins, or in flat foreland layers in a compressional pre-folding state (see pp. 113–116, 119) are likely released by a single slip event. Although the release of "locked-in" stresses may primarily cause tension jointing, it may also generate shear joints.

Similar to fully developed faults, the spacing of parallel shear joints increases with the extent of the strength reduction and the rate at which it takes place in relation to the applied shearing; the spacing also increases with the length of the shear joints and the thickness of the jointed layer. But since shear softening might not be completed during a single slip event, the post-peak shear joints may be more closely spaced than interspersed parallel faults, especially if the latter are longer than the shear joints.

Pre-peak shear bands are formed while the loaded rock is still in a *hardening* state, and can still sustain further increases in load. This implies that the "driving" shear stress (τ_{max}^{int} in Fig. 6.17A) exerted by the surroundings upon the boundaries of a "pre-peak" shear band is not at its peak value. Although pre-peak shear bands form before the rock fails under the applied load, pre-failure shear bands of the post-peak type cannot be excluded since softened shear bands may form in patterns (Fig. 6.10C) that would still allow the rock to support a further raise in external load.

In a theoretical analysis, Rudnicki and Rice (1975) determined the possible onset of pre-peak shear banding (Fig. 6.14) in a homogeneous elastic/frictional plastic (i.e. pressure-sensitive) material under a load that would also be compatible with a field of uniform straining (e.g. simple or pure shear). Interestingly, the pre-peak shear bands only occur in a field of plane, or almost plane, strain, and have normals perpendicular to the σ_{II}axis. (In contrast, axially symmetric loading, as commonly applied in rock-mechanics laboratories, permits shear band formation only in a post-peak state.) The theoretical analysis is corroborated by computer simulations of the formation of pre-peak shear bands in frictional plastic materials (Figs. 6.12, 6.13), and by experiments with dry sand (e.g. Fig. 6.11).

As shown in P. Cundall's (1990) numerical analyses, shear bands may even form in ideal elastic/frictional plastic materials, i.e. materials without any inelastic changes in material shear strength. The shear localization is then produced by a kind of "stress-state softening" which reduces the band-parallel normal stress ($\sigma_{\|}'^{int}$) inside the band (Fig. 6.16A).

How a pre-peak shear band is actually formed at the "bifurcation" of the stress solution is explained by the dilational "bookshelf" mechanism (Fig. 6.17A and insert). The mechanism operates by repeated formation and subsequent small rotations of transverse en-échelon slip elements. The collective rotation of the elements results in a component of band-parallel shear, some thickening (dilation) of the band, and a decrease of the internal band-parallel normal stress ($\sigma_{\|}'^{int}$). Thus, in contrast to faulting, the simple or quasi-simple shearing of the pre-peak band is not achieved by band-parallel slippage or flow ("card-deck" mode).

Remarkably, the kinematics of the mechanism is determined by a single material parameter – the angle of internal friction φ. The angle ω between shear band and slip elements is 90° – $\varphi/2$, the range of possible band inclinations to the external σ_I-direction is determined by Eqs. 6.7 and 6.7a, and the orientation of the conjugate pre-peak shear bands is uniquely determined by the construction in Fig. 6.19, provided the loading stress $\sigma'_I{}^{ex}$ at band formation (bifurcation) is given (or assumed) as a fraction of the peak load σ'^{ex}_{max} (at a lateral confining stress $\sigma_{III}'^{o}$).

Because of the lack of material softening, pre-peak shear bands should be rather transient phenomena, and in contrast to proper faults, be immobilized by the slightest changes in directions and magnitudes of the principal stresses that activated the bands. This is also illustrated by the dilational bookshelf mechanism of pre-peak shearing, which is immobilized when the external load is raised beyond the level at shear band formation (Fig. 6.20). Naturally, a raise of pre-peak loading will generate new shear bands at different inclination angles.

As a consequence of the stress-sensitivity of pre-peak shear bands, one should expect the bands to frequently form complex crisscross patterns. The observation of such patterns might be of some help in the, not very promising, attempts to locate pre-peak shear bands in the field.

Appendix to Chapter 6

The possible orientations of pre-peak shear bands

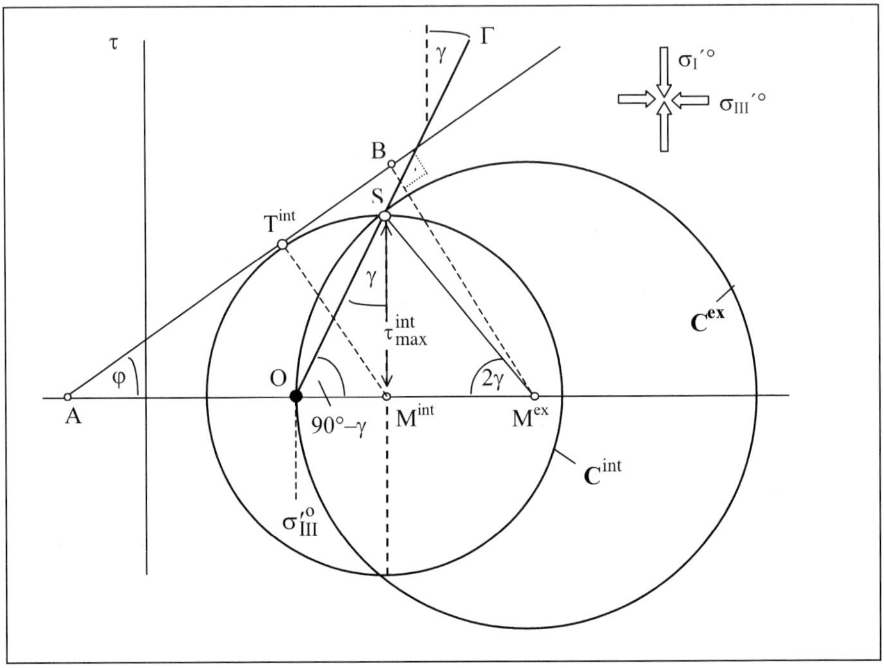

Fig. 6.21. Determination of the γ-range in Mohr diagram (after N. Tschierske)

The range of possible inclinations γ of a pre-peak shear band Γ to the external σ_I-direction is limited by the condition that the circle C^{ex} of the stresses outside the band must not intersect the limit lines. Hence, the radius OM^{ex} of C^{ex} in the above figure must not exceed the distance from the center M^{ex} to the limit lines:

$$OM^{ex} \leq M^{ex}B \tag{6.8a}$$

Since, in the triangle OSM^{ex} the sides OM^{ex} and $M^{ex}S$ are equal, the angle at M^{ex} is 2γ. Therefore, with reference to the figure:

$$OM^{ex} = M^{ex}S = \tau_{max}^{int} / \sin 2\gamma \quad \text{and} \quad M^{ex}B = AM^{ex}.\sin\varphi \tag{6.8b}$$

where

$$AM^{ex} = AM^{int} + M^{int}M^{ex} = \tau_{max}^{int}\left[\frac{1}{\sin\varphi} + \cot 2\gamma\right] \tag{6.8b}$$

Inserting the Eqs. 6.8b and 6.8c into condition 6.8a gives

$$\boxed{\sin 2\gamma + \sin\varphi.\cos 2\gamma \geq 1} \tag{6.7}$$

Chapter 7

Joints in Faulting and Folding

The formation of tectonic faults and folds in the brittle crust is commonly associated with a conspicuous increase in the frequency and variety of joints. The analysis of such fracture systems is complicated by the fact that some joint sets were forerunners of the faulting or folding, while others accompanied the formation of the structures or were generated afterwards. Therefore, in this chapter, rather than trying to disentangle specific fracture patterns in faulted or folded rocks in the field (and getting ourselves entangled in controversies) we shall attempt to review the various joint sets as they may form in succession during the development of fault- and compressional fold structures.

Joints and faults. We begin with joints that are associated with tectonic faults. First, let us recall that, apart from the very special case of slippage on a pre-existing plane of weakness, we regard tectonic faults in the brittle regime as narrow shear bands which are generated by a gradual process of shear concentration and localization in a region (the "proto-fault" zone) of pre-damaged rock. Accordingly, we shall first consider the fracture structures which are successively generated during the development and the activity of a tectonic fault. There are: the swarms of fractures from which tectonic shear bands evolve; the fracture structures initiated by incremental fault slips; and the subsidiary joint and fault structures that adapt a sliding fault block to fault curvatures, or accommodate the stretching of layers dragged along a fault.

After that, we shall turn to the effects which a fault may have on a set of joints that were *not* generated by the activity of a fault. This will include the opening of healed tension joints by the reactivation of a fault; the occurrence of strike-slip faults within sets of parallel joints; and the perturbation of joint sets by pre-existing faults.

Pre-faulting fractures. In the brittle regime, the pre-faulting damage consists in pervasive fracturing, mainly by tension cracks which may vary in length from the order of grain size to a mesoscopic scale. Evidence for this is found in tectonic shear zones which have stopped at an early stage of growth, and is abundantly illustrated by detailed laboratory studies of shear failure in rocks ranging from loose granular sediments to highly cohesive hard rocks. In FBR (Fig. 3.11) we illustrated the formation of a pre-failure swarm of tension fractures in axially loaded quartzite samples using photographs from the classical paper by D.K. Hallbauer, H. Wagner and N.G.W. Cook (1973); here, we reprint in Fig. 7.1, two fault maps of axial sections of triaxially loaded samples of Westerly granite from a paper by D.E. Moore, R. Summers and J.D. Byerlee (1990; in: H.-P. Rossmanith (ed.) *Mechanics of Jointed and Faulted rocks*, Balkema, Rotterdam, pp. 345–352). In both cases, a fault was formed inside a complex array of fractures. But, as the authors state, "no noticeable difference in fracture density was observed between samples for which the experiment was terminated at the initial failure and ones for which post-failure shear was permitted". Thus, in the main, the fractures were formed prior to the fault. Further, note that the amount of fracturing decreases gradually with increasing distance from the faults.

The formation of a "fault" in the experiment is a highly complex process involving the interlinking of small-scale tension cracks that are subparallel to the σ_1-axis, with inclined cracks. Various micro-mechanical models have been constructed to elucidate the damage processes that lead to shear failure in samples of hard rocks. The majority of models postulate, that the actual interlinking is established by tensional "wing" cracks which emanate from pre-existing inclined cracks under shear (see p. 20, Fig. 2.10A; p. 92, Fig. 4.29). However, we reiterate that the formation of wing cracks of longer than, say, the grain size, is severely impeded by the effective confining pressures that prevail in the Earth's crust. A numerical 2D simulation of the micro-damage process that did not involve the wing-crack mechanism, and thus was capable of modelling the effect of the confining pressure, was presented by D.A. Lockner and Th.R. Madden (1991; J. Geophys. Res. *96*, pp. 19623–19642). The authors studied the growth of populations of horizontal, vertical and diagonal (45°) cracks under biaxial compressive loading, and found that under low confining pressures, the crack growth is predominantly parallel to the vertical maximum load, but at higher (horizontal) confining pressures, the growth of diagonal mode II cracks dominates.

As a detailed review of micro-mechanical damage models would by far exceed the scope of this book, we restrict ourselves to a very schematic illustration in Fig. 7.2 of the basic concept of the various damage models. In the figure, vertical arrows indicate an axial compressive load, black lateral arrows indicate a confining pressure, while, alternatively, open lateral arrows represent a boundary tension. In Fig. 7.2A, continuous shear fractures are formed by the interlinking of axially oriented small-scale tensile fractures with oblique sliding fractures with the same sense of right-lateral or left-lateral displacement.

Another mode of formation of macro-shear fractures is illustrated in Fig. 7.2B, where small beams, bounded by axial tension cracks, are arrayed en-échelon and slightly rotated by shear stress, the rotation being accommodated by slip along the tension cracks. The reader will recognize the dilational "bookshelf" mechanism which was proposed in Chapt. 6 (Figs. 6.16 and 6.17) to explain the formation of pre-peak simple shearing bands. Eventually, the rotating beams, initially connected with the surrounding rock, are bent and torn loose from their surroundings by tension fractures induced by the bending.

Next, let us complement the experimental development of a shear zone with a field case study: the development of strike-slip faults in the dolomites of the Sella Group in the central area of the Dolomites of Northern Italy, as described and interpreted in an interesting paper by Pauline N. Mollema and Marco Antonellini (1999; J. Structural Geology *21*, pp. 273–292). Figure 7.3 reproduces a map of an en-échelon array of joints with cross joints. The systematic parallel joints are hairline joints parallel to the σ_I direction, which is related to the direction of the compression caused in the Neogene by the Alpine orogeny. According to the authors, the deformation of the area always occurred in regimes of moderate to low confining pressures under a burial depth of less than 1 km. The joint pattern concentrates into a zone which represents a preliminary stage in the development of a right-lateral strike-slip fault, and shows a certain similarity with the experimental fracture pattern in Fig. 7.1.

In spite of this similarity, we find it difficult to infer from experiments the dimensions and fracture density of pre-faulting damage zones at a field scale. We here encounter the same problem as when dealing with the spacing of cleavage joints (pp. 92–94): the conditions of the rock-mechanical experiments do not allow a direct transposition of the experimental results to the geological field. Even if the rock material were the same in the experiment and the field, the dilation associated with the fracture damage is accommodated by thickening of the test sample in the experiment, but, in general, is impeded in the field by lateral constraints on the fractured rock. Secondly, the experimentally produced shear zones are limited in length by the sample size, and may, therefore, only represent a single element in the chain of separate shear bands that are interlinked into a fault in the field.

G. King and C. Sammis have outlined the process of fault formation in brittle rocks (1992; PAGEOPH *138*, No. 4(S), pp. 611–640) in a scenario that starts with small-scale

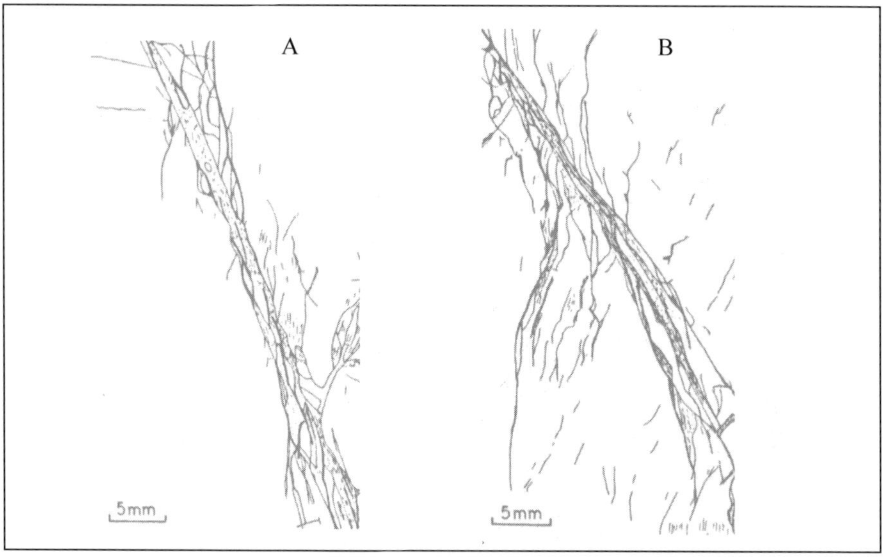

Fig. 7.1. Maps of fractures in dry samples of Westerly granite under vertical load: A) sample slid stably under a confining pressure of 83 MPa; B) sample slid in stick-slip mode under 485 MPa confining pressure. Note, that the subsidiary fractures formed prior to the continuous faults (from D.E. Moore, R. Summers and J.D. Byerlee (1990); in: H.-P. Rossmanith (ed.) *Mechanics of Jointed and Faulted Rock*, Balkema, Rotterdam, pp. 345–352)

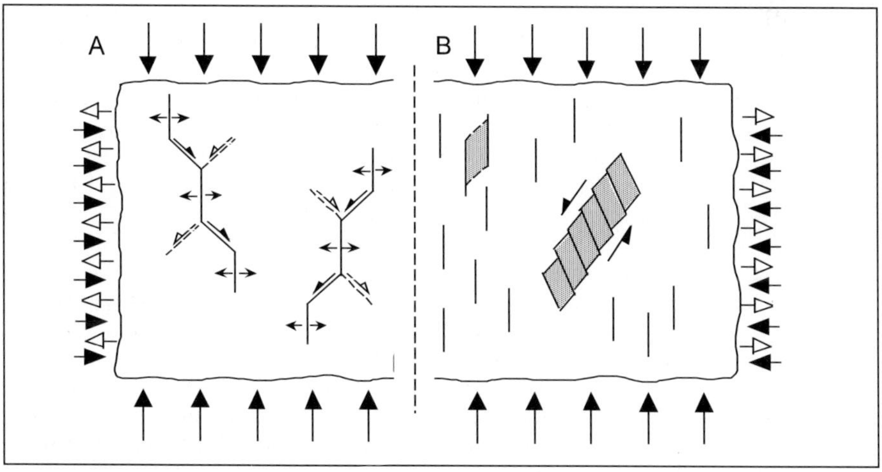

Fig. 7.2. Idealized concept of the incipience of macroscopic shear fractures in the brittle regime illustrating the basic role of small-scale tension fractures, statistically aligned along the σ_I-axis (open lateral arrows indicate boundary tension, black lateral arrows represent confining pressure). A) Oblique shear fractures formed by interlinking of axially oriented tensile fractures and inclined pre-existing fractures under shear; B) "bookshelf"-type shear band of tension fractures

fracturing concentrated in irregularly distributed clusters. The clustering develops at increasing scales. Larger clusters, when sufficiently close, coalesce into isolated shear bands at a somewhat larger scale. The localization of shearing into these bands or patches in a cohesive rock is, to a great extent, accomplished by locally intensified fracturing at various scales inside a narrowing zone. Thus, we expect each shear band to have its own "seam" of fracture-damage. Eventually, a compound fault is formed by the interlinking of shear bands in a step-wise manner. This results in a further increase in the damage seam that surrounds the newly formed compound fault.

We also expect the intensity and width of the damage seam to increase with increasing brittleness of the rock, since a greater brittleness involves higher and steeper local stress drops (cf. Fig. 1.9B,C) to accompany small-scale fracturing and, thereby causing the emission of stronger pressure/tension pulses (noticed as a stronger acoustic signal) that may trigger additional fracturing.

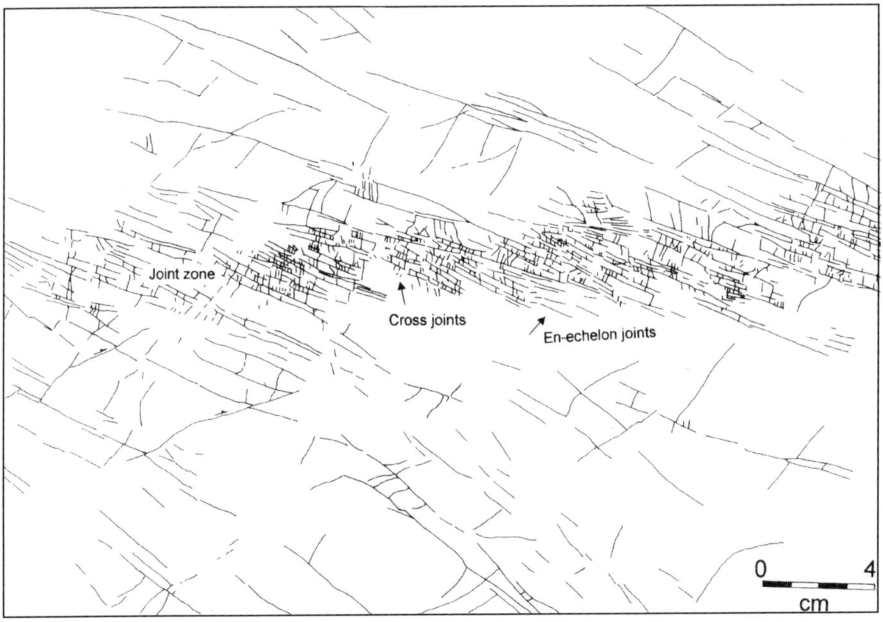

Fig. 7.3. Map of a zone of en-échelon and cross joints accommodating dextral shear in the Sella dolomites of northern Italy (latitude 46°31'N, longitude 11°51'E). The longer parallel (systematic) joints strike parallel to the direction of the maximum compressive stress (from Mollema and Antonellini 1999; J. Struct. Geology *21*, pp. 273–292, Fig. 4)

Fracturing in the tip region of a growing fault. In a fully developed fault, the surrounding damage zone is only partly the result of pre-faulting fracturing, since additional fracture damage accrues during the growth of the fault in length and shear displacement. An initial source of this syn-faulting damage lies in the tip region of a growing fault. Let us, therefore, consider first a perfectly planar fault which is completely enclosed by uniform rock ("blind" fault). As the fault grows in its own plane, either in a succession of seismic jerks or in episodic stable sliding, strains develop in the surrounding area of an advancing fault tip which, in general, are far too high to be accommodated by elastic deformation. Hence, at least part of the straining in the tip regions of propagating brittle faults is accommodated by brittle fracturing of the rock. This elastic/frictional plastic "process zone", attached to the propagating fault tip, leaves an additional halo of macroscopically fractured rock around the propagating fault. This is very schematically indicated in Fig. 7.4A, together with the obvious fact that, the material is compressed on the advancing side (+) of the leading edge and stretched on the receding side (–), at least as long as the growth of the single blind fault is not obstructed by other faults, joints or lithological barriers.

The growing fault causes perturbations of the stress field along the fault and in the tip regions. These perturbations may generate subsidiary faults, tension fractures or extension (cleavage) fractures. This has been more extensively discussed in Sect. 7.2 of FBR. Here, we restrict ourselves to the most essential features of fracturing by near-tip stress changes. At some distance from the fault, the "remote" or "far field" stresses may be considered as uniform, with the trajectory of the maximum compressive stress σ_I at a constant acute angle with the dip line of the fault, as indicated in Fig. 7.4B. But the σ_I-direction will be deflected along the fault because the fault material has a lower elastic stiffness and a lower shear strength than the adjacent rock, which limits the shear stress carried by the fault. Note, that if the fault were a frictionless plane, the σ_I trajectories would terminate tangentially on the advancing (compressive) side of the fault, and perpendicularly on the receding (extensional) side. Under realistic conditions of fault friction, the angular deflections of the σ_I trajectories will be less severe, as is also confirmed by various elastic analyses.

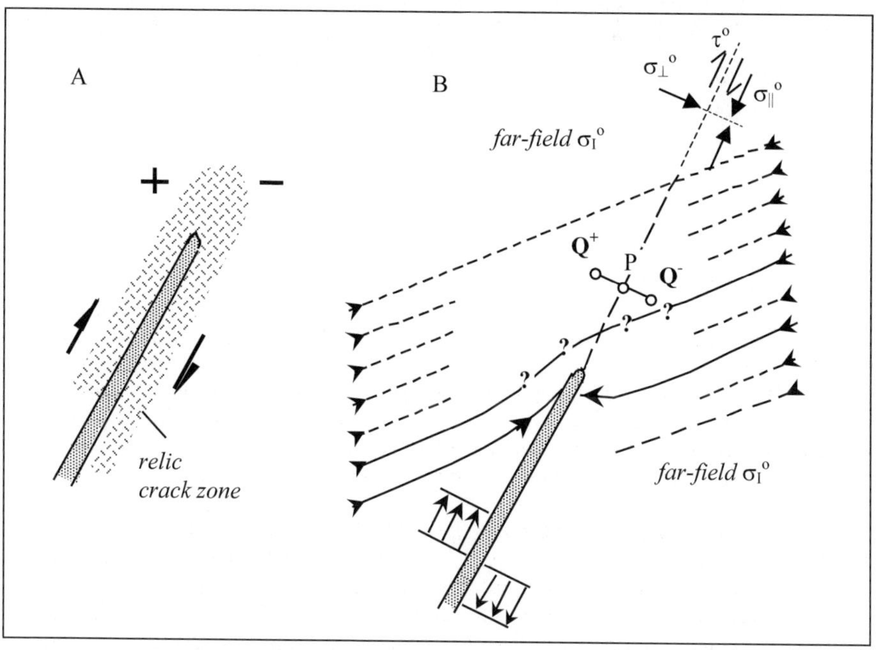

Fig.7.4. Planar section of the highly schematised near-edge region of a "blind" fault propagated by two-sided displacement (plane strain):
A) fracturing in the tip region causes a fracture halo in the displaced wall rock;
B) angular deviation of the σ_I direction near the fault tip

At this point, it should be noted that elastic analyses may be legitimately applied to the wall rock *behind* the propagating fault tips since shear softening of the fault material there has reduced the differential stress in the wall rock below the limit state, and thus reinduced an elastic state of stress. Quite differently, the near-tip region *ahead* of a propagating fault is, at least partly, in a frictional plastic state and cannot be adequately dealt with by elasticity theory. Nevertheless, it may be inferred from the elastic deflections of the σ_I trajectories behind the propagating fault tip (Fig. 7.4B), that the trajectories do not cross the inelastic near-tip region undeflected, but instead swing around the fault tip in a sigmoidal fashion, as tentatively indicated in Fig. 7.4B. Naturally, the curvature of the trajectories decreases – probably fairly rapidly – with increasing distance from the fault tip, where the stresses approach the far-field state. As indicated in Fig. 7.4B, we characterize these stresses by the components $\sigma_\perp^o$, $\sigma_\parallel^o$, τ^o, normal and tangential to the prolongation of the fault plane.

Despite the analytical complications due to the frictional plastic state in the tip regions, some insight into the development of the near-tip stresses can be gained in a fairly elementary way. Let us consider two points Q^+ and Q^- in Fig. 7.4B, which are positioned symmetrically with respect to the prolongation of the fault plane. And let us further assume that the fault propagates unhindered into a reasonably uniform and effectively unbounded rock mass. The relative displacements along the fault then produce strains and stresses at Q^+ and Q^- in mirror symmetry with respect to the plane of the fault. We relate the stress changes in Q^+ and Q^- to the effective remote stresses in the Mohr diagram of Fig. 7.5A, where the fault trace is plotted as a horizontal line which is supposed to correspond, in real space, with the dip direction of a normal or reverse fault, or with the strike of a strike-slip fault. The friction angle φ is that of the intact rock. In response to the relative displacement on the fault, the mean normal stress at Q^+ on the advancing side is increased to $\sigma'(Q^+) = \sigma'^o + \Delta\sigma'$, and decreased at Q^- on the receding side to $\sigma'(Q^-) = \sigma'^o - \Delta\sigma'$. The centers of the corresponding Mohr circles are thus displaced in opposite directions, as indicated in Fig. 7.5B. Similarly, the fault-parallel normal stress $\sigma_\parallel'^o$ is increased by $+\Delta\sigma_\parallel'$ at Q^+ and decreased by $-\Delta\sigma_\parallel'$ at Q^-. The changes in the fault-normal stresses $\sigma_\perp'$ at Q^+ and Q^- (not indicated in Fig. 7.5B) are then implicitly defined, since the stress points $\sigma_\perp'$ and $\sigma_\parallel'$ on the σ'-axis lie symmetrical with respect to the centers σ'^- and σ'^+ of the Mohr circles. It is fairly obvious that $|\Delta\sigma_\perp'|$ is considerably smaller than $|\Delta\sigma_\parallel'|$, because $\sigma_\perp'$ merely changes in response to the change in $\sigma_\parallel'$.

Finally, the changes in normal stresses are accompanied by an increase $\Delta\tau$ in the fault-parallel shear stress τ. For symmetry reasons, this increase is the same at Q^+ and Q^-, and the total fault-parallel shear stress at these points is raised to $\tau^\circ + \Delta\tau(Q^\pm)$. (Note, that $\Delta\tau(P) > \Delta\tau(Q^\pm)$ at the midpoint P in Fig. 7.4B). Clearly then, the stress changes $|\Delta\sigma_\parallel'|$, $|\Delta\sigma_\perp'|$, and $\Delta\tau$ increase with increasing displacement on the fault, and eventually one of the two stress circles will touch the limit line. Evidently, this is the circle associated with point Q^- on the receding side of the fault.

Admittedly, our argumentation is of a qualitative nature, but one can easily verify that the asymmetrical behaviour of the near-tip stresses still persists when the values of the remote stresses and the strength parameters are varied over a wide range. We therefore conclude that, *the plastic limit state at the leading edge of a fault propagating into a mechanically uniform and unbounded material is first induced on the receding side.* Note, however, that this asymmetry is due to the pressure sensitivity of the shear strength and does not occur in cohesive rocks with pressure-insensitive strength ($\varphi = 0$), as would exist in the transition region from brittle upper crust to ductile lower crust.

In Fig. 7.5A,B the remote stresses is sufficiently compressive to allow the stress circle at Q^- to remain in the compressive part of the stress plane. Hence, in the limit state, subsidiary faults, or shear fractures, should develop on the receding side of the near-tip region. Quite differently, at shallow depth, and in regions of highly overpressured pore fluids, the compressive remote stress $\sigma_{III}'^{o}$ in Fig. 7.5A, and the diameter of the stress circle, may be sufficiently small to allow the fault displacement to render the principal stress $\sigma_{III}'^{-}$ at Q^{-} tensile and make it eventually equal to the tensile strength of the rock ($\sigma_{III}'^{-} = -T_o$). Hence, tension joints will form on the receding side of the fault tip. Such tension fractures may branch off the fault as "wing cracks", as considered in Fig. 2.10A, if the remote

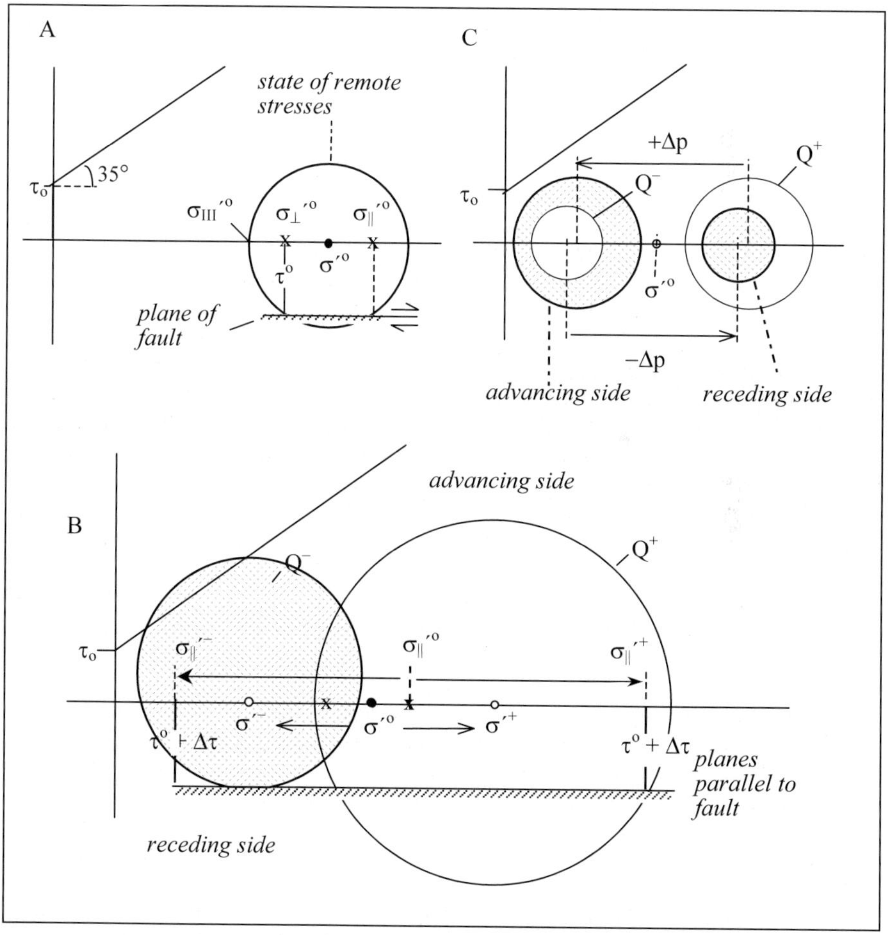

Fig. 7.5. Mohr diagrams of stresses induced near the leading fault edge by incremental shear displacement on the fault:
A) Mohr diagram of remote stress components referred to the plane of the "blind" fault.
B) Changes of the mean normal stress $\sigma' = (\sigma_{\perp}' + \sigma_{\|}')/2$, the fault-parallel stress $\sigma_{\|}'$, and the shear stress τ at advancing (+) and receding (–) sides of the fault plane. The limit state (shaded) is first attained at the receding side (see text for discussion).
C) Exchange of effective stress states due to increase (decrease) in pore pressure on the advancing (receding) side of the fault plane (see text for discussion)

loading is approximately uniaxial ($\sigma_{III}'^{o} \approx 0$, and the growth of wing cracks is, therefore, not suppressed by a compressive $\sigma_{III}'^{o}$. It should be mentioned that various aspects of the near-tip fracturing have already been elucidated in experimental studies by J.-P. Petit and M. Barquins (1988; Tectonics *7*, No. 6, pp. 1243–1256), and by M. Barquins and J.-P. Petit (1992; J. Structural Geology *14*, pp. 893–903).

Ideally, in the case where the fault grows in a mechanically fairly uniform material, not hindered or affected by any obstacles, such as material discontinuities or boundaries, we should, therefore, expect tension or extension joints, or shear joints, to form on the *receding* side of the near-tip region. Along a growing blind fault, these fractures should then leave a fracture seam, preferentially on the receding side of the fault.

However, in reality, near-tip fracturing and, in particular, en échelon tension joints, may also occur on the *advancing* side of a fault, and we should, therefore, briefly venture into the possible reasons for this deviation from the ideal case. First, note that we have tacitly assumed that the pore pressure is uniform throughout the near-tip region. But, we may envisage that slip or even creep on a fault may cause a transient rise in pore pressure at the advancing (compressive) side of the near-tip region, and a decrease in pore pressure on the receding (extensional) side. The consequences of the changes in pore pressure are illustrated in Fig. 7.5C. The unshaded circles represent the changes in effective stresses induced at the mirror points Q^+ and Q^-, when the pore pressure would remain the same at Q^+ and Q^-. If the pore pressure is raised by $|\Delta p|$ at Q^+, and reduced by the same amount at Q^-, the positions of the stress circles are interchanged, as shown by the shaded circles in Fig. 7.5C. Considering the (shaded) advancing-side circle in the figure, and recalling the conditions for cleavage jointing in Chapt. 2 ("Extension or cleavage fracture"), it appears quite feasible that cleavage joints may form on the advancing side of the near-tip region. The reader may also notice, that by choosing a lower value of σ'^{o} and a smaller diameter of the unshaded circle at Q^+, the pore pressure rise may even shift the advancing-side stress circle sufficiently far into the tension domain that (hydraulic) tension fractures may be initiated.

Another mechanism that changes the mirror symmetric development of the strains and associated stresses with respect to the fault plane, and thereby, possibly promotes fracturing on the advancing side of the tip region, was discussed in FBR (Sect. 7.2): the condition of mirror symmetry of the near-tip stresses may be violated when the growing fault approaches a boundary, such as may be imposed by a layer of stiffer rock, a free surface, or a fault or fracture plane. This is schematically illustrated by Fig. 7.6, reproduced from FBR. As long as the boundaries of the rock mass are far enough away from the fault, and thus do not affect the slip-induced near-tip changes in stress and strain, these changes depend on the *relative* displacement on the fault, and not on the way the displacement is actually distributed on either side of the fault, when referred to a reference frame at *rest*. (Note, that the distribution of the same relative displacement in A and B only differs by a uniform fault-parallel translatoric motion which, in the absence of interfering boundary reactions, would not affect the stress field.) If, however, confining boundaries of stiffer rocks (e.g. fixed basement, graben walls, or faults) interfere with the propagation of a fault and affect the mirror image symmetry of the near-tip stresses, the development of these stresses is then no longer controlled by the relative displacement on the fault, but by the displacements on either side of the fault, as referred to the same reference frame at rest.

If movement is suppressed on one side of a fault, as schematically illustrated for a near-tip fault segment in Fig. 7.6C, the slip-induced strains and associated stress changes will predominate on the opposite side of the tip-region. The result is then a tendency for near-tip fracturing to shift towards the advancing side of the parent fault.

This is best seen by considering the extreme situation, where the wall rock opposite the advancing side is completely immobilized by "rigid" boundaries. The shear stress (τ) exerted upon the immobilized block across the fault is then balanced by the opposite shear stress on the boundaries, and, therefore, cannot affect the fault-parallel normal stress $\sigma_{\parallel}'^{\circ}$ in the immobilized block. On the advancing side (+), however, the increase $\Delta\tau$ in shear stress required to advance the fault tip is provided by an increase $\Delta\sigma_{\parallel}'^{\circ}$ of the fault-parallel normal stress. This entails an increase $\Delta\sigma_{\perp}'^{\circ}$ in the fault-normal stress when the displacement normal to the fault plane is suppressed. Here, we assume that the advancing material can move laterally away from the fault plane without being hindered by boundary reactions; hence, $\Delta\sigma_{\perp}'^{\circ} = 0$. Consider then, in the Mohr-diagram Fig. 7.7, the circle C° as representing a state of effective stresses as may typically exist prior to fault progression at a relatively shallow depth, or under high overpressures of the pore fluid. One can then easily envisage that, in a certain area around the progressing fault tip, the stress changes on the advancing side may establish a stress condition such as the one represented by circle C^1, which may be conducive to cleavage fracturing. Also, indicated in the figure, is the direction of the maximum compressive stress ($\sigma_I'^{(1)}$), which includes the acute angle γ with the direction of the displacement on the fault plane. Hence, the trace of the cleavage fractures in the σ_I, σ_{III}-plane would include the angle γ with the parent fault.

Interestingly, movement on the fault may even induce tensile fracturing in the immobilized block. Although, as said before, $\sigma_{\parallel}'^{\circ}$ in the immobilized fault block is not affected by the increment $\Delta\tau$ of the shear stress on the fault, and the free boundary of the moving block allows the fault-normal stress $\sigma_{\perp}'^{\circ}$ to remain constant, the increase in fault-parallel shear stress affects the principal stresses inside the immobilized block. The corresponding stress circle may then enter the tension regime, as shown by the (dashed) circle C^2 in Fig. 7.7. The tensile fractures on the immobilized side of the advancing fault would then be parallel to the $\sigma_I'^{(2)}$-direction in the figure.

Also, recall that the circle C° in the figure was chosen in accordance with stress states associated with faulting at a relatively shallow depth or under high overpressures of the pore fluid. At higher levels of effective compressive stresses, the circle C° would be greater and, on the σ'-axis, centred more to the right than shown in Fig. 7.7. The circle C^1,

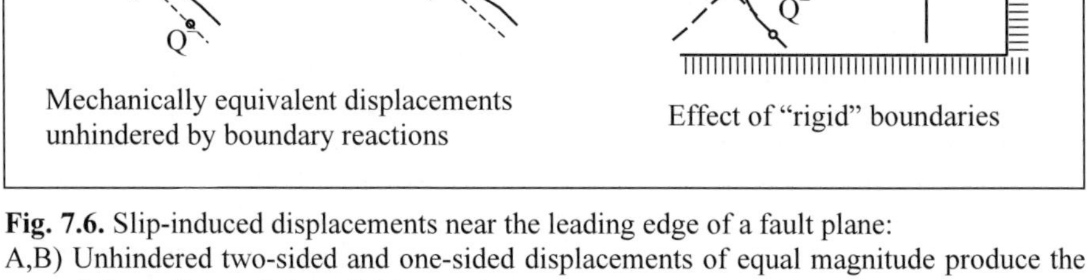

Fig. 7.6. Slip-induced displacements near the leading edge of a fault plane:
A,B) Unhindered two-sided and one-sided displacements of equal magnitude produce the same deformation and stress fields.
C) One-sided displacement and rigid confinement on the opposite side cause an accumulation of stress increments and strains on the advancing side

which refers to the increments of $\sigma_{\|}'$ and τ that allow the fault to advance, may then touch the Coulomb limit lines, indicating the onset of subsidiary faults or shear joints near the moving tip, that branch off the parent fault towards the advancing side.

At this point in the discussion, it is good to recall that our assessment of the fracture-inducing stresses only applies to the "near-tip" area of faults, that is, the small region where the growing fault overcomes the cohesive strength of the rock. In these frontal regions, the shear stress must exceed the shear stress required to maintain slip elsewhere on the fault. As seen in Fig. 7.7, this excess shear stress $\Delta\tau$ is the key parameter in our analysis of the stress perturbations. (In this case, $\Delta\tau$ is considered as constant, while in an earlier problem concerning the delamination of jointed layers (Chapt. 4, p. 77), the corresponding parameter $\delta\tau_c$ was assumed to vary linearly across the frontal decohesion zone.) Here it is assumed as constant. Clearly, the fracture damage by movement on a plane fault can only be produced at fault locations where $|\Delta\tau| > 0$.

This is primarily the case in the near-tip regions of a growing fault. But, the fracturing condition $|\Delta\tau| > 0$ may also be fulfilled at other fault locations, when, as is usual on tectonic faults, slip or creep does not occur along the whole fault at once, but in separate patches whose locations vary in the course of successive movements on the fault. At the edges of these patches, $|\Delta\tau|$ will be greater than zero, and fractures, similar to those at moving fault tips, may be generated.

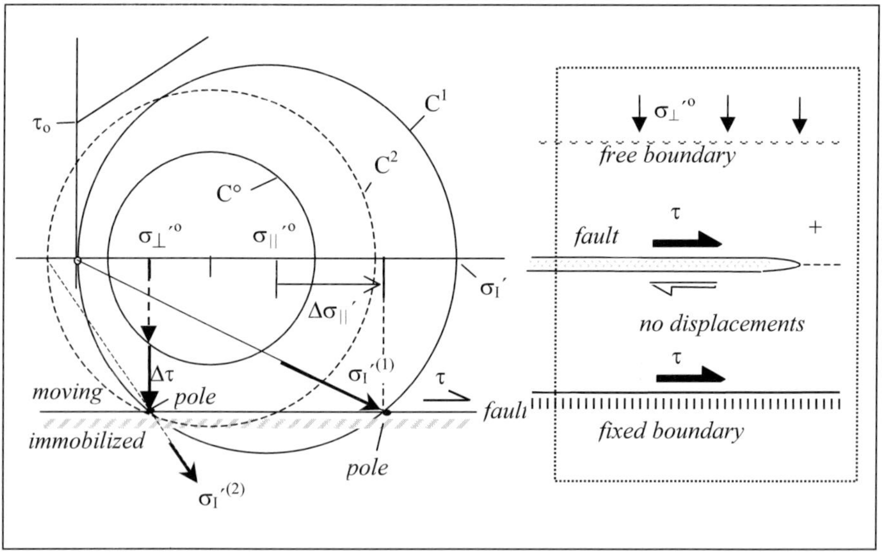

Fig. 7.7. Mohr diagrams of presumptive stress states prior to faulting (C°) and during fault progression, when boundary confinements restrict the displacements to the area on the advancing side of the fault (see insert on the right). The circle C^1 represents the stresses in the advancing block, which is free to deform perpendicularly to the fault, and remains loaded by the fault-normal stress $\sigma_\perp'^o$ that already acted before the advancement of the fault tip (see insert). The (dashed) circle C^2 represents the stresses on the immobilised side of the advancing fault. The position and diameter of the circle C° conform to the stress state prior to the fault progression at a relatively shallow depth, or under high over-pressures (see text)

The trends of the asymmetric development of tension- and cleavage fractures, and secondary faults, in near-tip regions are very schematically illustrated in Fig. 7.8. When the parent fault propagates unhindered into an effectively unbounded rock mass, tension fractures, or subsidiary faults and shear joints, develop on the receding side (A). Subsidiary faults and fractures formed on the receding side are deactivated as the parent fault propagates, and are arranged in a single-sided array behind the propagating fault tip. On the other hand, cleavage fracturing or subsidiary faulting is shifted to the advancing side (B) by the blockade of absolute displacement and deformation behind the advancing fault tip on the receding side. Since, in an absolute sense, the whole displacement takes place in the wall that moves with the fault tip, the splay faults or cleavage fractures should remain located near the moving fault edge on the advancing side of the parent fault. Furthermore, in a low effective stress regime, tension joints may also form on the immobilized side of the propagating fault tip, as was concluded from the position of the dashed circle C^2 in Fig. 7.7. These joints are indicated by the set of dashed lines in the lower part of B.

As was argued above, cleavage fracturing on the advancing side of a moving fault tip (lower figure in Fig. 7.8B) requires not only that the displacements on the opposite side are suppressed, but also, in general, that the material on the advancing side is allowed to move laterally away from the fault plane without, or only by slightly, increasing the fault-

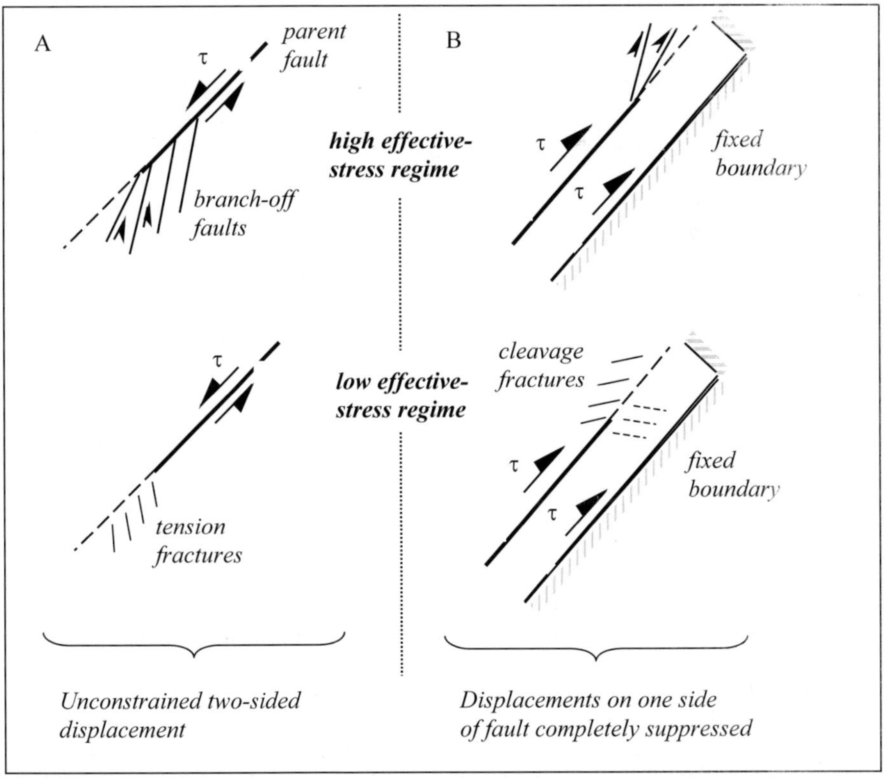

Fig. 7.8. Simplistic view of secondary faults and joints in near-tip regions on receding (A) and advancing side (B) of a propagating fault

normal stress $\sigma_\perp'$. These conditions may be fulfilled in various types of faults in near-surface layers or in overpressured layers; for instance, in thrust faults where the compressed material of the advancing side can expand towards the free surface while the footwall is at rest; or in normal faulting on a slope when the footwall does not move with respect to a reference frame at rest, but where the hanging wall expands in the down-slope direction. One may also envisage similar situations in strike-slip faulting, when displacement is confined to one side, for instance, by a graben boundary or an obliquely striking fault.

Certainly, our qualitative arguments are in need of rigorous poro-elastic analyses which, in particular, should include the *inertial* forces induced in the solid matrix and in the pore fluid caused by a sudden slip on a fault. The inertial effects also manifest themselves in stress pulses that travel along the fault on which the seismic slips occurred, and possibly generate fractures in the vicinity of the fault. Neville Price (in N.J. Price and J.W. Cosgrove 1990; loc.cit., pp. 150–159) analyzed the stress pulses (in terms of total stresses) and found that they rapidly fall off in magnitude with the distance from the slip locus. Price, therefore, concluded that fractures generated by seismic stress pulses should remain confined to relatively narrow zones on either side of the seismic fault.

Rock deformation along faults. Sedimentary rocks may be deformed near an active fault in many ways. Some typical fault-related deformation modes in extensional regimes are schematically illustrated in Fig. 7.9: A) Layers may be flexured and transected by minor normal faults before being cut by a major fault (after F. Lehner and W.F. Pilaar 1974, Shell Research Report). B) Layers are stretched by frictional drag on a fault plane, or C) deformed into "roll-over" structures when rotationally sliding along a listric normal fault (λιστρον = shovel). D) Strata in the hanging wall of normal faults (and in the reverse sense in thrust faults) are also extended in the strike direction as the dip-slip dies out along the strike, as shown in the block diagram. E) The diagram shows two downwarped layers, separated by a thinner weak layer, in the hanging wall of a normal fault. The flexure shape of the layers is controlled by the competent thick layer 2. The thinner layer 1, mechanically decoupled from the control unit by the weak interlayer, is stretched over its full length. In the brittle and semi-brittle regime, the deformation at a macro-scale is accommodated by minor faults and/or tension and extension joints, as sparsely marked in the diagrams. But, whereas the role of subsidiary faults in accommodating fault-imposed deformations is fairly well understood (see e.g. FBR, Chapt. 6), the when and where of an accommodation by jointing, or even the type of operating joints, are still much less known (notice the question marks in Fig. 7.9).

Although joint sets and minor normal faults are common features in competent layers which are extended in the vicinity of a fault (e.g. Fig. 4.16), the phenomenon poses several questions. First, consider a flat-lying layer which is extended by bending in the vicinity of a major normal fault, say, in one of the modes shown in Fig. 7.9. Then consider, in the Mohr diagram of Fig. 7.10, the critical stress states for tensile failure and shear failure in the layer, and compare the largest possible limit circle for tension failure and the smallest limit circle for shear failure (i.e. normal faulting of the type in Fig. 1.8D, p. 5). Whereas the difference in σ'_{III} between the two limit states is of little concern, because it would tend to vanish when tension joints are formed, the difference in σ'_I is considerable (amounting to almost $5T_o$, if $\tau_o = 2T_o$ and $\varphi = 35°$). Since, in general, in an extensional regime σ'_I differs little in magnitude from the overburden stress σ'_v, it appears impossible that both the tension joints and the minor normal faults can form together in the same layer

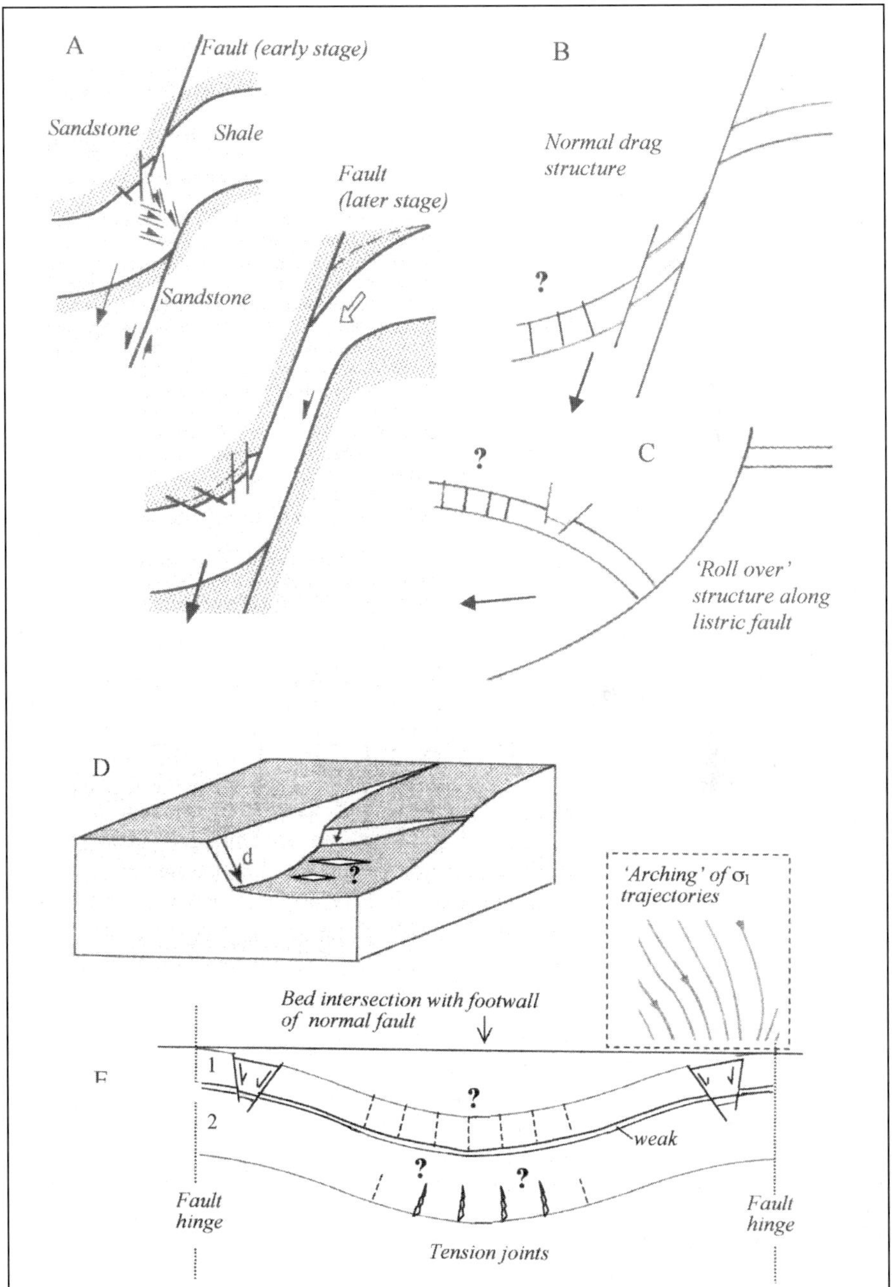

Fig. 7.9. Schematic diagrams of conjectural accommodation fracturing near normal faults (see text for comments)

of the hanging wall of a major normal fault. In fact, the tension jointing is restricted to layers under a small effective overburden stress, whereas faulting requires higher effective overburdens. (But note, that this restriction does not imply that a fault cannot approach the free surface; in a cohesive material the fault will steepen near the surface by changing into a dilational or hybrid tension-shear fault of the type shown in Fig. 1.8C (p. 5), and possibly terminate in subvertical tension joints.)

Although tensile fracturing and normal faulting are mutually exclusive processes, this does not exclude that, in the same layer, tension joints and normal faults may form in succession, when the magnitude of the effective stress σ'_I changes drastically with time, or joints and normal faults form simultaneously in separate parts of a layer along which σ'_I varies greatly in magnitude. The first situation could arise in downwarped strata when, after high overpressuring of the pore water has reduced σ'_v ($\sim \sigma'_I$) to values smaller than $3T_o$, and thereby allowed the formation of tension joints, the pore pressure is again "normalized" by drainage, facilitated by the joints; the associated increase in effective overburden stress may allow the formation of minor normal faults in the extending layer.

The second situation, where joints and faults form concomitantly in separate parts of a layer, might exist in strata downwarped between the hinges of a major normal fault, as shown schematically in Fig. 7.9E. Above a downbent layer, the σ'_I trajectories are deflected from an initially vertical position towards the layer shoulders, as indicated in the insert in the figure, thereby shifting the overburden load from the central part to the shoulders of the layer. The arch-type redistribution of the σ'_I trajectories ("stress arching") above a subsiding block or a synclinal structure is well-known in soil mechanics and rock engineering (see also MTF I.3.3 and Fig. II.7-32, and FBR Fig. 7.33). It is understood that, for a given shear resistance of the overlying material, the unloading of the inner part of the downthrown layer is more effective with a greater thickness of the overburden and a smaller distance between the fault hinges. We, therefore, presume that at some stage in normal faulting the inner part of a downwarped and pervasively stretched layer (such as

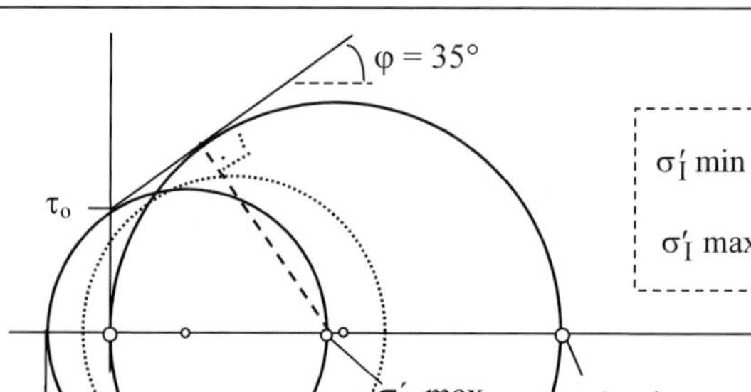

$$\sigma'_I \text{ min (fault)} = 2\tau_o \frac{\cos\varphi}{1-\sin\varphi}$$

$$\sigma'_I \text{ max (tension joint)} = 3T_o$$

Fig. 7.10. Mohr diagram of limit states of tension jointing and incipient normal faulting in the same layer, which shows the difference in magnitude of the maximum compressive stress that still permits the formation of tension joints and the minimum compressive stress required for incipient normal faulting (the dotted circle represents a stress state associated with the formation of a *dilational* fault – a hybrid extension-shear fracture; Fig.1.8C)

layer 1 in Fig. 7.9E) may be sufficiently unloaded to allow the formation of tension joints there, while minor normal faults (small grabens) are accommodating the extension of the marginal parts of the layer. We also note, that in thrusting (reverse faulting) the process should occur in reverse, with tension jointing at the margin and normal faulting in the inner part of an upbent layer.

So far, we have considered tension jointing and minor normal faulting in one and the same layer. What about two mechanically different layers at nearly the same depth of burial in a sedimentary sequence? Let us consider in Fig. 7.11A two horizontal sandstone layers, Sd 1 and Sd 2, which are separated by a weak clay layer. The sandstone layers differ in shear strength (τ_o, φ) as indicated in the Mohr diagram of Fig. 7.11B, but the

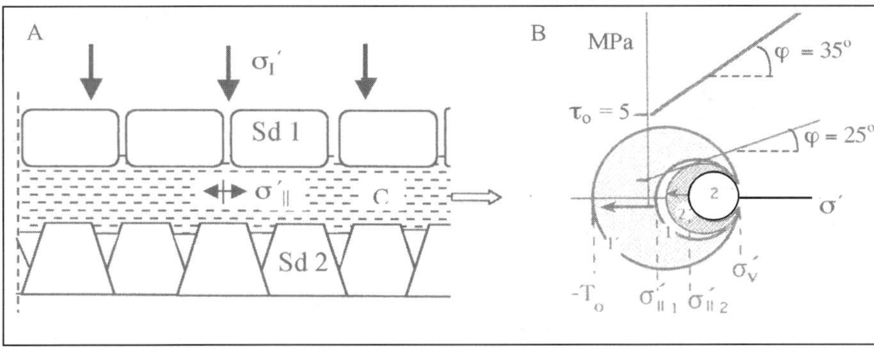

Fig. 7.11. Uniform extension of a horizontal multilayer generating tension joints and normal faults in different layers under the same overburden stress $\sigma'_I = \sigma'_V$.
A) Schematic picture of a jointed layer of competent sandstone (Sd 1), an intact interlayer of clay, C, and a sandstone layer (Sd 2) intersected by normal faults.
B) Mohr diagram with different strength conditions for Sd 1 and Sd 2 and corresponding limit circles 1′ and 2′

Fig. 7.12. Minor normal fault intersecting Liassic shale beds and continuing along tension joints in stronger sandstone beds (Lilstock, Bristol Channel, SE England)

overburden stress σ'_v may be taken as the same in all three layers. When the layers are extended together, it is possible that normal faults will be produced in one layer and tensile fractures in the other. This is demonstrated by the stress circles in the Mohr diagram, where the circles 1 and 2 represent the stress states in layer Sd 1 and Sd 2, respectively, before the extension; both circles run through the point σ'_v,0. The joint extension of the layers may change circle 1 into circle 1′ which intersects the σ' axis at $-T_o$, and at some stage of extension circle 2 may be altered into the circle 2′ which touches the limit line of the weaker material of layer Sd 2. Therefore, the extension of the layers will be accommodated in the stronger layer Sd 1 by tension joints, and by normal faults in the weaker layer Sd 2, as shown in Fig. 7.11A. Note that, in general, the two limit states will be attained at different stages of the extension of the whole. Also note again, that the formation of tension joints in Sd 1 is only possible as long as $\sigma'_v \leq 3T_o$, which at normal pore pressure and for the conditions in the figure, would imply a limiting overburden thickness of approximately 450 m.

A somewhat related phenomenon is shown in Fig. 7.12, where a small normal fault transects an alternating sequence of sandstones and shales by connecting fault segments in shale beds across tension joints in the stronger sandstone beds.

Healed joints opening concurrently with fault slip. The relationships between joints and faults considered above applies to the *formation* of joints that accompanies faulting. But one may also wonder how faulting might affect healed tension joints or veins which were generated *before* the onset of the faulting. Let us consider the problem in a simplified form as studied by R.H. Sibson (1981; Nature *289*, pp. 665–667). We assume a set of healed tension fractures or veins in flat-lying beds, which are aligned along the vertical σ_I (= σ_v) direction in a uniform stress field with a horizontal σ_{III} direction. Moreover, the layers are transected by a plane of weakness whose normal lies in the σ_I, σ_{III} plane; the trace (Γ) and dip angle (α) of the plane of weakness are shown in Fig. 7.13A. The chord of the stress circle in Fig. 7.13B is drawn parallel to Γ from the stress pole P, and, therefore, defines the normal and tangential stresses ($\sigma'_\perp$, τ) on the plane of weakness. Also indicated in the figure are the intact rock strength parameters τ_o, φ and T_o, a smaller tensile strength T_o* for the healed joints or veins, and the shear strength parameters τ_o* and φ* for the plane of weakness. In the figure, it is assumed that slip on the plane commences, which implies that the horizontal effective stress σ'_{III} is reduced until the end-point of the Γ-parallel chord arrives at the limit line for the plane of weakness. If it so happens that at this stage $\sigma'_{III} = -T_o$*, the healed tension joints will be opened.

Naturally, the opening of the joints at the moment of slip on the plane of weakness would appear as highly coincidental, were it not that slip-induced negative values of σ'_{III} exist through a certain overburden range, where σ'_{III} varies with the value of σ'_I. But, it is already clear from the diagram in Fig. 7.13B, that only relatively small values of σ'_I, corresponding with a shallow depth of burial or the presence of high overpressures, are compatible with negative values of σ'_{III}.

The range of σ'_I values which allows slip on the fault plane simultaneously with the opening of the healed tension joints is easily quantified by determining the coordinates $\sigma'_\perp(S)$, $\tau(S)$ of the intersection point S in Fig. 7.13B and using elementary geometry on the right-angled triangles in the figure. This gives:

$$\sigma_I'^* = \sigma'_\perp(S) + \tau(S).\tan\alpha \tag{7.1}$$

with

$$\sigma'_\perp(S) = \left(\tau_o^* - T_o^* \tan\alpha\right).\left(\tan\alpha - \tan\varphi^*\right)^{-1} \quad \text{and} \quad \tau(S) = \tau_o^* + \sigma'_\perp(S).\tan\varphi^*$$

Hence, a positive σ'_I*, implying a positive effective overburden stress σ'_v, is only possible if

$$\tau_o^* > T_o^* .(\sin\alpha.\cos\alpha + \sin^2\alpha.\tan\varphi^*) \quad (7.1a)$$

If, for example, $\varphi^* = 20°$ and $\alpha = 62.5°$, the cohesive shear strength τ_o* of the inactive plane of weakness would have to be greater than $0.7.T_o$*. With the effective overburden stress $\sigma'_v \sim \sigma'_I$*, slip and reopening of the joints require that

$$\sigma'_v = 3\,\tau_o^* - 2.1T_o^* \quad (7.1b)$$

Accepting, for the purpose of illustration, the values $\tau_o^* = 5$ MPa and $T_o^* = 1$ MPa, Eq. 7.1b predicts $\sigma'_v = 12.9$ MPa, which corresponds at normal pore pressures to a burial depth of approximately 860 m. Note, however, that overpressures would increase the depth level by the factor $0.6/(1 - \lambda)$, where $\lambda = p/\sigma_V$.

Now that the condition is established which must be fulfilled that slip on a normal fault and the opening of healed subvertical joints can occur together under the same effective overburden, we face a critical question concerning the *origin* of the normal fault that acted as a plane of weakness. To see the problem, consider the stress circle 1 in Fig. 7.14, which represents the effective stresses that generate a normal fault with the dip angle α of Fig. 7.13; all normal stresses are compressive, the pore pressure is normal ($\lambda = 0.4$), and the overburden stress $\sigma_V'^o$ is the maximum compressive stress. The stress circle is tangential to the straight Coulomb-limit line which, in the intact rock, defines the onset of faulting under all-compressive normal stresses. The formation of the normal fault causes a decrease in differential stress $\sigma_V'^o - \sigma_{III}'$, while the overburden stress $\sigma_V'^o$ remains constant, as indicated in the figure by the (shaded) circle 2. The fault plane is now a plane of weakness with the reduced strength parameters τ_o* and φ*. But, in order to satisfy the condition 7.1, the stress circle 2 would have to be brought into the position of the stress circle in Fig. 7.13. This is only possible if, *after* the formation of the fault, the effective

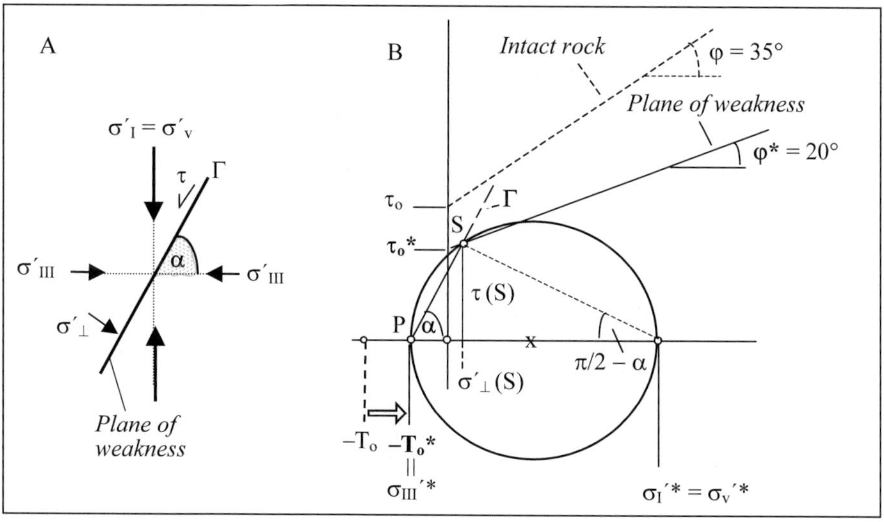

Fig. 7.13. Slip induced on plane of weakness by reduction of σ_{III}' at overburden ($\sigma'_v = \sigma'_I$). A) Trace of plane of weakness (Γ) in σ_I, σ_{III} plane; B) Mohr diagrams of stresses inducing slip on plane of weakness and opening healed joints of tensile strength T_o*

overburden is drastically reduced. This can only be achieved by overpressuring the pore fluid. The overpressuring will shift the circle 2 to the left. But, at the same time, the circle has to shrink (circle 3) in order to enter the tension region of the stress space. However, as was pointed out in Chapt. 3 (pp. 28–32), the shrinkage of the stress circle requires that lengthening of the layer in response to the rise in pore pressure is suppressed.

The shrinkage of the stress circle (Fig. 3.2B) is characterized by the parameter β, which is inserted into Fig. 7.14 from Eq. 3.1. Note, that under the strength conditions and the initial stress state (circle 1) assumed in the figure, the value β = 32 allows the Γ-chord of the shifted circle 3 to terminate on the limit line of the plane of weakness at the point S in Fig. 7.13B, thereby triggering slip on the fault. At the same time, the layer-parallel stress $\sigma_{III}'^*$ attains the value $-T_o^*$ and reopens healed tension joints or veins with a tensile strength T_o^*, or even generates new internal hydrofractures in a rock with this tensile strength. Clearly, the value of T_o^* depends on the shear strength parameters of the intact rock, the stress release by faulting, and the degree of weakening of the fault plane; but primarily, T_o^* varies with the thickness of the effective overburden.

The primary factor in this process is the overpressure of the pore fluid. Because of the low compressibility of the pore water, only very small amounts of pore water drainage are needed to drastically reduce the overpressure. As suggested by Sibson, the opening of vertical hydraulic tension joints may provide this drainage, and the resulting drop in pore

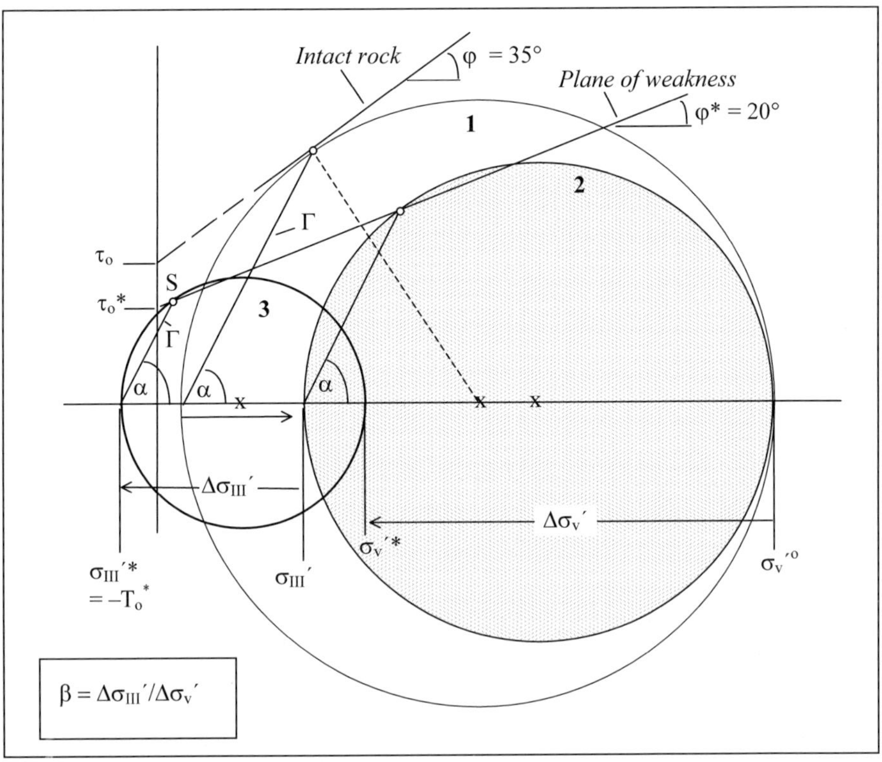

Fig. 7.14. Stress diagrams illustrating the history of a normal fault from incipience (1) and relief of differential stress (2), to the remobilization of the weakened fault plane by overpressuring of the pore fluid under *suppressed* horizontal extension (3)

pressure would immobilize the fault by increasing the fault-normal stress $\sigma'_{\perp}$. The deposition of hydrothermal minerals may then reheal and seal the joints or veins, and thereby stop the drainage and allow the overpressure to be restored to the level required for slip on the fault and reopening of the fractures. Obviously, one may think of the bleeding of overpressure, the rehealing of joints and the subsequent restoration of the overpressure as a repetitive process which results in intermittent slippage on a normal fault accompanied by the reopening of healed joints. The process may then be seen as a manifestation of John Ramsay's "crack-seal mechanism" (J.G. Ramsay 1980; Nature *284*, pp. 135–139).

Thus, the above process explains the possible coincidence of the opening of internal hydrofractures and the slip on a normal fault plane in highly overpressured rocks. But, it should be borne in mind, that the process neither implies that the fault slip is a prerequisite for the reopening of healed joints, nor that the fault slip must be accompanied by the reopening of healed hydrofractures.

A further point that should be remarked upon is the assumption that the layer is laterally confined and cannot extend in response to the increase in pore pressure, otherwise, the stress circle 2 in Fig. 7.14 could not be shifted into the tension domain. On the other hand, slip on the weakened normal fault in the stress state 3 would require some horizontal displacement of the adjacent rock, which might seem to contradict the postulation of a lateral confinement of the layers. However, the two requirements become compatible when the rock in the vicinity of the slipping fault contracts by the ejection of pore fluid along the fault plane. Nevertheless, the question remains why the geological boundary conditions have changed from a regime of horizontal extension that produced normal faults (states 1 and 2 in Fig. 7.14) to a regime of high overpressures under horizontal confinement. This question does not arise when the faults are *dilational*, as will be considered next.

A dilational normal fault, as shown in Fig. 1.8C, is formed by the interlinking, on a microscale, of tension fractures and shear fractures (as sketched in Fig. 7.2A). There, the tension fractures predominate in accordance with the minimum effective stress, being tensile on a macroscale. As a consequence, the interior of a dilational fault is more jagged and, at least initially, more porous and permeable than the surrounding rock; dilational faults, therefore, typically provide preferential conduits for fluid flow. This, in turn, produces a certain degree of cohesive strength, by mineralization of the fault material.

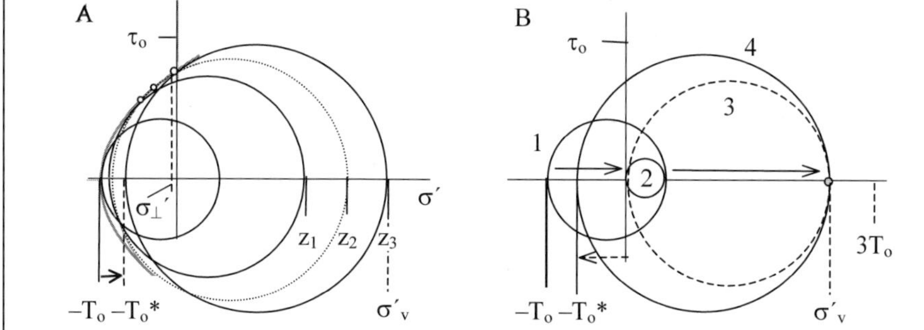

Fig. 7.15. Stress circles illustrating the origin of tension joints and dilational faults:
A) Stress conditions for dilational normal faulting at different depths ($z_1 < z_2 < z_3$); the stresses at depth z_3 open the healed joints of tensile strength T_o*.
B) Vertical tension jointing (1) and (2) with $\sigma'_{III} \Rightarrow 0$ at shallow depth; subsidence to a greater depth (3) and partial healing of the joints (4)

It is important to note, that the incipience of these faults is not controlled by a Coulomb-Mohr limit condition, since this exclusively applies to faulting when none of the effective stresses are tensile. In contrast, the critical stress circles for dilational faults have a curved envelope as shown in Fig. 7.15A (also see FBR, pp. 129–131), and the fault-normal stress $\sigma'_{\perp}$ is tensile. Assuming again, that the overburden stress σ'_v is the greatest compressive stress, the critical stress circles increase in diameter with increasing depth (z), but are limited by the circle whose point of tangency with the envelope has the abscissa $\sigma'_{\perp} = 0$. If we insert reasonable values of τ_o and T_o for a medium-strong rock, and allow for some variation of the envelope point on the τ axis, it transpires that dilational faulting at normal pore pressures is very likely restricted to a depth of less than 1 km. The same holds true for the formation of vertical tension joints.

Because the stress states that generate dilational normal faults are very similar to those which produce vertical tension joints, one can imagine how healed tension joints or veins may be opened by the formation of a dilational fault. Let us consider in Fig. 7.15B a stress state, represented by the stress circle 1, which produces an array of vertical tension joints in a horizontal layer of tensile strength T_o. As the parallel fractures are filled with pore fluid, the effective tensile stress is reduced to zero (circle 2), and remains zero under an accumulating overburden as long as the fracture walls remain separated (dashed circle 3). Under real conditions, the water-filled joints of the subsiding layer will acquire some tensile strength T_o* which allows a tensile stress to build up in response to further extension of the subsiding layer. When the tensile stress σ'_{III} reaches the value $-T_o$* (stress circle 4), the healed joints will reopen.

Now, let us go back to Fig. 7.15A, and consider the range of critical stress states for dilational faulting. At any depth z_i, a dilational fault will be formed if the tensile horizontal stress attains the value $\sigma'_{III}(z_i)$ that is uniquely related to the overburden load at this depth. If it so happens that $-\sigma'_{III}(z_i)$ equals the tensile strength T_o* of healed vertical joints at the depth z_i, the extension-induced $\sigma'_{III}(z_i)$ will simultaneously create dilational normal faults and reopen the healed joints. If $|\sigma'_{III}(z_i)| < T_o^*$ dilational faults may form, but without opening the tension joints. Conversely, joints may be reopened by extensional straining of layers under an overburden load which is still too small for the formation of a dilational fault (or any type of fault). When and where a concomitant reopening of joints and dilational jointing may actually occur will obviously depend on the interplay of the processes of sediment accretion and extension of the subsiding layers.

In any case, it transpires from the above discussion that the coincidental occurrence of reopened joints or veins and dilational normal faults is due to a very special state of stress which makes the coexistence of the two structures possible. In other words, there is no direct causal connection between the reopening of veins or joints and the activity of dilational faults. Hence, we cannot infer with confidence, the existence of normal faults from the observation of reopened and rehealed joints or veins, or vice versa. On the other hand, if both structures are observed at the same location, one may arrive at conclusions concerning the stress field and, possibly, the role of overpressured pore fluids.

Strike-slip faults parallel to joints. Faults can develop by the shearing of preexisting parallel joints. This has been convincingly demonstrated by Paul Segall and David D. Pollard in a paper on "Nucleation and Growth of Strike Slip Faults in Granite" (1983; J. Geophysical Research *88*, No. B1, pp. 555–568). The authors analyzed parallel fractures within granodiorite in two areas of the central Sierra Nevada, California. The fractures in the study areas form single sets of steeply dipping joints. In Fig. 7.16, a fracture map of one study area is reproduced from the paper. Some of the fractures were identified as dilational fractures and joints (without further specification as to whether the joints are of the tensional or cleavage type), and are mostly hairline in appearance. Other fractures were identified as small faults with left-lateral strike separation of up to 2 m, and a width of 1 to 10 mm or more. Also note that, in contrast to the regular spacing of joints in layered sedimentary rocks, the spacing of the parallel joints in the crystalline rock body is not uniform.

Obviously, the spacing mechanism in layered rocks considered in Chapt. 4 does not apply here. Instead, under horizontal tension, a vertical tension fracture grows in depth

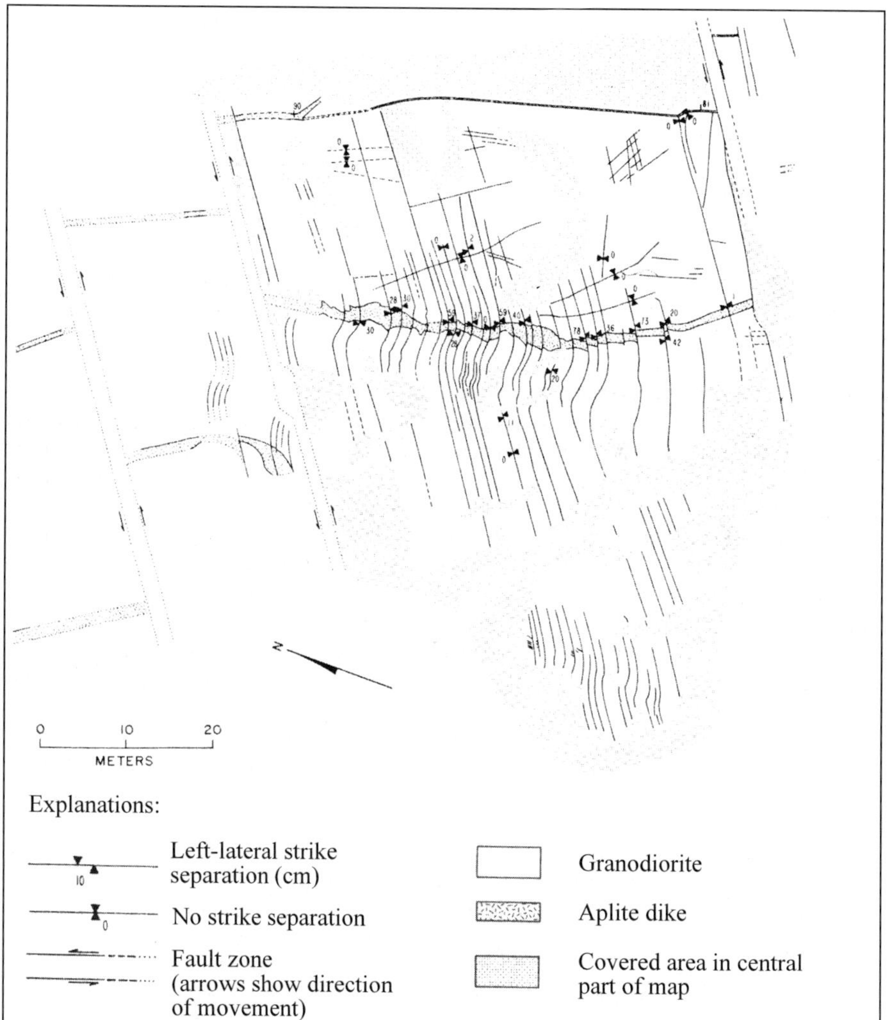

Fig. 7.16. Map of strike slip faults along pre-existing subvertical joints in granitic rocks of the central Sierra Nevada (Bear Creek area in the Mount Abbot quadrangle), California. The map is reproduced from P. Segall and D.D. Pollard (1983; J. Geophysical Research *88*, No. B1, p. 563; reproduced by permission of AGU) (see text for further explanation)

surrounded by a zone of elastic tension release and unhindered by the layer interfaces (A.H. Lachenbruch 1961; see reference on p. 73). Within this elastic halo or "shadow zone", a second fracture cannot grow. Hence, when a set of initially densely spaced shallow cracks is subject to a progressive horizontal extension, slightly deeper fractures will grow, while their tension release zones impede the growth of shorter fractures. As the process proceeds, the number of deepening fractures decreases, while their spacing increases. The process is very similar to the growth and spacing of parallel dyke-type intrusion fractures, which were considered in Chapt. 3 (pp. 34–38). Furthermore, it seems reasonable that a vertical fracture that grows in depth, also grows in horizontal length at the same time if it is not hindered by lateral barriers. Hence, in map view, longer tension fractures will be wider spaced than shorter ones (A. Nur 1982; J. Struct. Geol. *4*, No. 1, pp. 31–40).

The single set of dilational joints in Fig. 7.16 was formed when the region was extended, and left-lateral slip was imposed on the joints at a second stage of regional deformation that produced a counterclockwise rotation of the σ_I-axis through an acute angle from the strike of the joints. Thus, faults were nucleated on pre-existing joints, and proceeded along the joints even though the fractures were sealed with mineral fillings. However, the shearing did not continue into the intact rock as a single fault, coplanar with the parent joint, but was instead accommodated by secondary fracture structures in the two dilational quadrants at the receding side of the joint ends, as illustrated in Fig. 7.8A.

In the study area of Fig. 7.16, numerous strike slip fault zones (with a left-lateral strike displacement of approximately 10 m and a width of 0.5 to 2 m) also parallel the joints and small faults. These fault zones were formed by the linkage of adjacent smaller faults by secondary fractures. The linkage process was studied in detail by S.J. Martel (1990; J. Structural Geology *12*, No. 7, pp. 869–882). In essence, we think that the linkage of smaller faults and the transfer of the shearing displacement is accomplished by a dilational "bookshelf" mechanism which we have referred to before (see Fig. 6.17, and FBR Sect. 6.4). As schematically illustrated in Fig. 7.17, a sinistral strike-slip fault zone develops by left-laterally sheared joints or small faults stepping to the left between adjacent dilational tip zone quadrants (–), as was considered in Fig. 6.7. The shearing is transferred across the extensional offset areas by the "bookshelf"-type rotation of en échelon sets of secondary fractures. The inclination of the secondary fractures in the figure is chosen in qualitative agreement with Martel's observations, and implies that the shearing transfer is accomplished by the dilational mode of the bookshelf mechanism, schematically shown in the insert of Fig. 7.17. We expect the process to still work when the overlap of

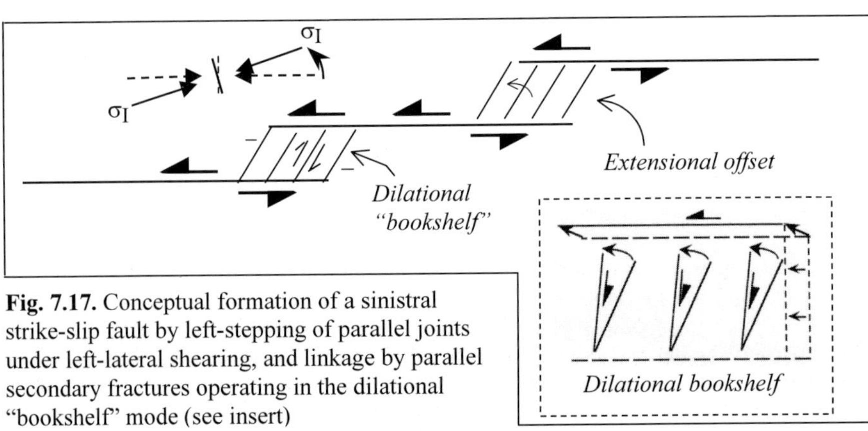

Fig. 7.17. Conceptual formation of a sinistral strike-slip fault by left-stepping of parallel joints under left-lateral shearing, and linkage by parallel secondary fractures operating in the dilational "bookshelf" mode (see insert)

the fault elements is smaller than in the figure, or even missing at all. We leave the question open as to whether the secondary fractures are "splay" fractures emanating from the fault tips, as assumed by Martel, antithetic Riedel shears, or fractures produced as tension joints in the extensional offset zone. Although at first sight one might be tempted to interpret the parallel joints in Fig. 7.16 as shear joints, Segall and Pollard firmly concluded, after closer inspection, that all of the fractures in the study area were originally formed as dilational joints, and none of them as shear joints. Nevertheless, in reality we also expect uniform stress conditions to exist which are conducive to the formation of parallel shear joints (pre-peak or post-peak), some of which, probably those of greatest length, develop into strike-slip faults.

Perturbation of joints by pre-existing faults. In the preceding sections on the reopening of healed joints and on joints turning into strike-slip faults, the faulting post-dated the formation of the joints. Now, we consider the reverse sequence, where jointing postdates the formation of strike-slip faults in horizontal, or nearly horizontal layers.

Let us begin with the simplest case of a straight vertical strike-slip fault (Andersonian type) which is not affected by other faults. We briefly recapitulate in Fig. 7.18 the stress states associated with the initiation, weakening and the eventual inactivation of the fault. The dotted stress circle in Fig. 7.18A represents the uniform stress system that initiated the fault, with the principal stresses $\sigma'_I{}^{(o)}$ and $\sigma'_{III}{}^{(o)}$ parallel to the layering. We consider the σ'-axis in the Mohr diagram as coinciding with the direction of $\sigma'_{III}{}^{(o)}$ in real space, and draw the chord of the stress circle from the "pole" at $\sigma'_{III}{}^{(o)}$, $\tau = 0$ to the point of tangency T with the limit line. The chord Γ then has the orientation of the strike line of the fault in real space (see Appendix), which includes the angle α with the $\sigma'_{III}{}^{(o)}$-axis. This is also schematically illustrated in Fig. 7.18B. We recall, that the formation of the fault is accompanied by a reduction in shear strength of the fault material, which causes a reduction of the stress difference $\sigma'_I{}^{(o)} - \sigma'_{III}{}^{(o)}$ to $\sigma'_I{}^{(1)} - \sigma'_{III}{}^{(1)}$, i.e. the diameter of the shaded circle in Fig. 7.18A. Thus, the active fault is a plane of weakness, whose shear strength is determined by a separate limit line defined by the reduced strength parameters $\tau_o{}^*$ and φ^*. Since the directions of the principal stress axes and the strike line Γ of the fault remain unchanged during the formation of the active fault, the point $\sigma'_{III}{}^{(1)}$, $\tau = 0$ is a stress pole of the shaded circle. Therefore, the Γ-line drawn through this pole defines the point where the stress circle intersects the limit line, and thereby determines the normal and tangential stresses ($\sigma_\perp'$, τ) on the fault plane.

Next, we assume (Fig. 7.18C) that the fault is inactivated by reducing the principal stresses to $\sigma'_I{}^*$ and $\sigma'_{III}{}^*$, which allow parallel tension joints to form. During the stress reduction, the directions of the horizontal principal stresses need not remain fixed, but may be turned through an angle ω around the vertical σ_{II}-axis, as illustrated in Fig. 7.18D. This angular change shifts the stress pole from the circle point $\sigma'_{III}{}^*$, $\tau^* = 0$ to the point shown in Fig. 7.18C, and the stresses $\sigma_\perp'{}^*$, τ^* on the inactive fault plane Γ are readily determined by the simple construction in the Mohr diagram.

The inactivated fault does not affect the stress field. The tension joints, therefore, propagate along the straight $\sigma'_I{}^*$-trajectories of the uniform stress field (Fig. 7.19A) either crossing the fault or being arrested by the weakness of the fault. When crossing the plane of weakness, a tension fracture will probably make a small side-step, as was discussed in Chapt. 4 (pp. 50–54).

Naturally, the assumption of a uniform shear stress along the fault plane will be rarely met in reality. Instead, one should expect the shear strength of the fault material to vary along the fault, and in some local areas even be insufficient to withstand the shear load imposed by the stress field. Let us, therefore, consider in Fig. 7.19B the extreme case of a strike-slip fault with a segment Σ on which the shear strength is zero. Since the segment

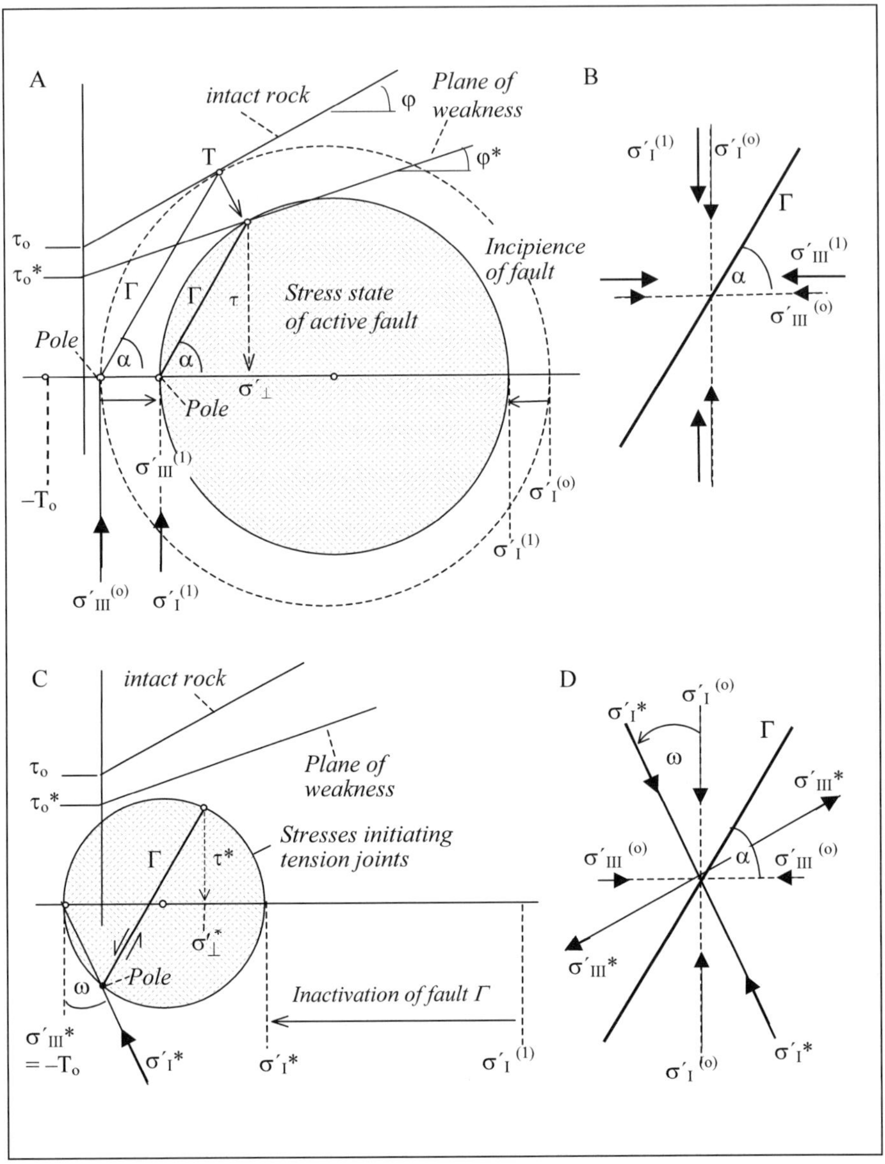

Fig. 7.18. Stress diagrams of the formation and inactivation of a vertical strike-slip fault:
A) Mohr diagram of the horizontal stresses (dashed circle) that initiate the fault (Γ) and are changed (shaded circle) by the weakening of the fully developed fault.
B) Diagram of the principal stress axes in the horizontal plane.
C) Inactivation of the fault by reduction of principal stresses to a limit state of tensile fracturing, possibly accompanied by a rotation (ω) of the horizontal principal axes; pole construction determines the normal and tangential stresses on the inactive fault plane.
D) Diagram of the rotated principal stresses

cannot carry a shear load, the deficit in the capacity of the fault to carry the imposed boundary load can only be compensated for by an increase in the shear force, on the part of the fault, downwards from the segment Σ. We assume, that the increased shear stresses remain below the limit line of the plane of weakness (Fig. 7.18C), and the fault, therefore, remains inactive. Since the fault segment Σ is free of shear stress, it is a principal plane, and on the fault segment the direction of σ'_{I} can only be normal or parallel to the fault. In fact both orientations occur, though on different quadrants of the segment as indicated in Fig. 7.19B. Clearly, the smallest principal stress σ'_{III} is parallel to the fault on the extensional part (–) on either side of Σ, while σ'_{I} is parallel to the fault on the compressed

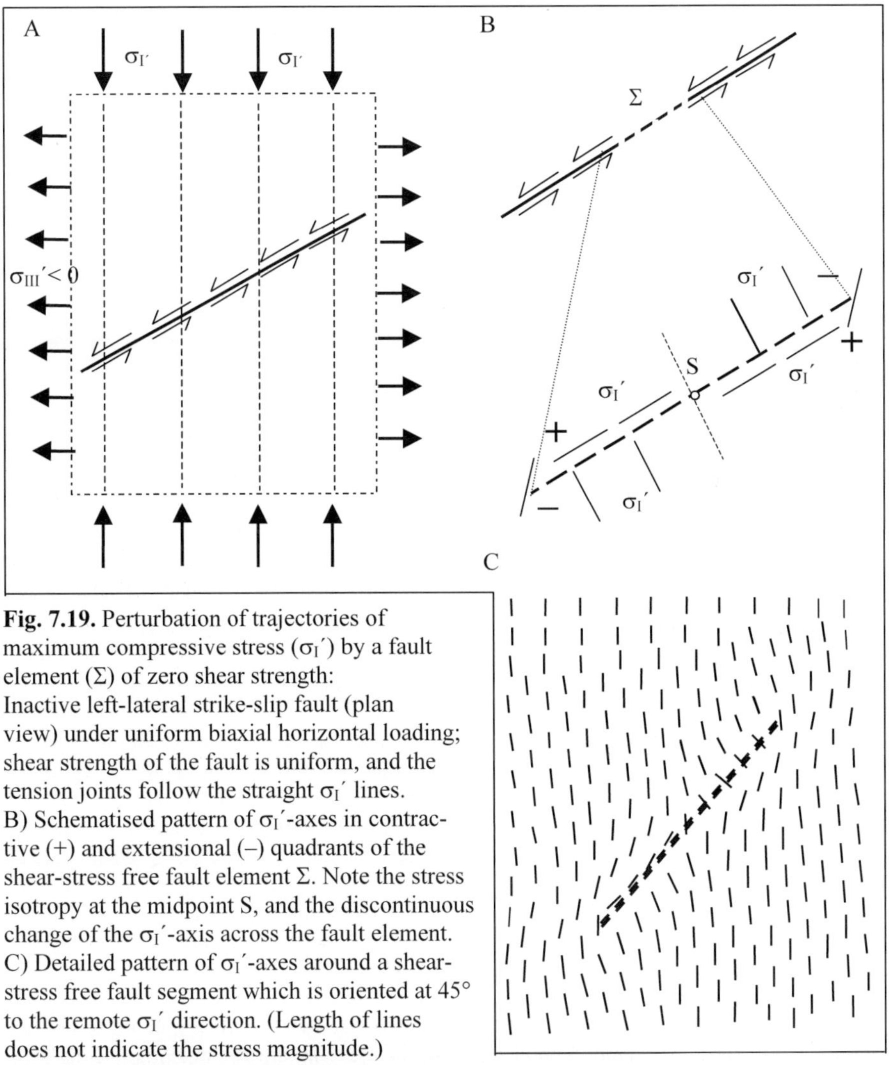

Fig. 7.19. Perturbation of trajectories of maximum compressive stress (σ_{I}') by a fault element (Σ) of zero shear strength:
Inactive left-lateral strike-slip fault (plan view) under uniform biaxial horizontal loading; shear strength of the fault is uniform, and the tension joints follow the straight σ_{I}' lines.
B) Schematised pattern of σ_{I}'-axes in contractive (+) and extensional (–) quadrants of the shear-stress free fault element Σ. Note the stress isotropy at the midpoint S, and the discontinuous change of the σ_{I}'-axis across the fault element.
C) Detailed pattern of σ_{I}'-axes around a shear-stress free fault segment which is oriented at 45° to the remote σ_{I}' direction. (Length of lines does not indicate the stress magnitude.)
The figure C is based on Fig. 8.6C of a paper by D. Pollard and P. Segall (1987); in: B.K. Atkinson (ed.) *Fracture Mechanics of Rocks*, Academic Press, p. 309)

part (+). The σ'_I trajectories which are straight and parallel further away from the fault in the uniform stress field, will curve smoothly into the σ'_I directions on the fault segment. This is illustrated in Fig. 7.19C by the direction field of σ'_I which was found by applying EFM to a fault segment that is oriented at 45° to the uniform σ'_I direction of the remote field. Also note, the stress isotropy ($\sigma'_I = \sigma'_{III}$) at the mid-point S in Fig. 7.19B, which is a center of symmetry of the perturbed field.

Further, note that the discontinuous change in the direction of σ'_I across the fault segment in Fig. 7.19 is only possible in association with a jump-like change of the fault-parallel normal stress ($\sigma_{||}$). This follows from a simple formula, shown in Fig. 7.20, and derived in the Appendix on the Mohr circle, which relates the acute angle Θ between the σ_I direction and the direction of the resolved shear stress τ on the fault plane, with the values of the stresses τ, $\sigma_{||}$ and $\sigma_{\perp}$. The equilibrium conditions only allow the fault-parallel normal stress $\sigma_{||}$ to discontinuously change across the fault; but a jump-like change in $\sigma_{||}$ would be accompanied by a discontinuous change of the associated fault-parallel longitudinal strains. This requires the decoupling of the fault-parallel strains on both sides of the fault plane by slippage on the fault. Thus, a diffraction of the σ_I trajectories on a fault plane can only be established during fault slip.

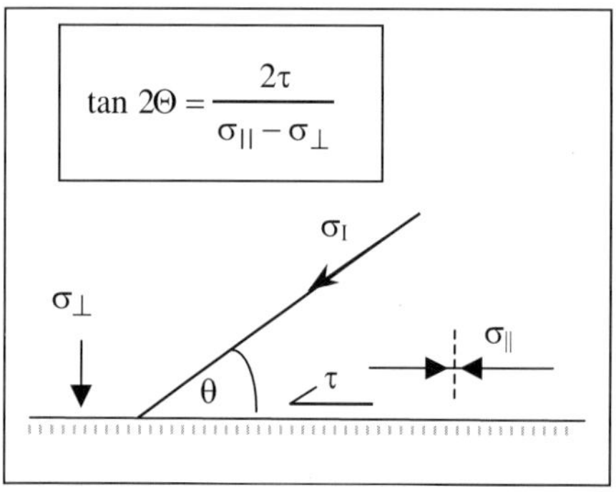

Fig. 7.20. Maximum principal stress direction and fault plane

This is quite different in the case of a very weak fault which, extremely simplified is considered as frictionless and cohesionless everywhere with the exception of a relatively small non-slip segment, representing a major asperity or a minor kink in the fault path. It is the non-slip element which allows shear forces to balance the constant remote boundary loads and to maintain the fault in a stable state, while outside this area the fault acts as a principal surface. This is schematically illustrated in Fig. 7.21A,B. The pattern of the principal stress trajectories were analyzed and simulated in photo-elastic experiments by K.D. Rawnsley, T. Rives, J.-P. Petit, S.R. Hencher and A.C. Lumsden (1992; J. Structural Geology *14(8/9)*, pp. 939–951), who used the model to interpret observed perturbations of joint systems near pre-existing faults on the U.K. coast.

Figure 7.21C is intended to show the essence of the model: near the fault, where it is free to slip, the σ_I trajectories must sweep into an orthogonal or a tangential position, whereas along the non-slip interval they should be inclined towards the fault as qualitatively indicated in the figure. Note that, contrary to the case of a non-slip fault with a slip element (Fig. 7.19), the principal stress directions may now jump from the normal to the tangential position across the fault outside the non-slip area, everywhere where slip actually occurs, thereby allowing the fault-parallel normal stress to change discontinuously across the fault plane. But it does not seem quite so simple to decide where along the shear-free fault the direction of σ_I will be normal and where tangential. In drawing Fig. 7.21C we assumed that the section shown is sufficiently far away from the ends of the "blind" fault so that the compressional or tensional effect of the tip regions can be neglected. On the other hand, a fault tip close to the non-slip region, might determine which of the two possible orientations the σ_I direction will attain.

Although the model of a frictional fault with a slip segment (Fig. 7.19) would often seem to be closer to reality than the model of a single frictionless fault kept in a stable state by an asperity or similar, the simplifying assumption of shear-free fault planes, even without any non-slip segments, appears apposite to the analysis of the stresses in systems of conjugate faults. It is easily seen, that a mosaic of wedges bounded by conjugate strike-slip faults of zero shear stress may be in static equilibrium under a biaxial remote load, with the σ_I-trajectories meeting the faults orthogonally or tangentially. Systems of tension joints between conjugate strike-slip faults have been studied by J.-P. Petit et al. (2000). A sketch of the joint traces is reproduced in Fig. 7.22A, and in Fig. 7.22B are compared to the result of a numerical analysis by St.J. Bourne and E.J.M. Willemse (2001), based on linear elasticity and zero shear stress on the faults. The calculated pattern compares quite well to the observed fracture pattern, except for the upper left corner and the right margin of the figure where faults or fault segments just beyond the limits of the map in Fig. 7.22A may exert an effect that is not accounted for in the calculation.

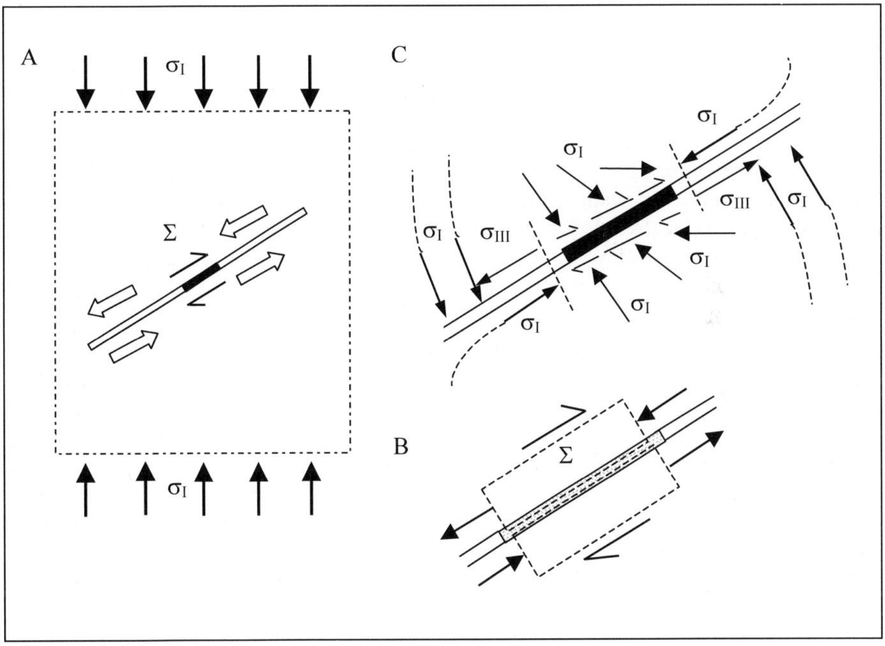

Fig. 7.21. Perturbation of trajectories of the maximum compressive stress (σ_I') by a non-slip element (Σ) in an otherwise frictionless and cohesionless fault:

A) Uniaxially loaded "blind" fault; open arrows schematically indicate the fault-parallel downward push and pull of the vertical boundary forces, which are balanced by a shear couple on Σ.

B) Balancing of fault-parallel forces by the presence of the non-slip element.

C) The maximum compressive stress (σ_I') is orthogonal or parallel to the friction- and cohesionless fault (dotted tails suggest further course of trajectories), but the σ_I'-axes are perturbed by the non-slip element, as schematised in the figure

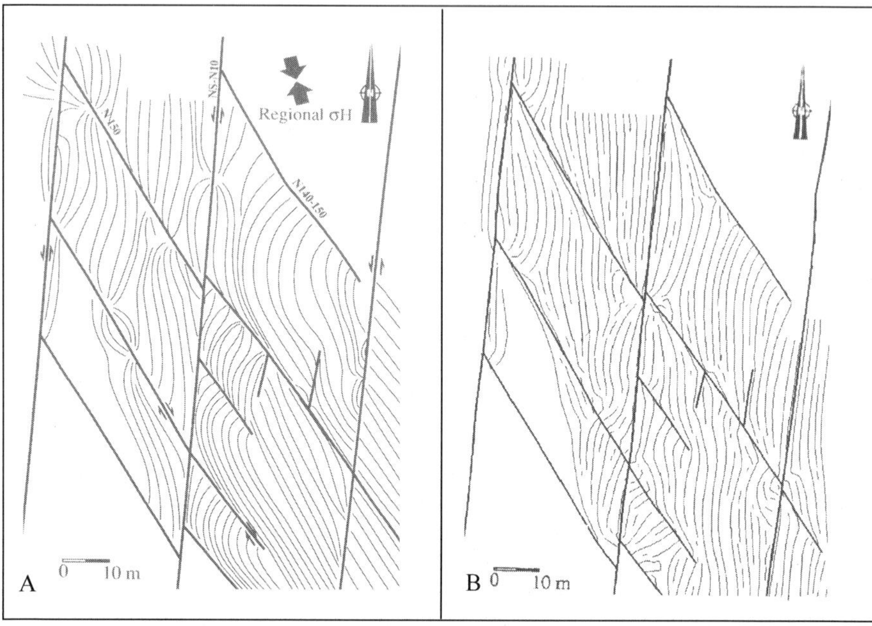

Fig. 7.22. Conjugate strike-slip faults (thick straight lines) and joint patterns (thin lines) produced in perturbed stress fields around faults (near Nash Point, Bristol Channel, UK).
A) Detailed sketch of observed fracture patterns by J-P. Petit, V. Auzias, K.D. Rawnsley and T. Rives (2000; in: F.K. Lehner and J.L. Urai (eds.): *Aspects of Tectonic faulting*, Springer)
B) Joint pattern calculated by three-dimensional numerical method based on linear elasticity and assuming zero shear stresses on the fault planes, by St. J. Bourne and E.J.M. Willemse (2001; J. Structural Geology *23*, pp. 1753–1770)

Joints in compressive folds. We now turn to tension and cleavage fracturing in compressive folding. Since the basic conditions and mechanisms involved in these processes have been addressed at various places in previous chapters, we content ourselves with summarizing the main results.

First, not all joint sets observed in a compressive fold need to be the result of the folding; some regular joint sets may form in flat lying layers while the compressive stress σ_I' is still building up towards the magnitude necessary for the onset of folding. As discussed in Chapt. 4 (pp. 88–90), a tectonic regime of this kind exists in the foreland of thrust and fold belts when overpressuring of the pore fluid by the foreland compression sufficiently reduces the smallest effective stress σ_{III}', orthogonal to the thrust direction, to allow the formation of vertical "internal hydraulic" joints of the tension or cleavage type. The joints will form a regional set, parallel to the σ'_I trajectories, and thus more or less at right angles to the later fold axis. Note, that the pattern of these cross-fold joints is not affected by the local curvature of the folded layers; but the opening of the individual joints is, of course, affected by the tension or compression (below the neutral surface in Fig. 5.12) due to the local curvature.

On the other hand, several sets of new joints are generated by the process of folding itself. There is, firstly a jointing mechanism that is still related to the pre-stresses in the flat lying foreland sediments. As argued in Chapt. 5 (p. 119), with reference to Fig. 5.10 (p. 116), a part of the layer-parallel compressive stress normal to the later fold axis is locked-in by intergranular cementation. In folding, the layer-parallel compressive stress is reduced, not only in the hinge region, but also in the reasonably straight fold limbs. This puts the intergranular cementing material under tension which, on a macro-scale, represents a maximum tensile stress along the trajectories of the former maximum compressive stress. The relief of this tension may produce "longitudinal" tension joints parallel or slightly out of parallel to the hinge line of the fold; the latter situation is to be expected if, during folding, the hinge line of a layer is diverted from orthogonality with the direction of the horizontal pre-folding compression.

Another set of hinge-parallel tension joints is generated by the extension along hinge-normal arcs (in the yz-planes of Fig. 5.12B). In contrast to the longitudinal joints that are caused by the release of locked-in compression, the tension joints generated by the curvature of a layer are expected to be located in or near the hinge region, where the spacing of the joints decreases with the spatial increase in curvature, as predicted by the Hobbs' model (Chapt. 4, pp. 63–64). Moreover, in a thicker layer the tension fractures may not completely transect the layer because the fractures are stopped inside a lower part of the layer that is put under compression by the bending (Fig. 5.12).

While these joints are caused by the bending of layers, hinge-parallel tension joints may also be generated in straight fold limbs of a flexural-slip fold. As the individual layers slide over each other towards the crest of the fold, as sketched in Fig. 7.23, they exert a shear stress (τ) on the interfaces, thus putting the layers (locally) under simple shear. This changes the directions and magnitudes of the principal stresses ($\sigma_I^{\circ\prime}$, $\sigma_{III}^{\circ\prime}$) that would exist in a layer if the interlayer friction were zero ($\tau = 0$). Considering, in Fig. 7.23, plane strain in a vertical cross-section at right angles to the hinge line of the multilayer fold, the axes of $\sigma_I^{\circ\prime}$ and $\sigma_{III}^{\circ\prime}$ lie in the plane, and are either normal or tangential to the frictionless interfaces, or the stresses are isotropic.

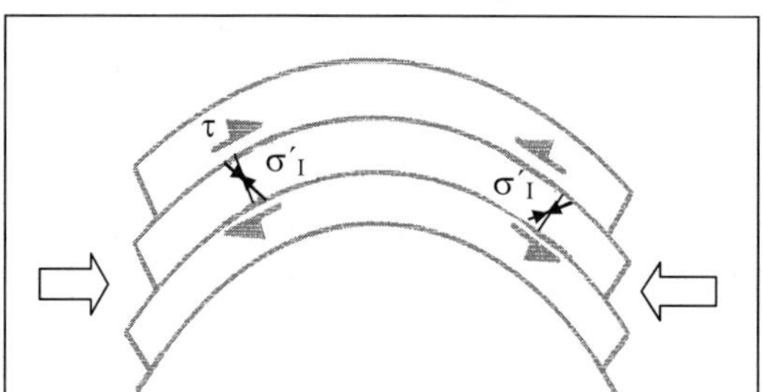

Fig. 7.23. Interlayer shearing in flexural-slip folding induces a rotation of the principal stress directions which would have existed in the case of zero interlayer friction

In the latter case, the superposition of the interlayer shear stress establishes different principal stresses ($\sigma_I' \neq \sigma_{III}'$) which include the angle of ±45° with the interface normal. When the imposed shearing reduces σ_{III}' to $-T_o$, tension joints form parallel to the 45° direction of σ_I'. This result may also easily be obtained by means of Mohr diagrams, as shown in Fig. 8.3 of the next chapter. There, it is also demonstrated that the imposed shearing can only produce tension joints if the effective stresses prior to shearing are smaller than $2T_o$.

In general, however, $\sigma_I^{\circ\prime}$ differs from $\sigma_{III}^{\circ\prime}$ in the frictionless situation. It can then be shown, either in Mohr diagrams or analytically (Appendix, Eq. A.6), that if $\sigma_I^{\circ\prime}$ is normal to the frictionless interface, the σ_I'-axis is rotated through less than 45° from the normal to the interface (as indicated in Fig. 7.23), but through more than 45° if $\sigma_I^{\circ\prime}$ is parallel to the frictionless interface. Hence, in the example of Fig. 4.3, $\sigma_I^{\circ\prime}$ would have been directed normal to the interface.

Finally, compressional folds are commonly transected by joint sets normal to the hinge line, which, together with hinge-parallel joints, may form a fundamental joint system. There have been two mechanisms identified as possibly generating cross-fold joints: first, the regional pre-folding stresses in an overpressured foreland, as mentioned above; and secondly, the hinge-parallel stretching due to the upwards convex curvature of the hinge line. Since this curvature is greatest near the axial end regions of the fold (see the example of a "whale back" shaped fold in Fig. 1.6), one should expect the highest frequency of cross joints in these regions. Nevertheless, the cross joints may be distributed over most of the fold because the axial end regions migrate outwards as folding progresses.

It is often suggested, that the suppression of the "anticlastic" bending of the buckled layers (pp. 118–119, Fig. 5.12A) would produce the tension required for the formation of the cross joints. But, although this factor may contribute to the tensional effect of the hinge-line curvature, in itself it cannot provide the necessary tension, even if stress reductions by inelastic deformation processes are not considered.

Summary of joints in faulting. In the brittle regime, the initiation, development and activity of a tectonic fault is accompanied by fracture damage in the rock surrounding the fault. Several phases can be distinguished in the build-up of the damage zone. Starting with the pervasive fracturing by tension fractures that precedes the concentration of shearing into narrow shear bands.

- ***Pre-faulting fractures*** (pp. 153–156). Rock mechanical experiments show that a fault forms by a complex process of the growth and interlinking of small-scale tension cracks, which – in a statistical sense – are aligned parallel to the direction of the macroscopic maximum compressive stress (see Fig. 7.1).

  At low effective confining pressures, the interlinking of tension fractures is thought to be established by tensional "wing" cracks which emanate from the tips of inclined tension cracks under shear. But at higher confining pressures, as are more common in the Earth's crust, the interlinking may involve mode II fractures or "bookshelf"-type arrays of tension or shear fractures, as illustrated in Fig. 7.2A,B. A small-scale field example of the bookshelf-type interlinking mechanism in a preliminary stage of fault development is shown in Fig. 7.3.

  It is difficult, if not impossible, to infer from rock-mechanical experiments the dimensions and fracture densities of pre-faulting damage "seams" in the field, due to the differences in boundary conditions and the experimental limitation in sample size. The latter does not allow observation of a later phase of fault development where separate shear bands, say of sample size, coalesce into a compound fault, embedded in a wider damage zone.
- ***Tip region fracturing*** (pp. 157–164). In a brittle rock, the stress perturbation near a propagating fault tip generates fractures of various kinds. Under ideal circumstances, that is, in the case where the fault propagates unhindered inside a mechanically uniform rock, the displacements along the fault produce stresses and strains in strict mirror image symmetry with respect to the fault plane and its straight prolongation. This allows the conclusion (Fig. 7.5) that tension- or shear fractures will form and branch off the fault on the *receding* side of the near-tip region. In a regional regime of low effective stresses, the fractures will be of the tensile type; but the fractures are of the shear type in a regime of higher effective stresses (see the schematic illustration in Fig. 7.8A). As the fault grows, the fracturing in the near-tip region will leave a halo of macroscopically fractured rock along the receding side of the fault.

In contrast, near-tip fracturing may be shifted to the *advancing* side when the mirror symmetry is broken by: (*1*) an increase in pore pressure on the advancing side, and decrease on the receding side of the near-tip region (Fig. 7.5C); or (*2*) when, on one side of the fault, the displacement is suppressed (Figs. 7.7 and 7.8B). In the latter case, it seems possible that, in regimes of low effective stresses, tension joints are also formed on the immobilized side (see circle C^2 in Fig. 7.7 and the dashed joint traces in Fig. 7.8B). The fracture halo will develop in unison with the moving fault edge.

- ***Rock deformation along faults*** (pp. 164–167). Rock layers are deformed by faulting in many ways, either before they are cut by a fault or while they are displaced along a fault. In extensional regimes, normal faults cause extensions in neighbouring layers of sedimentary rock, which give rise to tension joints or secondary normal faults; typical cases are schematically shown in Fig. 7.9.

  Tension joints and faults cannot form simultaneously at the *same* location (Fig. 7.10), but they may form in succession; for instance, in a layer downwarped between the hinges of a normal fault (Fig. 7.9E), when overpressures in the layer are drained by jointing, thereby increasing the effective overburden load to a stage conducive to normal faulting of the stretched layer. On the other hand, joints and faults may form at the same time in *separate parts* of the downwarped layer, when "arching" of the σ'_I trajectories above the downbent layer shifts the overburden load from the central part to the shoulders of the layer.

  Furthermore, when, at nearly the same depth of burial, two layers (for instance, two sandstones) of different shear strength are extended together, normal faults may be generated in the weaker layer and tension joints in the stronger one (Figs. 7.11 and 7.12).

- ***Healed joints opening concurrently with fault slip*** (pp. 168–171). It has been shown (Fig. 7.13) that slip on a normal fault and the opening of healed subvertical joints or veins can only occur together if a simple relation (Eq. 7.1) between the effective overburden stress, the tensile strength of the healed joints, and the shear strength parameters of the fault prior to slip, is fulfilled. This implies that: (*1*) the inactive fault must have a cohesive shear strength greater, or only a little less than the tensile strength of the healed joints, and (*2*), the thickness of the effective overburden is severely limited by the strength parameters of the inactive fault and the healed joints (Eq. 7.1b).

  But how does one explain the coexistence of a normal fault and a set of healed tension joints? It is readily seen (Fig. 7.14) that, after the formation of a normal fault under effective compressive stresses, high overpressures are required to allow the formation of vertical tension joints while leaving the weakened fault in a state of impending slippage. The opening of the internal hydraulic joints may rapidly reduce the overpressure, and thus immobilize the fault plane by increasing the fault friction. The open joints are sealed by mineral deposition, which stops the drainage of pore fluids; if the overpressure is restored, the fault can slip again, accompanied by a reopening of the joints or veins. If repeated, the process would be a manifestation of Ramsey's "crack-seal" mechanism.

  The question remains as to why the geological boundary condition should have changed from a horizontal extension conducive to normal faulting under effective compressive stresses, to a regime of high overpressures under horizontal confinement. This problem does not arise when the normal faults are not Coulomb-type (with the critical stress circles lying completely in the compression half of the stress plane), but are *dilational* type (with critical stress circles touching a curved limit line in the tension half of the stress plane, Fig. 7.15A).

Because the stress states that generate dilational normal faults are very similar to those which produce tension joints, it is not difficult to see how healed tension joints or veins may be opened by the formation of a dilational normal fault (Fig. 7.15B). However, the coincidence of the formation or reopening of joints and the formation or activity of dilational normal faults requires a very special state of stress, depending not only on the strength properties of the rock and the healed joints or veins, but also on the interplay of the accretion of the effective overburden and the extension of subsiding layers.

Hence, although it is understood that dilational faults can reopen healed joints, opened joints and dilational faults may also exist together, without one being the cause of the others.

- ***Strike-slip faults parallel to joints*** (pp. 173–175). Pre-existing subvertical joints (of tension or cleavage type) may be sheared in a stress field with a rotated subhorizontal σ_I-axis. In the field case (Fig. 7.16) studied by Segall and Pollard (1983), the sheared joints terminate in secondary fracture structures (schematically shown in Fig. 7.7A); and small faults of the sheared-joint type are linked into larger strike-slip faults by secondary fractures operating in the dilational "bookshelf" mode (Fig. 7.17).

  The spacing of vertical tension fractures in a crystalline rock under horizontal tension may grow in depth unhindered by mechanical interfaces; their spacing will, therefore, differ from the spacing in layered sedimentary rocks, as discussed at length in Chapt. 4. With greater depth of penetration, the joints will be longer in the strike direction, and surrounded by a wider halo of elastic tension release, resulting in a wider spacing.
- ***Perturbations of joints by pre-existing faults*** (pp. 175–180). The path of tension joints which formed in horizontal layers *after* the formation and inactivation of a straight strike-slip fault, may be affected by the fault. Consider first, a uniform biaxial horizontal stress system which generates the joints and is compatible with the existence of an inactive fault with uniform shear strength (Fig. 7.18C and D, and Fig. 7.19A). As the inactive fault does not affect the stress field, the joints will propagate along the straight σ_I trajectories of the uniform field, either crossing the fault or being arrested by the weakness of the fault.

  In the field, the shear strength will, most likely, vary along the fault and perturb the principal stress trajectories in some neighbourhood of the fault, thereby distorting the pattern of the joints. Two extreme cases of strength variation are dealt with to obtain an understanding of the perturbed joint patterns observed near faults in the field: (*1*) a "blind" fault with a shear-stress free element (Σ) and uniform shear strength everywhere else (Fig. 7.19B); and (*2*) a fault free to slip everywhere, except on a non-slip element (Σ) that keeps the fault in a stable state (Fig. 7.21A).

  In case (1), the joints will smoothly curve into the σ'_I directions on the slipping fault segment, which are normal on one half of the segment, and tangential on the other half, changing between the two directions across the slipping fault element (Fig.7.19B,C). In case (2), the joints curve near the fault into directions normal to the fault on one side of the non-slip element, and tangential on the other side, with a continuous change of directions along the non-slip element established by the balancing shear couple. Where the fault actually slips outside the non-slip element, the σ_I directions may jump across the fault plane from the normal to the tangential position (Fig. 7.21). Model (2) may come close to reality in a mosaic of wedges bounded by conjugate strike-slip faults (Fig. 7.22).

Chapter 8

Échelon Joints and Veins

In the preceding chapter, we considered how tension joints form on the receding side of the near-tip region of a propagating fault (Fig. 7.8A), and array en-échelon one-sidedly behind the advancing fault tip. More commonly, échelon arrays of dilatant fractures (i.e. tension and cleavage fractures, hydraulic intrusion fractures, gashes or veins), form independently of faults or sheared joints. Examples, of the kind shown in Fig. 8.1, are frequently exposed on denuded bedding planes of sedimentary rocks. Even in town, when strolling over a pavement of rock slabs, one may stumble upon the phenomenon. However, despite their abundance, the mechanical origin of échelon fractures is still the subject of debate. Several mechanisms have been proposed to explain the phenomenon; it transpires that "more than one mechanism can produce en échelon veins", as J.E. Olson and D.D. Pollard aptly remarked (1991; *The initiation and growth of en échelon veins*, J. Struct. Geol. *13*, No. 5, pp. 595–608). In this last chapter, we will examine these processes.

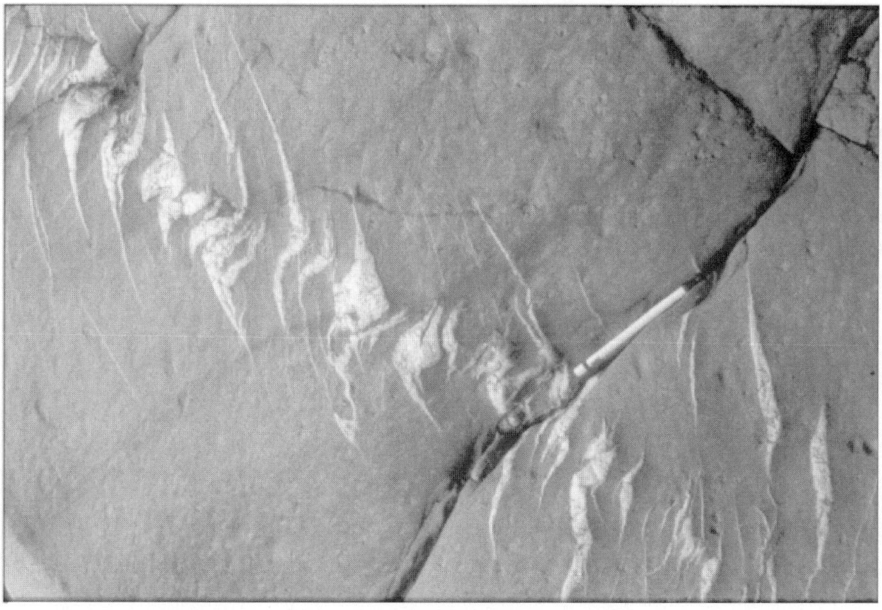

Fig. 8.1. Sigmoidal en échelon veins in slightly metamorphic sandstone (middle Miocene); Lushan formation, "Backbone Range" near Taimali, Taiwan (courtesy Prof. F.-J. Brosch) (see later explanation by Figs. 8.4 and 8.5)

En échelon cracks in shear zones. First, it is readily understood that échelon tension joints can be produced in weaker rocks which are sheared by stiffer ("competent") boundary rocks. This occurs, for instance, in grabens with horizontal movement of the graben boundaries (Fig. 8.2A), in the straight limbs of a flexural slip fold (Fig. 8.2B), and in the shearing of sediments by strike-slip faulting in stiffer substrata (Fig. 8.2C). Considering parallel boundaries, the shearing is of a quasi-simple mode if the distance between the boundaries is allowed to accommodate shearing dilatation. Typically, the simple or quasi-simple shearing imposed by the moving boundary rocks enforces a rotation of the σ_I direction inside the sheared material. This is shown in the drawings in Fig. 8.3A,B,C for the case that, prior to shearing, the effective stress state is isotropic in the plane of the shear couple. The Mohr diagram in Fig. 8.3D is included as a reminder of the fact that the effective normal stresses perpendicular ($\sigma_\perp'$) and parallel ($\sigma_{||}°'$) to the imposed shear direction must remain smaller than roughly twice the tensile strength T_o of the rock in order to permit tension fractures to be formed by the simple shearing. In contrast, if the effective stresses are at a higher level, such as is the case in the main part of the brittle crust of the Earth, simple shearing will produce Coulomb shears (so-called Riedel shears R and R′), as indicated in the Mohr diagram in Fig. 8.3E.

Note that, prior to shearing, in Fig. 8.3, the stresses in the plane of the shear couple are assumed to be isotropic, and the stresses $\sigma_\perp'$ and $\sigma_{||}'$ remain equal during the shearing. Only under these conditions does the σ_I' direction include the acute angle $\Theta = 45°$ with the imposed shearing direction (Fig. 8.3D). If $\sigma_{||}' > \sigma_\perp'$, the angle Θ is smaller than 45°; and greater than 45°, if $\sigma_\perp' > \sigma_{||}'$ (see Appendix, Eq. A.6). But in any case, prior to fracturing, the shear-induced angle Θ is smaller than 90°. Thus, although a 45° orientation of shear-induced tension joints may serve as a convenient reference orientation for the display of the basic mechanisms of tension jointing in a shear zone, the joints may actually form at a quite different orientation.

Once parallel tension fractures are formed in the shear zone, a further increase of the boundary shear stress may rotate the joints; because $\Theta < 90°$, the rotation must be dilational in style (Figs. 6.17 and 7.17). If unhindered by "rigid" boundaries, the fractures may grow in length, while the central part is passively rotated and opens up permitting a vein to develop, as is nicely shown in Fig. 8.1. As long as the fracture-bounded rock slices ("lithons") are still connected to the adjacent rock, the end parts of the advancing fractures propagate in mode I fashion, thus following the local σ_I direction. The joints, therefore, develop a sigmoidal shape (Fig. 8.1), as is further explained in Figs. 8.4 and 8.5.

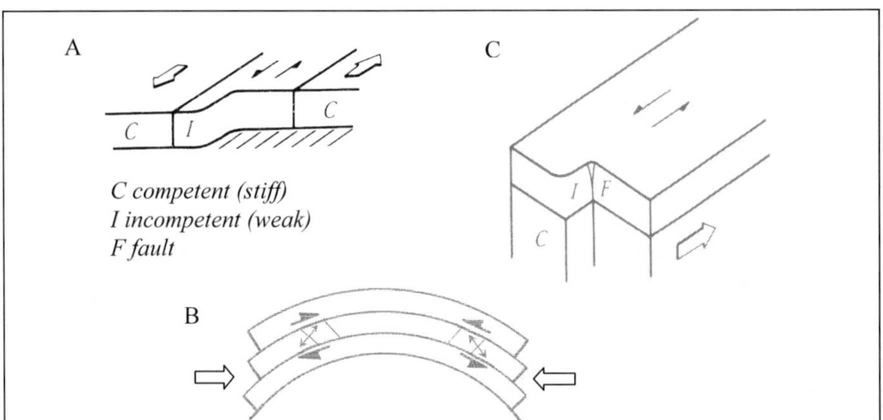

Fig. 8.2. Schematized geological processes of quasi-simple shearing:
A) horizontal shearing between lateral boundary blocks (e.g. "graben shoulders");
B) shearing in flexural-slip folding;
C) overburden sheared by strike-slip faulting in the "basement"

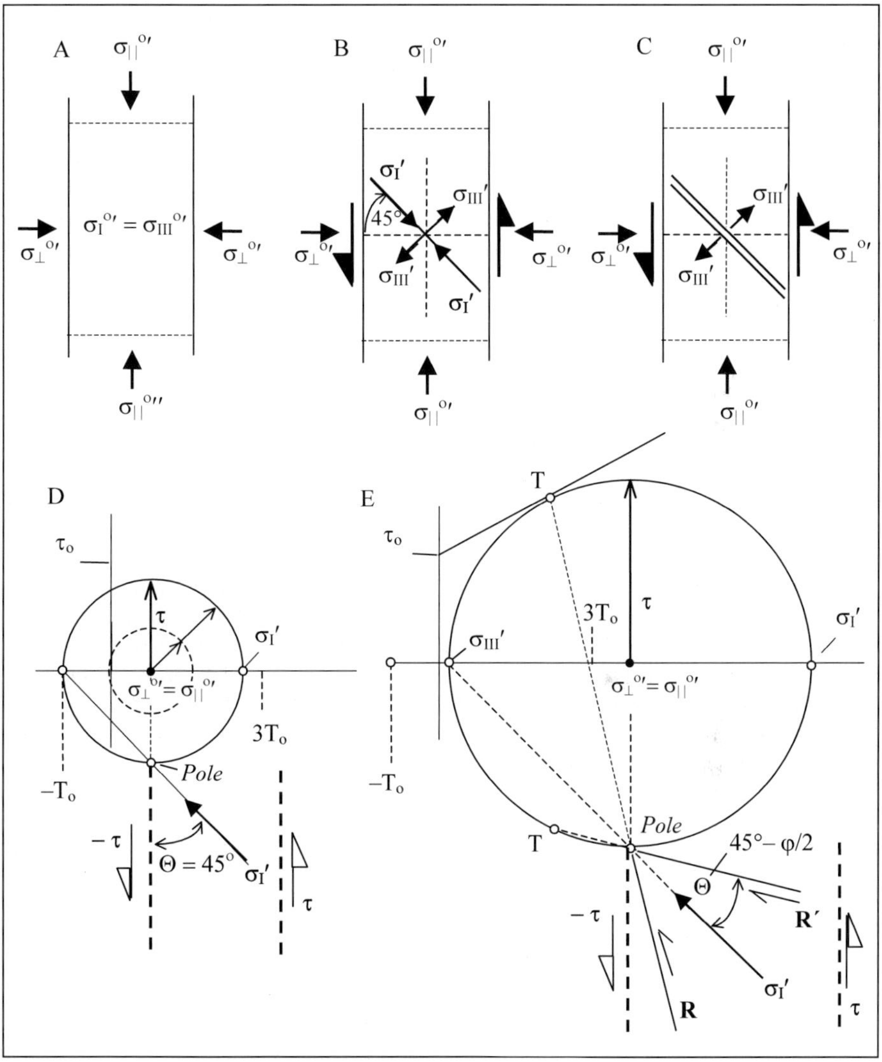

Fig. 8.3. Stress rotation by a shear couple (plan view of shearing arrangement):
A) Isotropic state of stress ($\sigma_\perp{}^{o\prime} = \sigma_{||}{}^{o\prime}$) before onset of shearing.
B) Elastic quasi-simple shearing ($\sigma_\perp{}^{o\prime}$ = const.) establishes the direction of σ_I' at 45° to the external shear direction, if $\sigma_{||}'$ also remains unchanged ($\sigma_{||}' = \sigma_{||}{}^{o\prime}$).
C) When $\sigma_{III}' = -T_o$ (and the material in the σ_I', σ_{III}'-plane is still isotropic) tension joints form parallel to the 45° direction of σ_I'.
D) Mohr diagram determining the 45° direction of σ_I', and indicating that simple shearing can produce tension joints only in regimes of rather low effective stresses: $\sigma_\perp{}^{o\prime}$, $\sigma_{||}{}^{o\prime} < 2T_o$.
E) Mohr diagram illustrates how even very moderate effective compressive stresses $\sigma_\perp{}^{o\prime}$, $\sigma_{||}{}^{o\prime}$ prevent tension fracturing by simple shearing, and promote instead the formation of synthetic (R) and antithetic (R′) Riedel shears

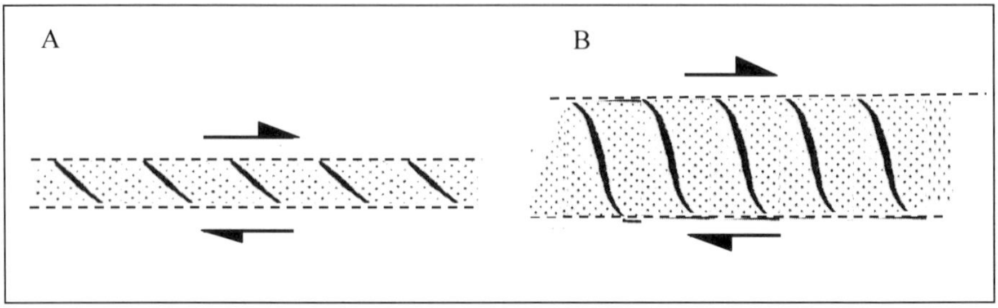

Fig. 8.4. Dilational development of an en échelon array of tension joints in a growing shear zone: A) initial state; B) advanced state with sigmoidal veins

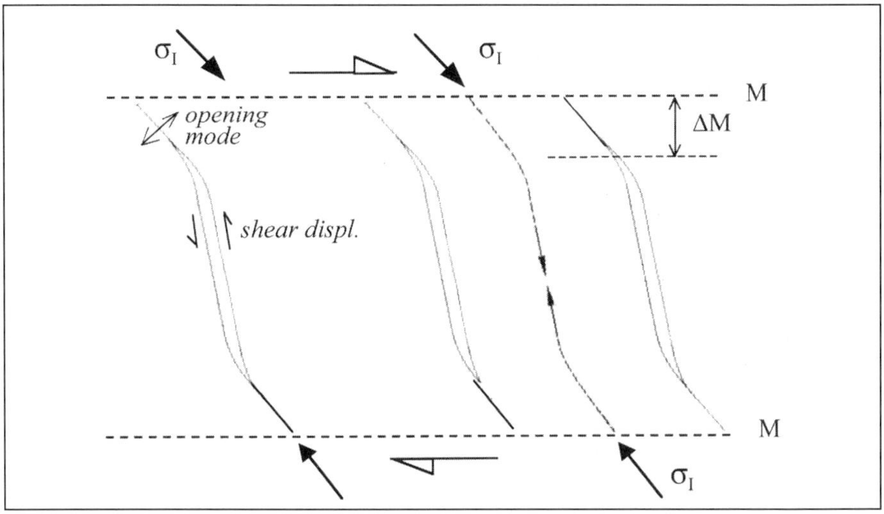

Fig. 8.5. Echelon tension joints rotated in dilational mode; opened central parts of fracture planes and associated σ_I trajectory (conjectural)

In the absence of separate σ_I' direction indicators, such as pressure solution seams, it may be difficult, in practice, to decide whether rotated en échelon joints were formed as tension (or cleavage) joints, or as shear fractures. This ambiguity stems from the range of possible inclination angles Θ in the σ_I direction, mentioned above, which does not exclude the formation and subsequent rotation of antithetic Riedel shears (the R′-shears in Fig. 8.3E). During rotation, the R′ shears may also attain a sigmoidal shape by propagating at an angle $(45° - \varphi/2)$ to the σ_I direction. Like tension fractures, the shear fractures will open up in dilational mode when rotated (Fig. 8.5), and may, thereby, even develop into veins.

When the shearing of a well-defined narrow band of en échelon joints proceeds, the friction along the opened, fluid-filled, middle part of the rotated fractures is lost, and the σ_I trajectories are deflected, as indicated in Fig. 8.5. In addition, the normal stress $\sigma_\perp'$ on the shear boundary is likely to increase as a consequence of the dilation which accompanies the rotation of the fracture-bounded "lithons".

Therefore, as stated by the author (G. Mandl 1987; Tectonophysics *141*, pp. 277–316): "sophisticated analyses are not needed to recognize that slender rock slices will hardly be capable of carrying the eccentric load imposed from outside on the shear band. The shear stress that acts upon the band margin M (Fig. 8.5) cannot be transmitted across the band of open fractures and would have to be balanced inside the boundary zones ΔM by an extra normal stress acting parallel to the band boundary M. Since this stress would increase linearly in the direction of the imposed shearing in order to balance the resulting shear force, it seems evident that arrays of open fractures with slender lithons would have to be rather short and arranged in a patchy fashion. We may thus infer that long arrays of tension or shear fractures would be associated with relatively short open fractures, i.e. fractures in which the open length would hardly amount to more than two or three times the fracture spacing."

Following this line of reasoning, it appears likely that during the rotation of en échelon fractures a state is reached where the marginal zone of the shear band experiences a reduction in shear strength by small-scale ruptures, thereby allowing shear localization along the band margin, which possibly develops into a fault parallel to the shear zone. The normal fault in Fig. 8.6 may serve as an example of a fault cutting through an échelon array of quartz-filled fractures. Although, in this case, the fractures are found to pre-date the actual formation of the fault, in similar cases, these "feather (pinnate) joints" may have formed on the receding side of the near-tip region of a growing fault (see Fig. 7.8A).

Since the shear zones considered were enforced parallel to the direction of the maximum shear stress, parallel to the kinematically imposed simple shearing (Fig. 8.2), the shear bands differ in orientation from Coulomb's slip direction. Therefore, a fault along such a kinematically controlled fracture band, is *not* a Coulomb- (or Andersonian) type fault. It shares the non-Andersonian aspect with the class of "simple-shear faults" (see e.g. FBR, Sect. 8.2) which are enforced by the movement of "rigid" boundary blocks in regimes of all-compressive effective stresses.

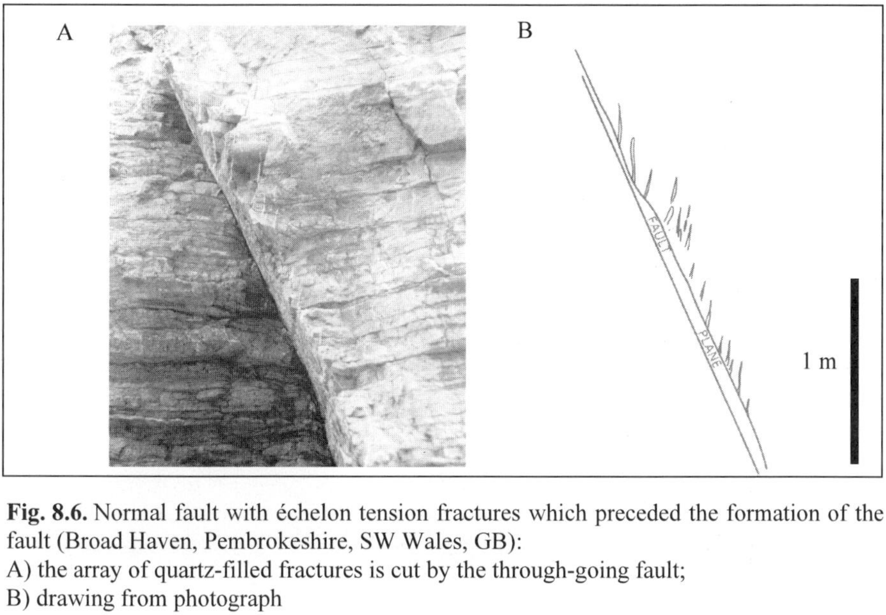

Fig. 8.6. Normal fault with échelon tension fractures which preceded the formation of the fault (Broad Haven, Pembrokeshire, SW Wales, GB):
A) the array of quartz-filled fractures is cut by the through-going fault;
B) drawing from photograph

But the mechanisms of shear localization into a continuous fault may differ widely between a compressional stress regime conducive to faulting, and a stress regime that generates échelon tension or cleavage fractures. Consider first, the limit state of stress in Fig. 8.3E, where prior to the application of the shear couple (τ, $-\tau$) the stresses $\sigma_{\perp}^{\,o\prime}$ and $\sigma_{||}^{\,o\prime}$ are equal. The development of a continuous strike-slip fault is a process of successively forming Coulomb-type faults (for reference see MTF, pp. 78–80, and FBR, Sect. 8.2). The essence of the process is illustrated in Fig. 8.7, where a first set (1) of Riedel shears, operating synthetically with the external shear direction, causes the material to be compressed in the overlapping region of neighbouring faults. This causes both an increase in magnitude of σ'_{I}, and its closer alignment with the direction of the Riedel shears (1). Once the critical state is reached in the overlap regions, new faults (2)

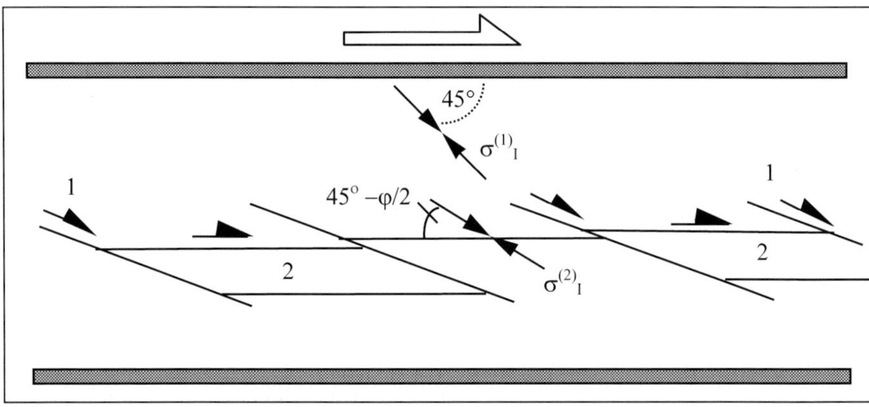

Fig. 8.7. Development of a continuous "simple-shear fault" (in plane strain) by the interlinking of synthetic Riedel shears (1) into shear lenses (schematic). The second set (2) of synthetic faults is associated with a reoriented σ_I between parallel Riedels (see text)

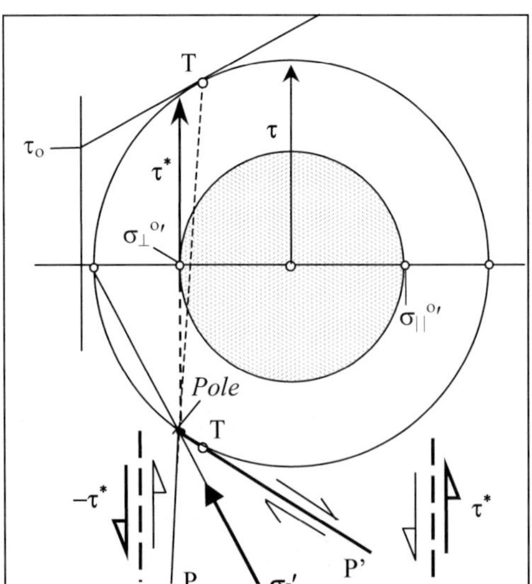

Fig. 8.8. The formation of antithetic faults (P′) by a shear couple ($\tau*$, $-\tau*$) when $\sigma_{||}^{\,o\prime} > \sigma_{\perp}^{\,o\prime}$.
Note, that the required external shear stress $\tau*$ is smaller than the "driving" shear stress τ if $\sigma_{||}^{\,o\prime} = \sigma_{\perp}^{\,o\prime}$ (Fig. 8.3E). Arrayed en échelon, the antithetic P′ shears could accommodate the external shear in a similar dilational bookshelf mode as the tension fractures in Fig.8.5

will develop at a smaller angle to the external shear direction. As a result, a series of shear lenses is formed parallel to the external shear direction. Inside the lenses, new faults may form at even smaller angles to the external shear direction, concentrating the shearing in an ever-narrowing band of cross-connecting Coulomb-type shears that constitute a continuous fault.

However, in contrast to this development which is commonly observed in conventional shearing devices, the external shear may also be accommodated by antithetic Riedel shears in dilational bookshelf mode, similar to, and perhaps undistinguishable from, rotated échelon tension fractures (Fig. 8.5). This is pointed out by the pole construction in the Mohr diagram in Fig. 8.8, which shows that the formation and subsequent dilational rotation of antithetic Riedel shears (P′) is possible in the case where the normal stress ($\sigma_{||}{}^{o\prime}$) parallel to the imposed shear direction is markedly greater than the normal stress ($\sigma_{\perp}{}^{o\prime}$) perpendicular to it. Under these conditions, we expect it to be much more likely that the external shear is accommodated by rotating antithetic shears rather than by non-rotating synthetic Riedel shears (Fig. 8.7). This is simply because the required "driving" shear stress $\tau*$ in Fig. 8.8 is smaller than the shear stress τ in the corresponding Fig. 8.3E.

In view of the range of possible orientations which échelon tension fractures and shear fractures can have in common, it is often difficult to decide whether the observed échelon fractures and veins were generated as tension fractures, cleavage fractures, shear fractures, hybrid extension-shear fractures (Fig. 1.8), or sheared tension fractures. Note, that even veins need not necessarily originate from tension or extension fractures, since similar to the opening of tension joints in Fig. 8.5, shear fractures in cohesive rocks also open up under a dilational bookshelf-type rotation. Thus, understandably, the mechanical history of échelon joints and veins in the field are a matter of quite some debate in the literature. For a review of relevant papers, the reader is advised to consult J.E. Olson and D.D. Pollard (1991; loc.cit.), or Terry Engelder's article "Joints and shear fractures in rocks" (in: B.K. Atkinson (ed.) (1987) *Fracture mechanics of rocks*, Academic Press).

Differentiation between a tension- and a shear-fracture origin of en échelon veins is easier when pressure solution seams occur together with the veins, as considered in Fig. 8.9. The example shows conjugate bands of échelon quartz veins and pressure solution seams in a sandstone bed. The dark solution seams are orthogonal to the direction of $\sigma_I{}'$ during the activity of the solution process, and the quartz veins are at right angles to the pressure solution seams. From this, and from the fact that the veins show very little sign of rotation, one may conclude that the veins have originated from tension or extension (cleavage) cracks. Therefore, on both sides of the line bisecting the angle between the conjugate bands, the $\sigma_I{}'$ trajectories must somehow curve into the direction of the veins, as is tentatively indicated in the figure by dashed lines.

A. Beach, in studying similar structures (1977; Tectonophysics *40*, pp. 201–225), concluded that the veins developed as hydraulic fractures under high pore-fluid pressures, and received the mineral infill from neighbouring pressure solution seams. Assuming that the pressure solution does not significantly raise the local fluid pressure, the opening of the fractures is attributed to a pervasive overpressure. Thus, with reference to Chapt. 3, the veins originated as *internal* hydraulic fractures (Fig. 3.1A) which, under elastic confinement, should form as "hairline-cracks" (p. 34). Obviously, this is in sharp contrast to the large vein dilations that are actually observed. However, a closer inspection of hand samples and thin sections revealed that the dilation of the veins is generally the result of John Ramsay's "crack-seal" mechanism (J.G. Ramsay 1980; Nature *284*, pp. 135–139), which implies a repetitive sealing of cracks by the deposition of quartz, and the opening of new cracks, adjacent to the sealed ones, by a periodic rising and falling of the high pore pressure.

In general, the rotation of principal stress axes under simple shearing allows the development of échelon fracture bands of just one orientation at a time. In contrast, the simultaneous formation of equally well developed conjugate bands, either singly or in pairs, as shown in Fig. 8.9, indicates that the structures were formed in a stationary stress field, such as that caused by a biaxial straining of bed-parallel extension and shortening. Similar to the formation of conjugate shear joints in a field of non-rotating principal stress axes, which was discussed in Chapt. 6 (Fig. 6.7), the occurrence of conjugate fracture bands immediately raises the question of how the bands originate. Asked more specifically: Did the pressure solution seams and tension (or extension) joints form before, or after, the formation of a shear zone? What controls the angle between the conjugate fracture bands? Can Coulomb-Mohr's theory of fault orientation be applied (see FBR, Sect. 4.3), or should the fracture bands be considered as dilational faults (Fig. 1.8C), or as something different? And, did the structures develop in a uniform or a locally perturbed stress field? A perturbation might be caused, for instance, by pressure solution at an isolated location, such as the intersection of the conjugate bands show in Fig. 8.9; in passing, it may be noted that the removal of material may force the σ_I-trajectories to "arch" over the "subsiding" intersection area of the fracture bands (see the discussion of "stress arching" in FBR, pp. 252–253).

Echelon fractures in pre-peak shear bands. Answers to the above questions may be found by interpreting échelon fracture bands as "pre-peak shear bands". This con-tinuum-mechanical concept was discussed in Chapt. 6 (pp. 134–149). There, it was explained how a simple-shear band can form under a compressive biaxial load by the operation of a

Fig. 8.9. Photograph of conjugate sets of échelon quartz veins and pressure solution seams in sandstone beds (Marloes sands, Pembrokeshire, SW Wales, GB). Note the orthogonal intersection of the dark weathered solution seams by the quartz veins; from this it is inferred that the fractures formed parallel to the local σ_I' direction. Dashed lines indicate conjectural σ_I-trajectories

dilational bookshelf mechanism of parallel slip elements (see Fig. 6.17A and insert). The mechanism can produce shear bands inside a loaded rock, which is still in a "hardening" state, and thus capable of sustaining yet higher loads. Although the bookshelf fractures in Fig. 6.17A are Coulomb-type *shear* joints, the concept of a pre-peak shear banding is easily adapted to the operation of parallel *tension* joints, or *cleavage* fractures, inside the band.

For this purpose, the Coulomb-Mohr limit condition inside the incipient band is replaced by the conditions of tension jointing or cleavage fracturing. Let us consider the formation of tension fractures inside an incipient simple-shear band (Γ) in the Mohr diagram in Fig. 8.10. The stresses inside the band are represented by the (shaded) circle C^{int} with $\sigma_{III}'^{int} = -T_o$ and $\sigma_I'^{int} = 3T_o$ which is the highest value that is still compatible with the onset of tensile fracturing. The circle C^{ex} represents the uniform compressive loading stresses outside the shear band. The principal stresses $\sigma_I'^{ex}$ and $\sigma_{III}'^{ex}$ act parallel to the bedding of the rock in the directions indicated in the upper right corner of the figure. The circle C^{ex} must intersect the smaller circle C^{int}, since the normal stress $\sigma_\perp'$ and the shear stress τ on the shear band must be the same both inside and outside the band boundaries, at least as long as the pore pressure is the same on both sides of the boundaries. Moreover, the intersection points S and S* have to coincide with the opposite vertices of the circle C^{int} to ensure that the band-parallel shear stress is the maximum shear stress τ^{int}_{max} inside the band, as required for simple shearing. In contrast to the continuity of $\sigma_\perp'$ and τ across the band boundaries, the band-parallel normal stress $\sigma_{||}'$ in Fig. 8.10 decreases discontinuously from $\sigma_{||}'^{ex}$ to $\sigma_{||}'^{int}$ across the band boundaries. Naturally, such a stress jump, although compatible with static equilibrium, would invoke an infinite shear stress at the band boundary, on either side of which the material is the same. Therefore, the discontinuous change of $\sigma_{||}'$ in the Mohr diagram should be understood as a continuous transition inside a narrow band margin. In fact, the abrupt decrease of $\sigma_{||}'$ from outside to inside the shear band is the main feature of the "stress softening" model which is applied here (see pp. 142–149).

Next, we consider the line Γ drawn through the points O and S of the circle C^{ex}. It represents the trace of the shear band in real space, where it includes an angle γ with the direction of the maximum loading stress $\sigma_I'^{ex}$. (Note that the point O is a stress pole of the circle C^{ex}, since the orientation of the element that is acted upon by the stress $\sigma_{III}'^{ex}$ in real space is known.) Since S also lies on the circle C^{int}, the intersection of the Γ-line with the lower part of the circle C^{int} serves as a pole of the stresses inside the band. The graphical construction then shows that the maximum compressive stress $\sigma_I'^{int}$ inside the band is directed at 45° to the shear band, in accordance with the maximum shear stress τ^{int}_{max} acting parallel to the band.

Thus, as a result of the difference in the band-parallel normal stresses $\sigma_{||}'$, the σ_I'-directions inside and outside the band are different. Consequently, close to the band, the σ_I-trajectories should bend into the $\sigma_I'^{int}$ direction; in fact, much more sharply than conjectured in Fig. 8.9. Since inside the band, tension fractures or cleavage fractures are generated parallel to the direction of $\sigma_I'^{int}$, the fractures are arranged in dilational "bookshelf" mode at the acute angle of 45° to the shear band, as indicated in the lower left corner of Fig. 8.10. Further note, that the joint-bounded rock slices of the band are still connected to the surrounding rock; this suppresses the tendency of the band to shorten when the slices are rotated (see insert in Fig. 6.17A). Consequently, the band-parallel stress $\sigma_{||}'^{int}$ is decreased, which is quite in line with the starting assumption $\sigma_{||}'^{int} < \sigma_{||}'^{ex}$.

Also note, that the conjugate band Γ*, that would be inclined towards the vertical $\sigma_I'^{ex}$-direction at the anticlockwise angle $-\gamma$, is constructed, in the Mohr plane of Fig. 8.10, by drawing the Γ*-line through O and the second point of intersection S* of the two circles, and again applying the simple pole construction (not shown in the figure). The two conjugate bands and fracture sets are then each others' mirror image with respect to the $\sigma_I{}'^{ex}$-direction in the σ_I, σ_{III}-plane. While the bands include the angle 2γ, the fractures of the conjugate sets include the angle $\pi/2 - 2\gamma$.

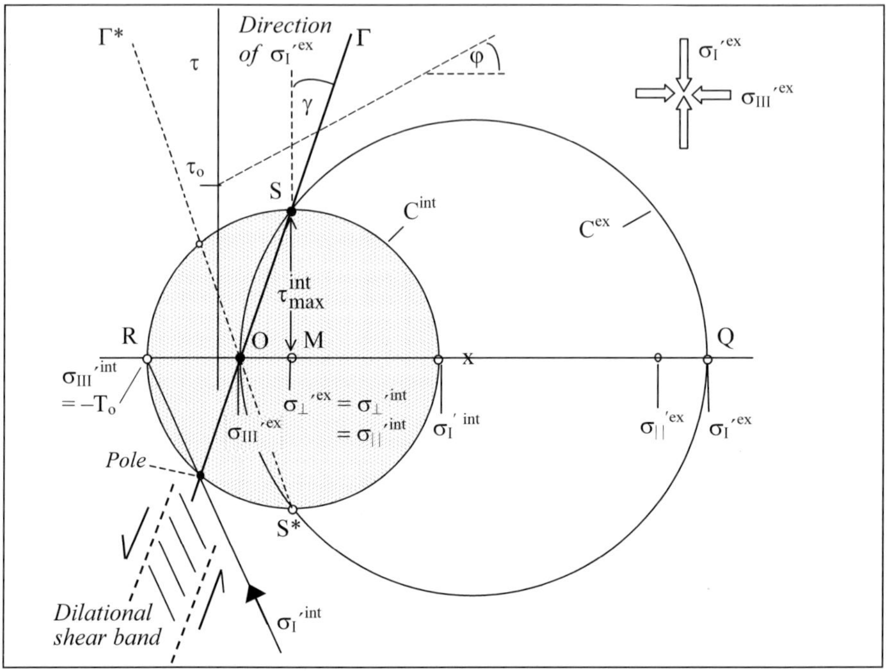

Fig. 8.10. Mohr Diagram of the pre-peak formation of a simple-shear band (Γ) under compressive biaxial load ($\sigma_I{}'^{ex}$, $\sigma_{III}{}'^{ex}$). The band-parallel shearing is produced by a dilational "bookshelf" mechanism of échelon tension fractures (see lower left corner). The shaded circle C^{int} represents the state of stress *inside* the shear band, with $\sigma_I{}'^{int}$ being assigned the greatest possible value ($3T_o$) compatible with the onset of tensile fracturing. The circle C^{ex} through O and S represents a homogeneous stress state *outside* the shear band (see text for further comment)

The inclination angle γ in Fig. 8.10 is found from the relations in the triangle Δ OSQ:

$$\tan\gamma = \frac{\overline{OM}}{\overline{SM}}, \quad \overline{SM}^2 = \overline{OM}.\overline{MQ}, \quad \overline{SM} = \overline{RM} \tag{8.1}$$

$$\tan\gamma = \frac{\overline{RM}}{\overline{MQ}} = \frac{(1+\lambda)T_o}{\sigma_I'^{ex} - \lambda T_o} \qquad (0 \le \lambda \le 1) \tag{8.2}$$

where λT_o is the abscissa of M.

The circle C^{ex} that intersect a circle C^{int} at S and S* is defined by the pairs of dimensionless principal stresses $\sigma_I{}'^{ex}/T_o$, $\sigma_{III}{}'^{ex}/T_o$ that satisfy the relationship

$$\frac{\sigma_{III}'^{\,ex}}{T_o} = \lambda - (1+\lambda)^2.\left(\frac{\sigma_I'^{ex}}{T_o} - \lambda\right)^{-1} \qquad (0 \le \lambda \le 1) \tag{8.3}$$

which is obtained by combining the relationships 8.1 and 8.2.

The stress parameters in this equation are further limited by the condition that the circle C^{ex} has to remain below Coulomb-Mohr's limit line to exclude the onset of faulting. In terms of the external principal stresses the limit condition is

$$\frac{\sigma_I'^{ex}}{T_o} = \frac{\sigma_{III}'^{ex}}{T_o}.k + 2\frac{\tau_o}{T_o}.\sqrt{k} \quad \text{with} \quad k = (1+\sin\varphi)/(1-\sin\varphi) \tag{8.4}$$

Hence, when taking $\tau_o = 2T_o$ and $\varphi = 37°$ as in Fig. 8.10, the restrictive condition for the circle C^{ex} becomes

$$0.4\frac{\sigma_I'^{ex}}{T_o} - 1.6\frac{\sigma_{III}'^{ex}}{T_o} < 3.2 \tag{8.5}$$

Note, that the radius of the stress circle C^{int} in Fig. 8.10 is chosen as equal to $2T_o$, which is already the greatest radius that is theoretically compatible with the formation of tension joints. This implies that the parameter λ, introduced in Eq. 8.2, attains the value 1. The limit circle C^{int} with $\sigma_I'^{int} = 3T_o$ will be intersected at the vertices by the growing C^{ex}-circles that maps the build-up of a biaxial load at a mean stress $(\sigma_{III}'^{ex} + \sigma_I'^{ex})/2 \geq 3T_o$. Some typical loads and the associated inclination angles, determined by the relationships 8.2 and 8.3 with $\lambda = 1$, are given below:

$\frac{\sigma_{III}'^{ex}}{T_o}$	0	0.1	0.2	0.3	0.4	0.5
$\frac{\sigma_I'^{ex}}{T_o}$	5.0	5.4	6.0	6.7	7.7	9.0
γ °	26.6	24.2	21.8	19.3	16.7	14.0

(8.6)

Note that, in view of the limit condition (Eq. 8.5), $\sigma_{III}'^{ex}/T_o$ can only attain values up to 0.57. Further, note that the stars mark the parameter values of the circle C^{ex} which is tangential to the vertex line, – the line through the vertices S of all limit circles C^{int}.

The table shows, that pre-peak shear bands which are produced at a mean stress $(\sigma_{III}'^{ex} + \sigma_I'^{ex})/2 \geq 3T_o$ are inclined at an angle γ which decreases with increasing maximum load $\sigma_{III}'^{ex}/T_o$. Quite differently, at a mean stress $0 \leq (\sigma_{III}'^{ex} + \sigma_I'^{ex})/2 < 3T_o$, biaxial loading produces pre-peak shear bands at the uniform inclination angle $\gamma = 22.5°$. This is readily seen by considering the C^{ex}-circles that are tangential to the vertex line of the C^{int}-circles, which is inclined at 45° to the σ-axis. Once a growing stress circle C^{ex} has touched the vertex line, and initiated a first shear band, any further increase in differential load will cause the stress circle to intersect the vertex line at two points. Hence, for the initial shear band to remain active, the internal stress circle would have to shift with placing its vertex at an intersection point. Moving the centre of the C^{int}-circle to the right implies an increase of the parallel stress $\sigma_{II}'^{int}$; this is clearly incompatible with the shortening tendency of dilational bookshelf-type shearing. Moving the circle to the left implies a reduction of the shear stress on the band. In general, this will also halt the rotational operation of the "bookshelf" fractures. Hence, it is rather unlikely that biaxial loading at $0 \leq (\sigma_{III}'^{ex} + \sigma_I'^{ex})/2 < 3T_o$ will generate pre-peak shear bands of tension fractures at inclination angles different from $\gamma = 22.5°$.

But this still does not answer the question as to how the shear banding started. Did the échelon fractures come first, and produce the localized reduction of the normal stress component, or was the stress reduction first, and localized the formation of the fracture array? Or did the two phenomena develop simultaneously in an intertwined fashion? It is this last mode of development that is corroborated by a fracture-mechanical model proposed by Olson and Pollard (1991; loc.cit.) for the development of en-échelon vein arrays. The essence of the model is outlined next, in qualitative terms.

Self-organization of en échelon fractures. Consider a rock with randomly located and oriented flaws. When a biaxial load is applied to the rock, micro-cracks will develop from flaws and align sub-parallel to the direction of σ_I. Each fracture is surrounded by a tensile stress release halo, which is widest at the middle part of the fracture and narrows towards the fracture tips. As the fractures grow in length, their halos also grow in size, and the fractures start to interact, with longer fractures suppressing the growth of shorter ones. It is clear, that the shape of the stress release halos allows neighbouring fractures to approach each other more closely (in terms of the orthogonal distance of the fracture planes) when the parallel fractures only partly overlap, as schematically indicated for the fractures 1–3 in Fig. 8.11. In numerical experiments, Olson and Pollard showed that the growth of neighbouring fractures is enhanced by mechanical fracture interaction for all fracture arrays with δ (see the figure) less than about 50°.

Thus, the model tells us that, in a random fracture population under a tension load, certain arrays of a few neighbouring micro-cracks will preferentially grow into macro-fractures in en échelon positions. However, we now face the question: will a rather short en échelon array, say, of three fractures, born by chance, grow into a longer array by incorporating additional fractures? In other words, is the formation of en échelon fracture sets a self-organizing process?

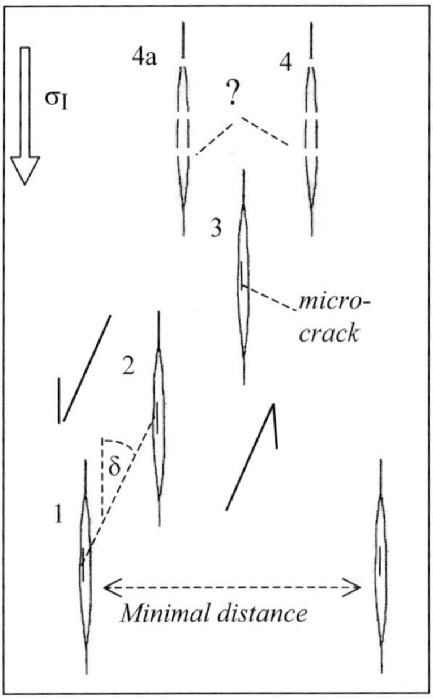

Fig. 8.11. The self-organisation of en échelon tension fractures and veins (see text for explanation)

The problem is illustrated in Fig. 8.11. Assume that there are, ahead of the en échelon array, micro-fractures in favourable growth positions. Then, a candidate fracture can grow into a macro-fracture on only one side of the free tip region at the end of the en échelon set (4 or 4a in the figure). Otherwise the distance between the new fractures would be too small. If it were fracture 4a, the growth of the en échelon ensemble would be stopped. Thus, a systematic growth of the en échelon set requires a stress asymmetry, with extra tension on the outer side of fracture 3.

One may safely assume that this tensional asymmetry is generated by the external shear couple that acts upon the fractured band of rock, and tends to rotate the rock slices in dilational bookshelf mode.

In Fig. 8.11, the parallel fractures are assumed as straight; but it is easily seen, as very schematically indicated in Fig. 8.12, that the dilatant fractures exert shear stresses upon each other, in the region of overlap. This may cause the growing fractures to curve towards each

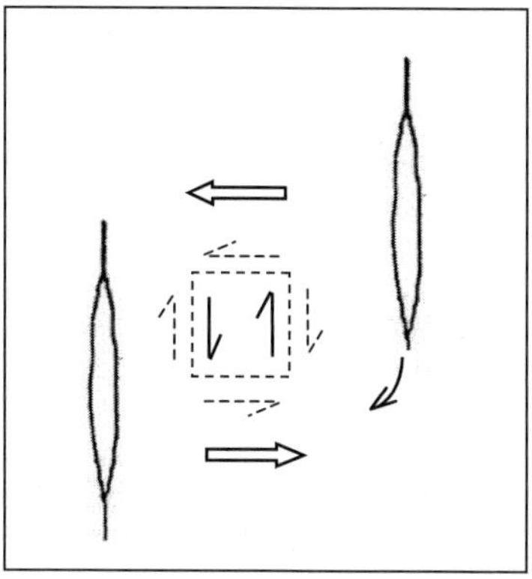

Fig. 8.12. Shearing interaction of neighbouring en échelon fractures (highly simplified)

other, as indicated in the right fracture. This may result in a sigmoidal shape of the en échelon veins; this was numerically simulated by Olson and Pollard (1991; loc.cit.), following earlier fracture-mechanical studies by D.D. Pollard, P. Segall and P.I. Delaney (1982; Geol. Soc. America Bull. *93*, pp. 1291–1303).

Recalling from the end of Chapt. 5 (Fig. 5.14E,F) how a remote compressive fracture-parallel stress tends to stabilize a straight fracture path, we would expect the sigmoidal deformation of neighbouring fractures due to shearing interaction to be suppressed by high remote compression parallel to the fractures. This, and other mechanisms which tend to stabilize the straight propagation of tension fractures were discussed by C.E. Renshaw and D.D. Pollard (1994; J. Struct. Geol. *16*, No. 6, pp. 817–822).

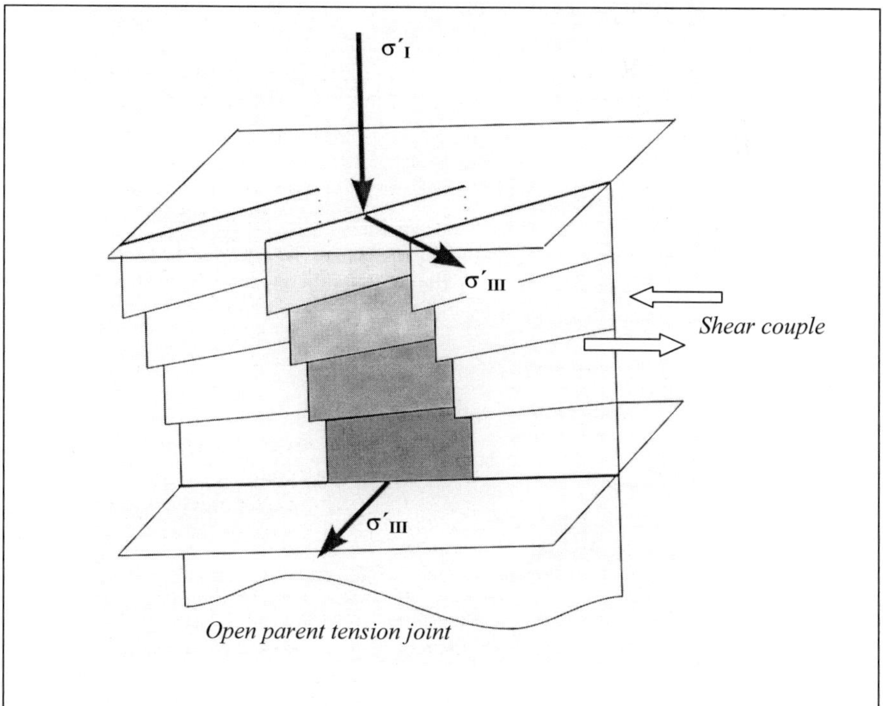

Fig. 8.13. Segmentation of an open vertical tension joint by a superimposed horizontal shear couple that rotates the horizontal σ_{III}- and σ_{II}-axes

The breakdown of parent cracks into dilatant échelon cracks. When a dilatant fracture, (be it an open tension fracture, a cleavage (extension) fracture, or a hydraulic intrusion fracture,) enters a region where the axis of the principal stress σ'_{III} has been rotated about the local σ'_{I} axis, the parent fracture will disintegrate into numerous en échelon fractures. This is schematically illustrated in Fig. 8.13. The mechanics of this particular mode of en échelon fracturing in rocks was studied in a fracture-mechanical analysis by Pollard, Segall and Delaney (1982; loc.cit.). Here, we shall circumvent the discussion of fracture-mechanical details by resorting to the "integrability theorem" of vector fields in differential geometry, which we have already employed to explain the segmentary development of faults (G. Mandl 1987; J. Struct. Geology *9*, No. 1, pp. 105–110; see also MTF, II.7.6, and FBR, 4.5.)

The theorem states, that a vector field $\vec{v}$ (satisfying all the required differentiability, etc.) will be normal everywhere to a set of smooth, non-intersecting surfaces if, and only if, the scalar product of the vector field with its own curl vanishes:

$$\vec{v}.\text{curl}\,\vec{v} = 0 \tag{8.7}$$

Written *in extenso* for a Cartesian x, y, z system, this condition is

$$v_x\left(\partial v_z/\partial y - \partial v_y/\partial z\right) + v_y\left(\partial v_x/\partial z - \partial v_z/\partial x\right) + v_z\left(\partial v_y/\partial x - \partial v_x/\partial y\right) = 0 \tag{8.7a}$$

Now, we identify the vector **Fehler! Textmarke nicht definiert.** with the unit vector **Fehler! Textmarke nicht definiert.** in the direction of the minimum principal stress σ_{III}, which we assume to lie in a plane z = const.; the unit vector $\vec{e}_I$ in the direction of σ_I is parallel to the vertical z-axis.

Since the z-component $e_{III\,z}$ of $\vec{e}_{III}$ vanishes, Eq. 8.7a reduces to

$$e_{III\,y}.\partial e_{III\,x}/\partial z - e_{III\,x}.\partial e_{III\,y}/\partial z = 0 \tag{8.7b}$$

in terms of the x and y-components of $\vec{e}_{III}$.

Introducing the angle α of the unit vector $\vec{e}_{III}$ with the (horizontal) x-axis, the components of the vector are $e_{IIIx} = \cos\alpha$ and $e_{IIIy} = \sin\alpha$. Inserting these expressions into Eq. 8.7b, and noting that α may vary in the z-direction to allow the direction of the horizontal minimum principal stress to vary with z as assumed in Fig. 8.13, Eq. 8.7b turns into

$$\frac{d\alpha}{dz} = 0 \tag{8.8}$$

This simple result states that surfaces orthogonal to the direction field of the minimum principal stress σ_{III} exist if, and only if, the normal to the plane through the σ_I- and σ_{III}-axes has the same direction everywhere. Note, that this condition only refers to the directions of the principal stresses, and not to their magnitudes. Since, as is generally accepted, tensile, cleavage or hydraulic macro-fractures must develop normal to the direction of σ_{III}, the fractures cannot form as coherent and continuous structures if condition 8.8 is not satisfied.

Without going into further mathematical details, it should be noted that the above condition also holds for curved σ_I- and σ_{III}-trajectories; as long as the two sets of trajectories lie in parallel planes or on smooth parallel surfaces, the macroscopic fractures propagating along a curved σ_I-trajectory (as in Figs. 4.26, 5.5, 7.22) will be continuous and smooth, at least as long as the fractures are not affected by material inhomogeneities or anisotropies.

A typical situation, where condition 8.8 is not satisfied, is schematically shown in Fig. 8.13, where a vertical parent fracture enters a region in which the σ_{III}-axis spirals around the vertical; this causes the parent fracture to break down into segments. Clearly, the differential-geometrical approach does not tell us what the width of the fracture segments may be in real materials; but it does suggest that twisted fracture segments themselves may be segmented in a "spiral staircase" mode. Naturally, the twisting of the fracture segments will stop when, ahead of the propagating fracture, the orientation of the σ_{III}-axis becomes constant.

M. Lagalli already recognized in 1930 (Z. Gletscherkunde *17*, pp. 285–301) that a three-dimensional variation of principal stress directions with depth caused the discont-inuity of crevasses in glaciers. In rocks, similar variations of principal directions with depth may result from a variety of tectonic situations. A. McGarr (1980; J. Geoph. Res. *85*, pp. 9213–9338) showed that, when σ_I is the overburden stress and the horizontal stresses vary linearly with depth, the horizontal principal stresses may, from the surface downward, monotonically change their directions. Thus, even in a fully compressive regime where no tension or cleavage joints were to be expected, sheetlike intrusions would break up into separate dykes or dykelets. Some similar tectonic situations were discussed in MTF (pp. 328–330, II.7.6) and FBR (pp. 147–149, 398–401). In general, the changes in the σ_{III} direction are associated with shear couples superimposed on a major stress field.

Interestingly, the break-up of a vertical joint into échelon dilatant fractures is also observed on the scale of hand specimens (see Pollard, Segall and Delaney 1982, loc.cit.). In this case, the break-up requires a rotation of the local principal stresses over a very short distance. What kind of geological conditions might produce the phenomenon? Convincing answers to this intriguing question seem to be scarce. Nevertheless, a few suggestions may be made.

Let us consider a horizontal two-layer system of sedimentary rocks consisting of a stiff layer (1) overlaid by a softer rock (2). The rocks are assumed to be mechanically isotropic in all layer-parallel directions (transverse isotropy), and are laterally confined during the accretion of the overburden. The effective overburden load causes horizontal normal stresses σ'^{o}_{h}, as indicated for a horizontal element in Fig. 8.14A. Naturally, these stresses differ between the two layers; they are smaller in the stiff rock than in the soft rock: $\sigma'^{o(1)}_{h} < \sigma'^{o(2)}_{h}$. The initial state of the horizontal stresses is changed by a uniform tectonic extension of the whole two-layer unit in the horizontal direction "1", without straining the layers in the horizontal direction "2". In addition, a simple shear is applied parallel to the horizontal direction "2", as schematically shown in Fig. 8.18B. The deformation reduces the horizontal normal stress in the "1" direction by $\Delta\sigma'_h$, and induces a horizontal shear stress τ_h, as indicated in the figure. Although the straining is the same in both layers, the induced stresses are different, with $\Delta\sigma'^{(1)}_{h}$ (stiff layer) > $\Delta\sigma'^{(2)}_{h}$ (soft layer). In contrast, the shear stress $\tau_h^{(1)}$ (stiff layer) will be greater than $\tau_h^{(2)}$ (soft layer).

The orientations of the horizontal principal directions are described in Fig. 8.14C by the angle Θ, which is easily determined by the formula shown in the figure. We assume that the rocks respond linear elastically to the uniaxial straining Δe_1, under the lateral confinement $\Delta e_2 = 0$, and to the simple shearing through the angle γ. The stresses in the formula in Fig. 8.14C may then be expressed in terms of strains by applying Hooke's law (Eq. 3.3 and $\gamma = \tau/G = 2(1 + \nu).\tau/E$). The orientation of the horizontal principal directions is then given by the simple relation

$$\tan 2\Theta = (1 - \nu)\frac{\gamma(\text{radians})}{\Delta e_1} \qquad (8.9)$$

Note, that the E-modulus dropped out of the formula, because it was assumed to be the same in both the extensional straining and the simple shearing operation. In general, the E-modulus is greater in unloading than in loading of the rock. Hence, if the shearing takes place after the extension of the rock, the shear modulus G would be related to a smaller E-value, and a multiplier smaller than 1 would have to be inserted into the right side of Eq. 8.9. Another uncertainty occurs due to the Poisson ratio ν, which is known to vary with the mean normal stress.

Notwithstanding these simplifications, the relation 8.9 provides a first impression of the extent to which the combined operation of extension and simple shearing may affect the orientation of the principal stress axes. Taking $\gamma = 1.5° = 0.026$ rad and $\Delta e_1 = 0.02$, and $\nu = 0.15$ for layer "1", and $\nu = 0.45$ for layer "2", Eq. 8.9 yields: $\Theta_1 = 24°$ and $\Theta_2 = 18°$. Thus, the principal axes would be rotated through only the modest angle of 6° about the vertical σ_1-axis, even though the parameter values were chosen for maximum effect.

However, Eq. 8.9 also suggests that the rotation angle may substantially increase if the shear angle varies in the vertical direction. This may be geologically achieved, for example, by horizontal shearing between "rigid" boundary rocks of an inclined contact with the sheared layers (Fig. 8.15). If then, in a vertical interval, γ changes from, say 1.5° to 0.75°, while the values of Δe_1 and ν remain the same as in the example before, the principal axes are rotated through the angle $\Delta\Theta = 14°$.

Similarly, the rotation angle is raised when the uniaxial straining differs between two layers. This may occur in the two-layer system if the straining of the unit as a whole continues after tension joints have been generated in the stiff layer "1". As discussed before (see Figs. 4.18 and 5.5), at the instant of jointing, the stresses and strains in layer "1" become highly non-uniform, and further straining of the unit is accommodated in layer "1" by the opening of the joints. Layer "2" is continuously, albeit non-uniformly

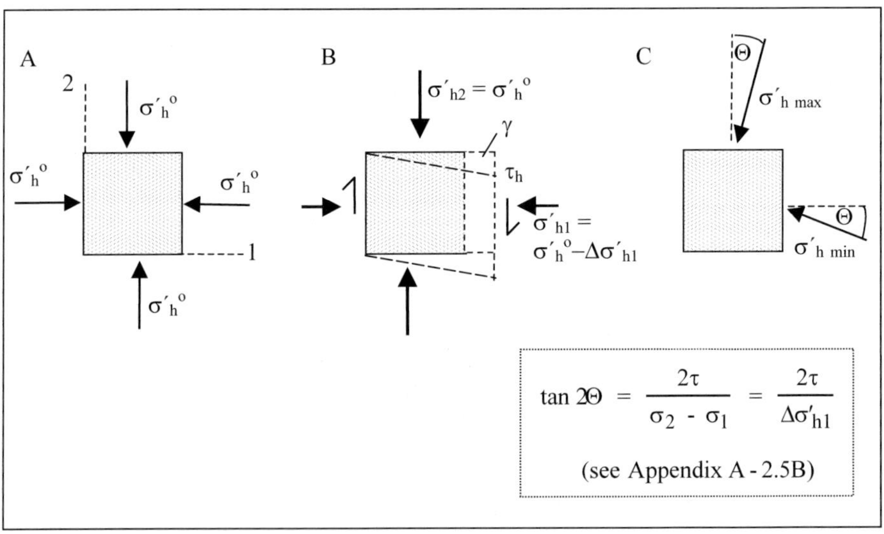

Fig. 8.14. Stresses in transverse isotropic horizontal layer of sedimentary rock:
A) horizontal effective stresses caused by overburden load;
B) uniaxial horizontal stretching and simple shearing through angle γ;
C) orientation of principal stress directions

$\vec{s}$ strained, with alternating extensional and contractive local deviations from the average strain $\Delta e_I^{(2)}$ (also see p. 66). Taking $\Delta e_I^{(2)} = 0.02$ for the mean value of Δe_I in layers "2", and $\Delta e_I^{(1)} = 0.005$ in layer "1" at the onset of jointing, and $\gamma = 1.5°$, $\nu_1 = 0.15$, $\nu_2 = 0.45$, as chosen before, Eq. 8.9 predicts a rotation angle $\Delta\Theta = 21°$. Hence, in the case where at the total horizontal strain $\Delta e_I^{(2)} = 0.02$ tensile jointing is also induced in layer "2", the new joints will strike at an angle of 21° with respect to the earlier formed joints in layer "1". But, although arrayed en échelon, the joints in layer "2" can hardly be considered as "break-up" structures of a parent tension fracture. However, a break-up would occur to a hydraulic intrusion fracture that enters layer "2" from the jointed layer "1".

We think that these examples provide an idea of how a layer-parallel uniaxial straining *in combination* with a transverse simple shear couple may rotate the σ_{III}-axis around the σ_I direction such as is required to break-up a dilational parent fracture into en échelon segments. Certainly, other geological settings will exist that may produce these kind of stress rotations; but detecting them must be left to the inventive reader.

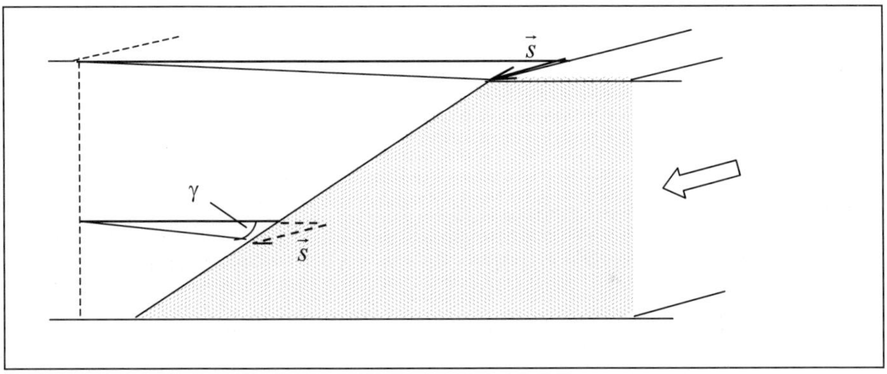

Fig. 8.15. Horizontal simple shearing along the inclined contact with a moving boundary block causes the shear angle to vary with depth

Summary of dilatant en échelon fractures. Despite the abundance of dilatant fractures (i.e. tension and cleavage fractures, hydraulic intrusion fractures, gashes and veins) in en échelon arrays in rocks (Figs. 8.1, 8.6, 8.9), and despite the similarity of these structures, they are the result of quite different mechanical processes. Not all of them are as yet sufficiently well-understood.

- ***Echelon cracks in shear zones.*** The formation of échelon fractures in sediments under simple or quasi-simple (i.e. dilating) shearing by stiffer boundary rocks (Fig. 8.2) is well understood. The movement of the parallel boundaries enforces a rotation of the σ_I direction through the angle $\Theta = 45°$ if the normal stresses parallel ($\sigma_{||}$) and normal ($\sigma_\perp$) to the external shear direction are equal; otherwise $\Theta < 45°$ if $\sigma_{||} > \sigma_\perp$, and $\Theta > 45°$ if $\sigma_{||} < \sigma_\perp$. Tension fractures form parallel to the rotated σ_I direction only if the effective stresses $\sigma_{||}'$ and $\sigma_\perp'$ remain smaller than about twice the tensile strength of the rock. At moderately higher effective compressive stresses the shearing promotes the formation of Riedel shears (Fig. 8.3), whose orientation is also affected by the difference between $\sigma_{||}'$ and $\sigma_\perp'$. Hence, in the absence of other indicators, the angular fracture orientation with respect to the external shear boundaries is not a reliable means of differentiating between tensile and shear fractures.

Further shearing rotates tension fractures in a dilational bookshelf mode causing the central fracture parts to open up and thus, allowing the infill of minerals, while the tips of growing fractures follow the local σ_I direction and give the fractures a sigmoidal shape (Figs. 8.1, 8.4, 8.5). Small scale rupturing in the marginal zones of a band of rotated fractures (Fig. 8.5) may then promote the localization of shear along a band boundary, which could possibly develop into a minor fault. The fault, enforced by the kinematic constraints of simple shearing, is not of a Coulomb- (or Andersonian) type, but rather comparable to the non-Coulomb type faults produced under "shear-box" conditions in all-compressive regimes (Fig. 8.7).

In an all-compressive regime, antithetic shears (P' in Fig. 8.8) are likely to be formed under simple shearing conditions when $\sigma_{||} > \sigma_{\perp}$; the shears may accommodate the external shear by rotation in a similar way to tension fractures, and even form veins by opening during rotation in the dilational bookshelf mode. The similarity in appearance may make it difficult, or even impossible, in the field to decide with certainty whether the en échelon fractures were formed as tension fractures or as shear fractures. The differentiation may become easier when pressure solution seams occur together with veins in an en échelon array, since they indicate the local orientation of the maximum compressive stress (Fig. 8.9).

- ***En échelon fractures in pre-peak shear bands.*** The central question about en échelon fractures is: how and why are they formed? To answer this question, at least to a great extent, it is proposed that the en échelon fracture bands are formed as shear bands in a rock under a compressional biaxial load, which is still below the limit load that the rock can sustain. While the rock material is, therefore, still in a "hardening" state, the uniformity of the stress field is perturbed by bands whose band-parallel normal stress $\sigma_{||}$ is smaller inside than outside ($\sigma_{||}'^{int} < \sigma_{||}'^{ex}$). This kind of "stress softening" (already discussed in Chapt. 6, pp. 142–148) is associated with the formation and rotation of en échelon tension (or cleavage) fractures inside the bands. Tensile fractures are produced, and rotated in the dilational "bookshelf" style by the external shear stress that acts on any band-parallel section in the rock (see the Mohr diagram Fig. 8.10, and again Figs. 6.16, 6.17A). This has two effects: the band-parallel stress $\sigma_{||}'^{int}$ is reduced, and the bands are sheared in a quasi-simple mode.

  There are two conjugate shear bands Γ and Γ^* inclined at $\pm\,\gamma$ to the external direction of the maximum compressive stress ($\sigma_I'^{ex}$). With reference to Fig. 8.10, the inclination angle γ is found to decrease with increasing σ'^{ex}_I (see Table 8.6) if $\sigma'^{ex}_I/T_o + \sigma'^{int}_{III}/T_o \geq 6T_o$, and to remain at 22.5° if $0 \leq \sigma'^{ex}_I/T_o + \sigma'^{int}_{III}/T_o \leq 6T_o$. The tension fractures in a shear band are generated at 45° to the band boundary; hence, near a band of en échelon tension fractures, the σ_I-trajectories must bend through an angle of 45° – γ towards the band. Furthermore, the pre-peak shear bands are immobilized by minor changes in the external stress system, which makes it unlikely that shear bands of this type develop into minor faults.
- ***Self-organization of en échelon fractures.*** The continuum mechanical model of pre-peak shear banding leaves the question open as to how the shear banding starts. A likely scenario is presented on page 194 (Fig. 8.11): Olson and Pollard (1991; loc.cit.) showed that, in a random population of micro-cracks, a few neighbouring cracks which, by chance, form a certain array, will preferentially grow under tension load into macro-fractures in the en échelon position. The preferential growth of the fractures is due to the mechanical interaction of partly overlapping parallel fractures.

But an en échelon array of a few macro-fractures, born by chance, can only develop into a longer array by incorporating additional fractures in a systematic way, which preserves the en échelon geometry of the ensemble. This is achieved if the near-tip stresses of the rearmost fracture of the array, to which the new fracture is to connect, are asymmetric with a greater tension on the side facing the new fracture. This stress asymmetry is induced by the external shear couple that acts on the compound of the initial en échelon fractures (see Fig. 8.11).

- ***The breakdown of parent cracks.*** A dilatant fracture (tension-, cleavage-, or hydraulic intrusion fracture) must disintegrate into numerous en échelon fractures when entering a region where the σ_{III} axis is rotated around the σ_I axis (Fig. 8.13). A simple straightforward proof of this statement is provided by the "integrability theorem of vector fields" (Eq. 8.7).

  In general, the changes in the σ_{III} direction are associated with shear couples superimposed on a major stress field. As a typical example, a horizontal multilayer of stiff and softer layers is uniaxially stretched, and subjected to in-plane simple shearing (angle γ) perpendicular to the direction of the extension (Fig. 8.14). A simple elastic analysis (Eq. 8.9) reveals that the contrast in Poisson ratios (ν) of superincumbent layers may rotate the horizontal σ_{III} axis around the vertical σ_I axis by a few degrees. Considerably greater angles of rotation are obtained when the shear angle γ changes in the vertical direction; this would be caused, for example, by an inclination of the contact between the shearing boundary rocks and the sheared multilayer (Fig. 8.15). Likewise, the rotation angle is raised when the extensional straining of the multilayer continues *after* tension joints have formed in a stiff layer.

The Stress Circle

O. Mohr's stress circle diagram is an extremely useful geometrical representation of the state of stress at a point. It is particularly convenient in problems of plane deformation, since the state of stress may then be visualized by a single stress circle.

Derivation. We consider in Fig. A.1 the equilibrium of forces on a prismatic volume element, whose mutually orthogonal faces are chosen so as to be acted upon by the principal stresses, σ_I, σ_{II}, σ_{III} (the associated vectors being $\underline{\sigma}_I$,...). The oblique face (shaded in the Fig. A.1) is acted upon by a normal stress $\sigma_\perp$ and a shear stress τ. In addition, the volume element experiences the gravity force $\underline{\gamma}\,dV$. As stated in the figure, equilibrium requires that the vectorial sum of all forces on the prismatic element vanishes. We divide $\underline{\sigma}_I$, $\underline{\sigma}_{III}$, $\underline{t}$ the equilibrium equation in the figure by the surface area $d\Sigma$ and let the prismatic element shrink towards a corner point. Since the ratio $dV/d\Sigma$ is proportional to a characteristic length of the prismatic element, it tends to zero as the element shrinks, and the gravity force disappears from the equation. Thus, the vectorial equilibrium equation for the infinitesimal prism becomes:

$$\underline{\sigma}_I.\cos\alpha + \underline{\sigma}_{III}.\sin\alpha + \underline{t} = 0 \tag{A.1}$$

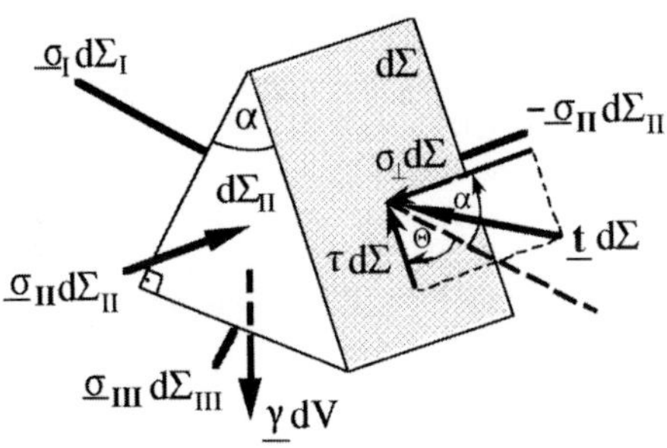

Equilibrium of forces:

$$\underline{\sigma}_I\, d\Sigma_I + \underline{\sigma}_{III}\, d\Sigma_{III} + \underline{\gamma}\, dV + \underline{t}\, d\Sigma = 0$$

where $d\Sigma_I = d\Sigma.\cos\alpha$, $d\Sigma_{III} = d\Sigma.\sin\alpha$

Fig. A.1. Forces on a prismatic element

The vanishing of this vector sum implies that any component of it vanishes as well. We consider two components: the first parallel to the exterior normal of the oblique side of the prism, and the second parallel to the shear stress on this side (Fig. A.2). The projections of the vectors $\underline{\sigma}_I$, $\underline{\sigma}_{III}$, $\underline{t}$ on the exterior normal of the oblique surface element are $\sigma_I.\cos\alpha$, $\sigma_{III}.\sin\alpha$, $-\sigma_\perp$ respectively, while the projections on the direction of τ are $-\sigma_I.\sin\alpha$, $\sigma_{III}.\cos\alpha$, τ. Hence, the componental equations of Eq. A.1 are

$$\sigma_\perp - \sigma_I.\cos^2\alpha - \sigma_{III}.\sin^2\alpha = 0 \tag{A.2a}$$

$$\tau - (\sigma_I - \sigma_{III}).\sin\alpha.\cos\alpha = 0 \tag{A.2b}$$

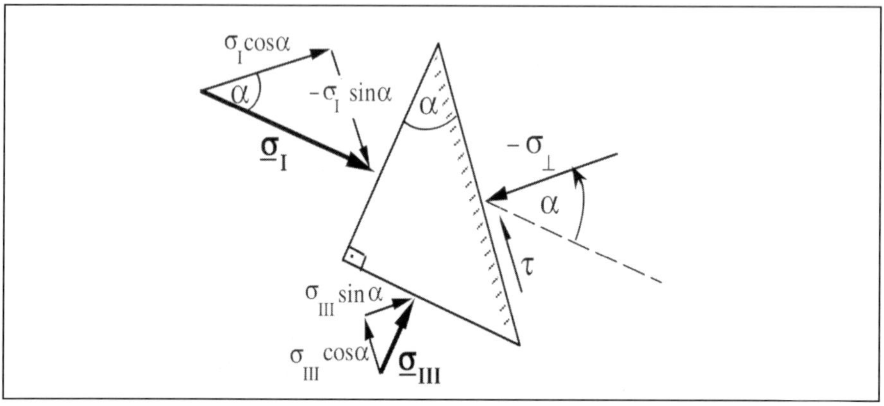

Fig. A.2. Components of stresses on prismatic element

The Eqs. A.2a,b may be rewritten by means of the well-known trigonometric identities $\sin^2\alpha = (1 - \cos 2\alpha)/2$, $\cos^2\alpha = (1 + \cos 2\alpha)/2$, $\sin 2\alpha = 2\sin \alpha.\cos \alpha$:

$$\sigma_\perp = \frac{\sigma_I + \sigma_{III}}{2} + \frac{\sigma_I - \sigma_{III}}{2}.\cos 2\alpha \tag{A.3a}$$

$$\tau = \frac{\sigma_I - \sigma_{III}}{2}.\sin 2\alpha\tau = \tag{A.3b}$$

These formulae are extremely useful, since they relate the normal and tangential components of a stress vector that acts upon a surface element, with the principal stress values σ_I, σ_{III} and the angle α which the normal of the element includes with the σ_I direction. But note that these relations only apply to surface elements which are parallel to the direction of the intermediate principal stress σ_{II}. Also note that the last term in Eq. A.3a attains a negative sign when the angle α is replaced by the complementary angle $\Theta = \pi/2 - \alpha$ (Fig. A.1).

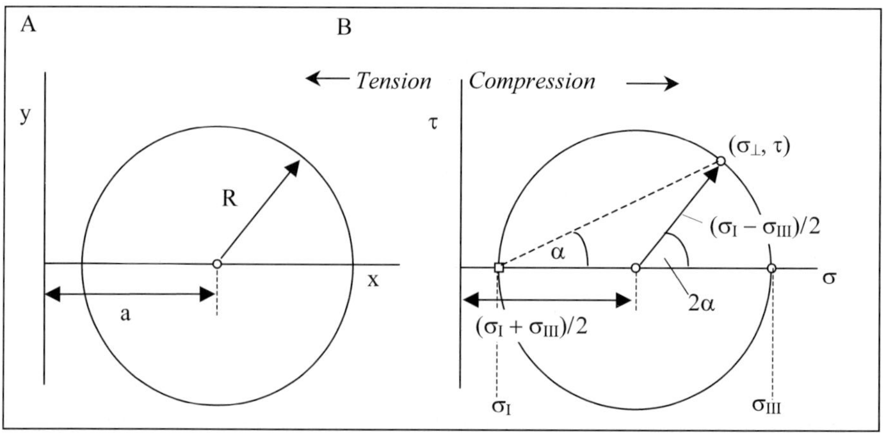

Fig. A.3. Mohr's circle representation for stresses acting parallel to the σ_I,σ_{III}-plane

The angle α can be eliminated from the Eqs. A.3a,b. Because of $\sin^2 2\alpha + \cos^2 2\alpha = 1$, rearranging terms in the equations, squaring and adding yields

$$\left[\sigma_\perp - \frac{\sigma_I + \sigma_{III}}{2}\right]^2 + \tau^2 = \left[\frac{\sigma_I - \sigma_{III}}{2}\right]^2 \tag{A.4}$$

Mohr's stress circle. If we compare Eq. A.4 with the equation

$$(x - a)^2 + y^2 = R^2$$

of a circle in a Cartesian x,y-coordinate system (Fig. A.3A), we notice immediately that Eq. A.4 represents a circle in a two-dimensional Cartesian frame with the coordinates $\sigma_\perp$ and τ (Fig. A.3B).

The "stress circle" has the radius $(\sigma_I - \sigma_{III})/2$, intersects the σ-axis at the points $(\sigma_{III},0)$ and $(\sigma_I,0)$, and is centred on the σ-axis at a distance of $(\sigma_I + \sigma_{III})/2$ from the origin. The coordinates $\sigma_\perp$, τ of a point on the circle represent the normal and tangential stress components of the stress vector $\underline{t}$ which acts on a planar element which is parallel to the intermediate principal stress σ_{II}. Since we count compressive stresses as positive, the corresponding stress points will lie in the positive half-plane of the diagram.

It is important to note that the radius vector to a stress point, associated with a planar element whose normal includes the angle α with the σ_I direction (Fig. A.2), includes twice that angle with the σ-axis. The corresponding relation $\tan 2\alpha = \tau /[\sigma_\perp - (\sigma_I + \sigma_{III})/2]$ in Fig. A.3B follows directly from the parameter equations (Eqs. A.3a,b) of the stress circle. Needless to say, α is not only the angle between the normals of two planar elements, but also the angle included by the elements themselves. Hence, when a planar element in real space is rotated anticlockwise around the σ_{II}-axis through an angle $\Delta\alpha$, the radius vector of the stress circle is rotated through $2\Delta\alpha$.

Note also that the chord extending from the stress point $(\sigma_{III},0)$ to a stress point $(\sigma_\perp,\tau)$ makes the angle α with the σ-axis, since according to an elementary geometrical theorem the angle extended from any point on the periphery of a circle over a given arc of the circle amounts to half the angle extended over the same arc from the center of the circle.

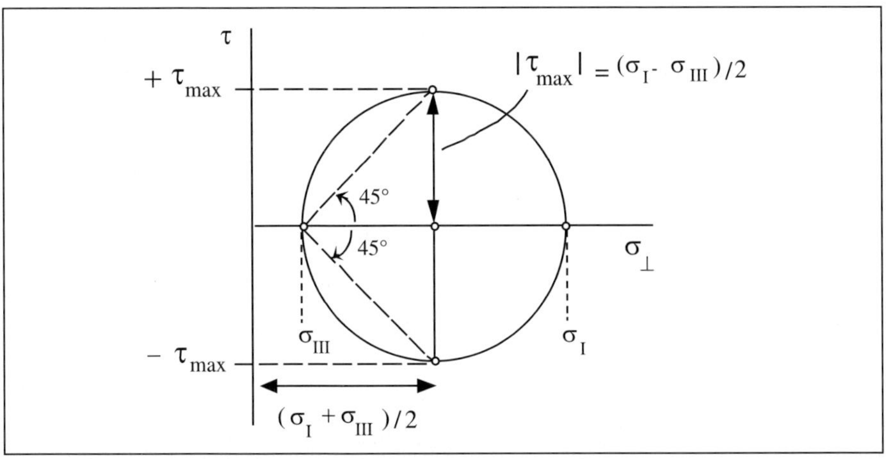

Fig. A.4. Maximum shear stress in the Mohr diagram

The Mohr circle is fully determined and easily constructed when the principal stress values σ_I and σ_{III} are prescribed, or when the normal and shear stresses that act on two arbitrary, mutually orthogonal planes, are known, in which case the associated stress points lie at the opposite ends of a diameter of the circle.

Maximum shear stress. From the Mohr Diagram (Fig. A.4) it is immediately apparent that maximum shear stresses τ_{max} occur on elements inclined at 45° to the σ_I-axis (i.e. $2\alpha = 90°$) since $\alpha = 0$ for the radius vector to the principal stress point $(\sigma_I,0)$. The normal stress acting on these elements is $(\sigma_I + \sigma_{III})/2$, while $\tau_{max} = \pm(\sigma_I - \sigma_{III})/2$. The two maximum shear stresses are equal in magnitude, as is required by the symmetric orientation of the two elements with respect to the σ_I-axis.

Stress relations. If the normal and shear stresses that act upon two mutually perpendicular planes which are parallel to the σ_{II} direction are known, the values of the principal stresses σ_I, σ_{III} and the maximal shear stress τ_{max} can be read from the Mohr Circle. These stresses may be also determined from the analytical expressions:

$$\left.\begin{matrix}\sigma_I \\ \sigma_{III}\end{matrix}\right\rangle = \frac{\sigma_1 + \sigma_3}{2} \pm \sqrt{\frac{(\sigma_1 - \sigma_3)^2}{4} + \tau_{13}^2} \qquad \text{(A.5a)}$$

$$|\tau_{max}| = \frac{\sigma_I - \sigma_{III}}{2} = \sqrt{\frac{(\sigma_1 - \sigma_3)^2}{4} + \tau_{13}^2} \qquad \text{(A.5b)}$$

which are derived by applying Pythagoras' Theorem to the shaded triangles in Fig. A.5A.

Similarly, the directions of the principal stresses σ_I and σ_{III} are easily determined with respect to a material element whose normal and shear stress are known. For the simple construction in the Mohr Diagram in Fig. A.5B the material reference element was chosen with the normal parallel to the x_3-axis of a Cartesian system (see insert of Fig. A.5B). The element is acted upon by the stresses σ_3 and τ_{31}. Hence, the radius vector to the point (σ_3,τ_{31}) on the stress circle includes the angle 2α with the radius vector to the point $(\sigma_I,0)$, i.e., twice the angle between the σ_I-axis and the x_3-axis in real space. The complementary angle 2Θ in the Mohr Diagram is then twice the angle between the σ_I axis and the x_1-axis. The shaded right-angled triangle in Fig. A.5B immediately provides the important analytical expression for the orientation of the σ_I-axis:

$$\tan 2\Theta = \frac{2\tau}{\sigma_1 - \sigma_3} = -\tan 2\alpha \qquad (\sigma_1 > \sigma_3) \qquad \text{(A.6)}$$

where σ_1 and σ_3 are the normal stresses that act upon the elements with normals parallel to the x_1- and x_3-axes of a Cartesian frame, and $\tau = \tau_{31} = \tau_{13}$ is the shear stress on these elements.

Note that the roles of the x_1- and the x_3-axes in Eq. A.6 are interchanged, if $\sigma_1 < \sigma_3$, and the angle Θ^* in $\tan 2\Theta^* = 2\tau/(\sigma_3 - \sigma_1)$ becomes the angle between the σ_I-axis and the x_3-axis.

Three-dimensional state of stress. So far, we have dealt with stresses that act upon material elements whose normals are perpendicular to the σ_{II}-axis Although we shall mostly deal with this type of elements, occasionally we may also consider elements with normals perpendicular to the σ_I-axis or to the σ_{III}-axis. The stresses of these elements map on one of the smaller circles in Fig. A.6. Without proof, we add that the stresses $\sigma_\perp$, τ which act on elements that are obliquely oriented with respect to all three principal stress directions map on the interior of the crescent-shaped region enclosed by the three principal stress circles.

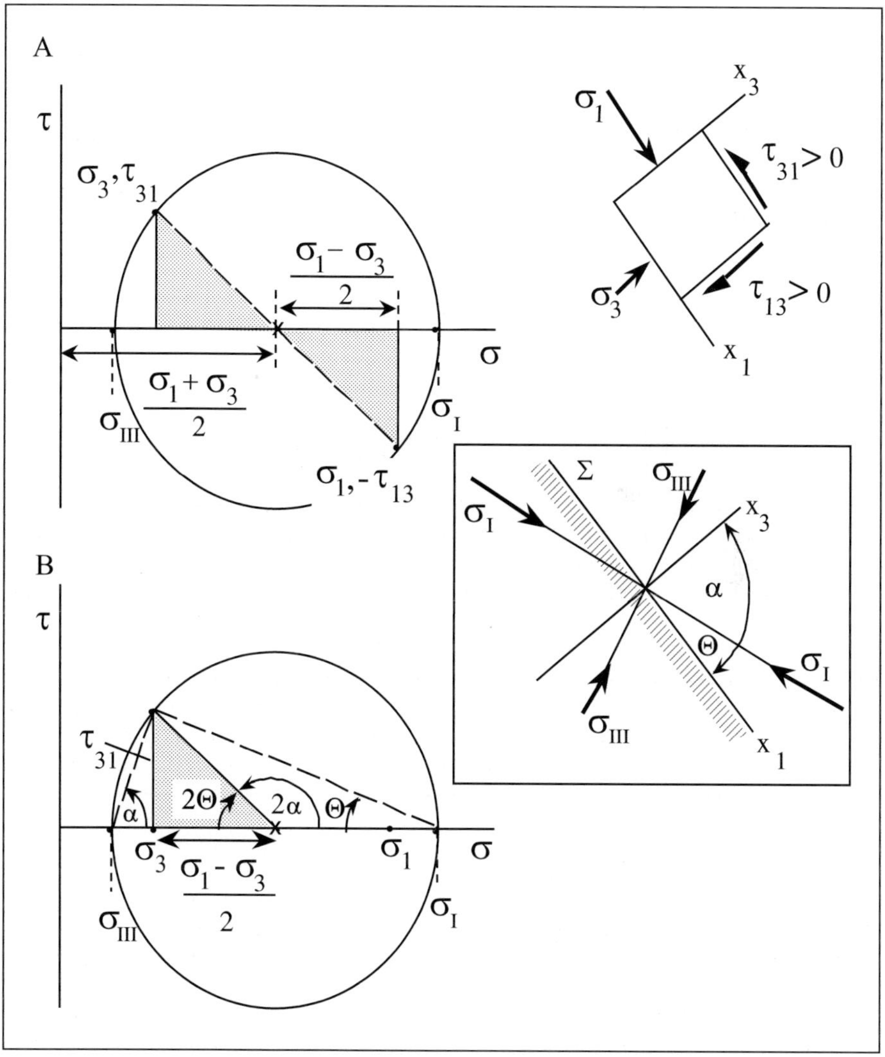

Fig. A.5. Stress relations in the Mohr Diagram:
A) principle stresses and Cartesian stress components;
B) orientation of principal axes in Cartesian reference frame (see text)

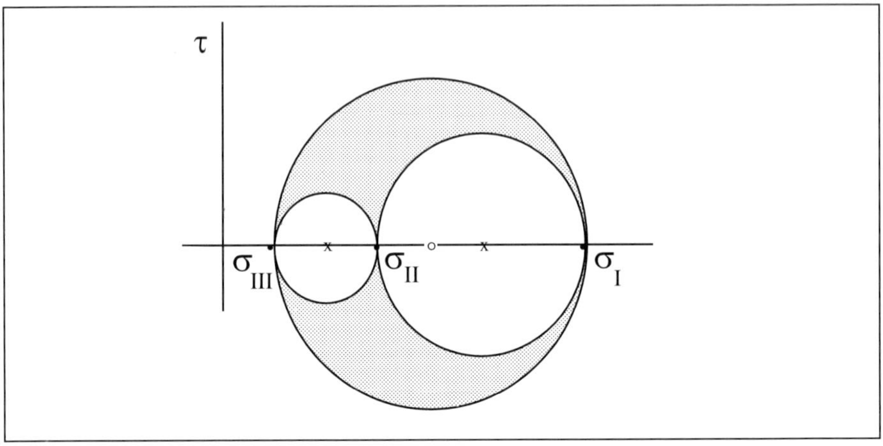

Fig. A.6. Mohr's representation of a three-dimensional state of stress

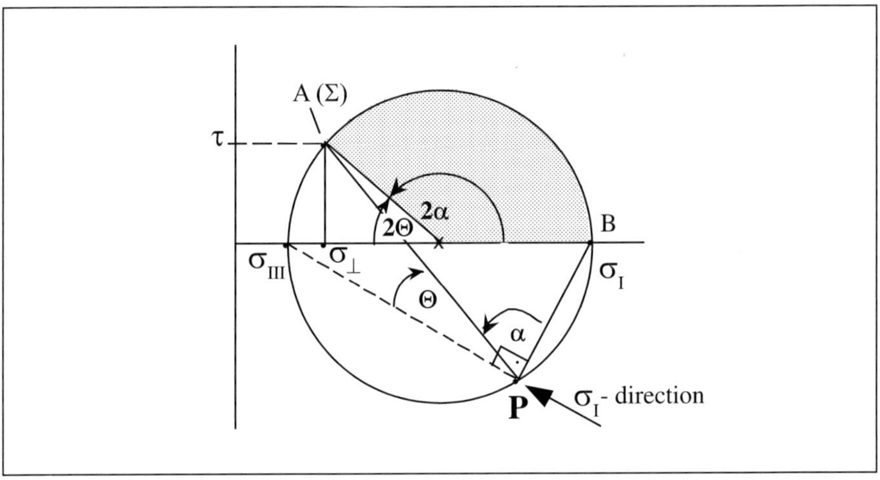

Fig. A.7. Construction of the σ_I direction by transposing the trace (AP) of a material element in the σ_I, σ_{III} plane (Σ in the insert of Fig. A.5) into the Mohr Diagram

The "pole" of the stress circle. There is an even simpler way of determining in Fig. A.5B the direction of σ_I with respect to the element whose normal and shear stresses are known. Consider in Fig. A.7 again the Mohr Circle and the stress point of Fig. 2.5B and draw the chord from this point (A) parallel to the trace of the associated material element Σ in the σ_I, σ_{III}-plane of real space. The chord meets the stress circle at a point P. Now connect P with the stress point (σ_I,0) at B. The angle α subtended by the chord AP and the line BP is then half the angle subtended by the radii from the centre to the ends of the arc AB. The angle α is, therefore, the angle which, in real space, the element that is acted upon by σ_I includes with the reference element. Since we have conveniently chosen the chord AP parallel to the trace of the reference element in the σ_I, σ_{III}-plane, the line segment BP is also parallel to the trace of the element acted upon by the maximum principal stress σ_I. Since σ_I acts perpendicularly on this element, the σ_I-axis is parallel to the chord drawn from the point (σ_{III},0) to P.

Naturally, the same procedure can be applied to determine the orientation or the stresses of any material element which is parallel to the σ_{II} axis, provided the Mohr Circle and a single reference element with known orientation and stress components are given. In Fig. A.8A,B the chords AP represent the traces of such reference elements, transposed from the σ_I, σ_{III}-plane into the Mohr Diagrams. If we wish to know the orientation of the material element that is acted upon by a given shear stress τ and a normal stress $\sigma_\perp$, all we have to do, is to connect the point P with the corresponding stress point B on the circle. This line is then parallel to the trace of the element in the σ_I, σ_{III}-plane. Conversely, if we wish to know the stresses that act on a given element, we draw from P the chord parallel to the trace of the element. The other endpoint of the chord represents the stresses on the element. The point P is referred to as the *pole* of the stress circle, because of its prominent role in this construction. The two diagrams in FigA.8 illustrate the difference between reference elements with sinistral and dextral sense of shearing. The shear stress acts towards the acute angle between reference element (AP) and σ_I direction.

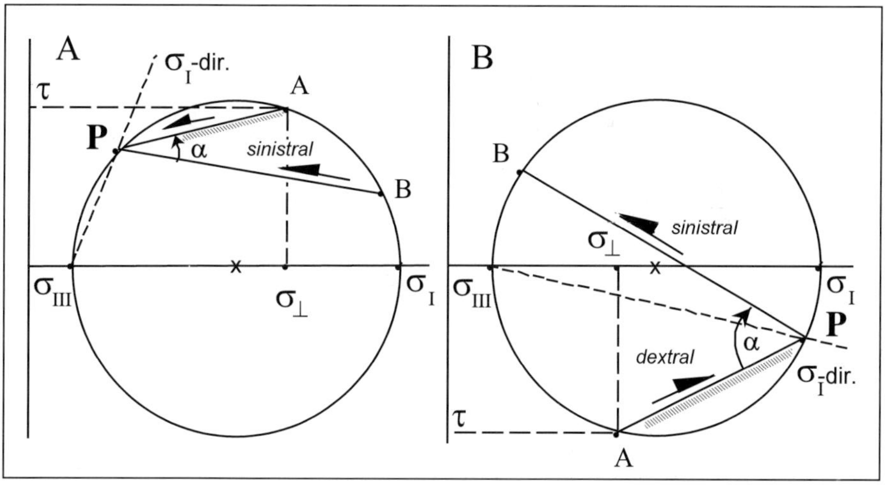

Fig. A.8. Pole of the stress circle; chord AP represents the trace of the reference element in the σ_I, σ_{III}-plane. Shear stress on reference element: A) sinistral; B) dextral

Examples. 1. The principal stresses depicted in Fig. A.9A may act parallel to any plane in space. In a vertical plane, for instance, $\sigma_I°$ could be the stress exerted by an overburden upon a horizontal plane, while acting in horizontal directions $\sigma_I°$ and $\sigma_{III}°$ may give rise to horizontal displacements along vertical fault planes. A Mohr Circle (Fig. A.9C) represents the stresses on the material elements with normals parallel to the σ_I, σ_{III}-plane. The chord of the circle, drawn parallel to the trace of the plane acted upon by $\sigma_I°$ in Fig. A.9A, determines the pole at ($\sigma_{III}°$,0). Since the two points coincide, the chord representing the $\sigma_I°$ direction by connecting the two points on the circle degenerates into the tangent shown in Fig. A.9C.

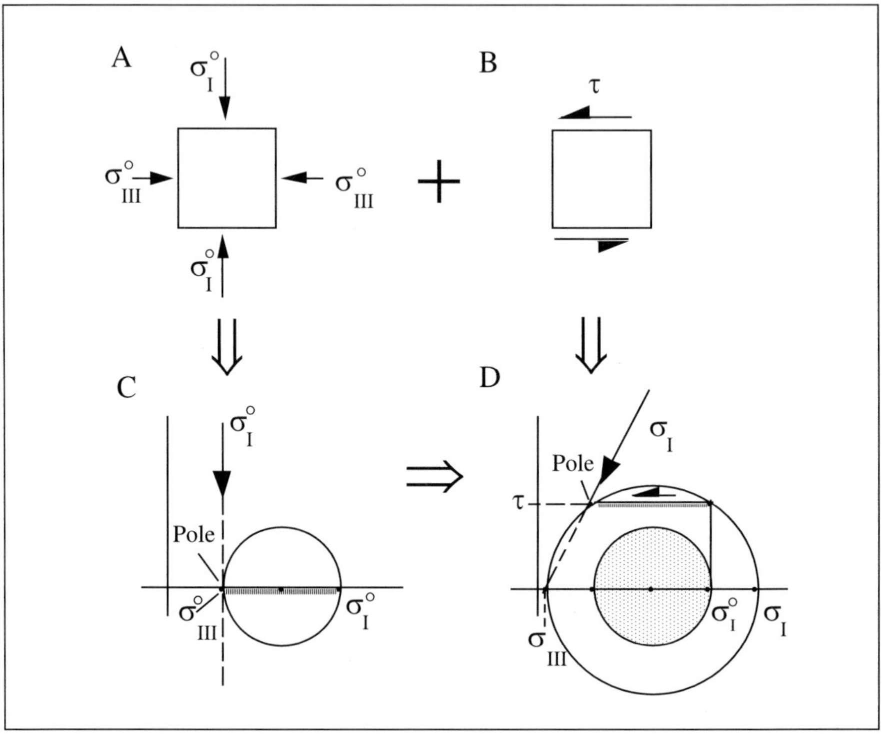

Fig. A.9. Change of σ_I-direction by superposition of a shear stress

Next, we assume that a shear stress τ is imposed by some tectonic simple shearing (Fig. A.9B), leaving the normal stress components on the orthogonal faces of the volume element in Fig. A.9A unchanged. Thus these normal stresses maintain the original values of $\sigma_I°$ and $\sigma_{III}°$, which implies that the center of the stress circle remains fixed. How are the principal stresses affected in magnitude and direction by the imposed shearing? The question is important, since incipience and orientation of tectonic faults and joints are determined by the principal stresses.

Figure A.9D shows again the original stress circle (shaded), but the point $(\sigma_I°,0)$ is no longer associated with the plane originally acted upon by $\sigma_I°$. Instead, the stress point $(\sigma_I°,\tau)$ is now associated with this plane, and completely determines the new stress circle, since the centre of the circle is not changed by the imposed simple shear. The orientation of the new maximum principal stress σ_I is found by the simple pole construction in Fig. A.9D.

2. In the next example we consider a normal fault in a horizontally layered rock mass with a horizontal surface. Figure A.10A shows the trace of the fault in a dip section, which also is the σ_I, σ_{III}-plane as the fault strikes parallel to the σ_{II}-axis. The vertical overburden pressure is the maximum principal stress σ_I, and the value of σ_{III} is assumed to be known, at least approximately, either from in-situ measurements or from shear strength data of the rock. We wish to determine the normal stress $\sigma_\perp$ and the shear stress τ that act

upon the fault plane. With σ_I and σ_{III} being known, we can draw the stress circle in Fig. A.10B. As in the preceding example, the point (σ_{III},0) becomes again the pole, from which we draw the chord parallel to the fault trace in Fig. A.10A. The coordinates of the point where the chord meets the circle represent the normal and tangential stress components on the fault.

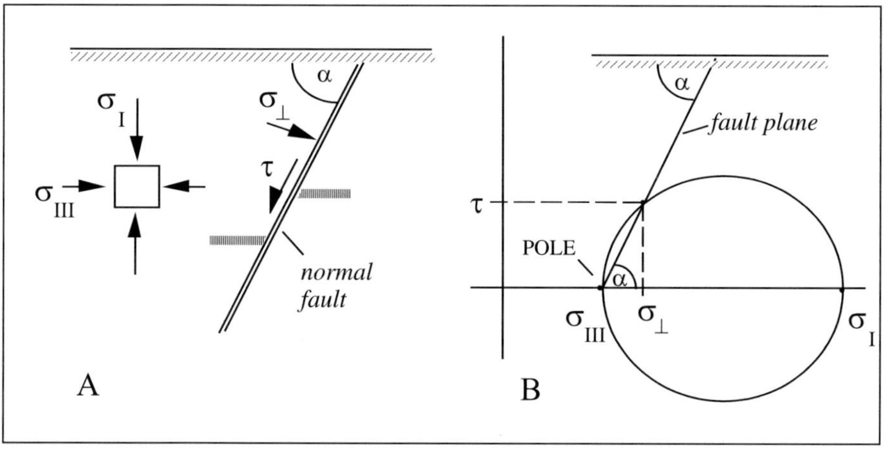

Fig. A.10. Normal and shear stress on a normal fault in a horizontally layered rock mass: A) dip section of the fault with principal stresses; B) determination of fault stresses in the Mohr diagram

References

Ashby MF, Hallam SD (1986) The failure of brittle solids containing small cracks under compressive stress states. Acta metallurgy 34:497–510

Atkinson BK (1987) Fracture mechanics of rocks. Academic Press, London

Bai T, Pollard DD (2000) Closely spaced fractures in layered rocks: initiation mechanism and propagation kinematics. J Struct Geol 22:1409–1425

Bankwitz P, Bahat D, Bankwitz E (2000) Granitklüftung – Kenntnisstand 80 Jahre nach Hans Cloos. Z Geol Wiss 28(1/2):87–110

Barquins M, Petit J-P (1992) Kinetic instabilities during the propagation of a branch crack: effects of loading conditions and internal pressure. J Struct Geol 14:893–903

Beach A (1977) Vein arrays, hydraulic fractures and pressure solution structures in a deformed flysch sequence, SW England. Tectonophysics 40:201–225

Bock H (1980) Die Rolle der Werkstoffkunde bei der mechanischen Interpretation geologischer Trennflächen. N Jb Geol Paläont Abh 160:380–405

Bourne SJ, Willemse EJM (2001) Elastic stress control on the pattern of tensile fracturing around a small fault network at Nash Point, UK. J Struct Geol 23:1753–1770

Brace WF, Bombolakis EG (1963) A note on brittle crack growth in compression. J Geophys Res 68(12):3709–3713

Cooke ML, Underwood CA (2001) Fracture termination and step-over at bedding interfaces due to frictional slip and interface opening. J Struct Geol 23:223–238

Cotterell B, Rice JR (1980) Slightly curved or kinked cracks. Int J Fracture 16:155–169

Cundall P (1990) Numerical modelling of jointed and faulted rock. In: Rossmanith HP (ed) Proceedings of the International Conference on the Mechanics of Jointed and Faulted Rock, Vienna, Austria. Balkema, Rotterdam, pp 11–18

Dyer R (1988) Using joint interactions to estimate paleostress ratios. J Struct Geol 10:685–699

Engelder T (1987) Joints and shear fractures in rock. In: Atkinson BK (ed) Fracture Mechanics of Rocks. Academic Press, London, pp 27–69

Engelder T (1993) Stress regimes in the lithosphere. Princeton University Press

Friedman M (1972) Residual elastic strain in rocks. Tectonophysics 15:297–330

Friedman M, Logan JM (1970) Influence of residual elastic strain on the orientation of experimental fractures in three quartzose sandstones. J Geophys Res 75(2):387–405

Gramberg J (1989) A non-conventional view on rock mechanics and fracture mechanics. Balkema, Rotterdam

Gretener PE (1983) Remarks by a geologist on the propagation and containment of extension fractures. Bull Ver Schweiz Petroleum-Geol Ing 49(116):29–35

Griffith AA (1925) Theory of rupture. In: Proc. 1st Int. Congress Appl. Mechanics, Delft. J. Waltman, Delft, pp 55–63

Gross MR (1993) The origin and spacing of cross joints: examples from the Monterey Formation, Santa Barbara coastline, California. J Struct Geol 15:737–751

Hallbauer DK, Wagner H, Cook NGW (1973) Some observations concerning the microscopic and mechanical behaviour of quartzite specimens in stiff, triaxial compression tests. Int J Rock Mech Min Sci 10:713–726

Helgeson DE, Aydin A (1991) Characteristics of joint propagation across layer interfaces in sedimentary rocks. J Struct Geol 13:897–911

Hobbs DW (1967) The formation of tension joints in sedimentary rocks: an explanation. Geol Magazine 104:550–556

Hoek E, Bieniawski ZT (1965) Brittle fracture propagation in rock under compression. Int J Rock Mech Min Sci 1:137–155

Holzhausen GR, Johnson AM (1979) Analysis of longitudinal splitting of uniaxially compressed rock cylinders. Int J Rock mech Min Sci 16:163–177

Josselin de Jong G (1959) Statics and kinematics in the failable zone of granular material. Dr. thesis, Techn. University Delft

Kármán von T (1911) Festigkeitsversuche unter allseitigem Druck. Z Ver dt Ing 55:1749–57

Kemeny JM (1993) The micromechanics of deformation and failure in rocks. In: Pasamehmetoglu et al. (eds) Assessing and Prevention of Failure Phenomena in Rock Engineering, Balkema, Rotterdam, pp 23–33

King G, Sammis C (1992) The Mechanisms of finite brittle strain. Pageoph 138(4):611–640

Lachenbruch AH (1961) Depth and spacing of tension cracks. J Geophys Res 66:4273–4292

Lachenbruch AH (1962) Mechanics of thermal contraction cracks and ice-wedge polygons in permafrost. Geological Society of America Special Paper 70

Lagalli M (1930) Versuch einer Theorie der Spaltenbildung in Gletschern. Z Gletscherkunde 17:285–301

Lawn BR, Wilshaw TR (1975) Fracture of brittle solids. Cambridge University Press

Lehner F, Pilaar WF (1974) Shell Research Report (internal)

Lehner F, Pilaar WF (1997) The emplacement of clay smears in syn-sedimentary normal faults: inference from field observations near Frechen, Germany. In: Moller-Pedersen P, Koestler AG (eds) Hydrocarbon seals. Norwegian Petroleum Soc Special Publ 7:39–50

Leroy Y, Ortiz M (1990) Finite-element analysis of transient strain localization phenomena in frictional solids. Int J Numer Anal Met Geomech 14:93–124

Lockner DA, Madden TR (1991) A multiple-crack model of brittle fracture, II. Time-dependent simulations. J Geophys Res 96:19643–19654

Mandl G (1987a) Tectonic deformation by rotating parallel faults: the "bookshelf" mechanism. Tectonophysics 141:277–316

Mandl G (1987b) Discontinuous fault zones. J Struct Geol 9:105–110

Mandl G (1988) Mechanics of tectonic faulting, models and basic concepts. Elsevier, Amsterdam

Mandl G (2000) Faulting in brittle rocks: an introduction to the mechanics of tectonic faults. Springer-Verlag, Heidelberg Berlin

Mandl G, Harkness R (1987) Hydrocarbon migration by hydraulic fracturing. In: Jones ME, Preston RMF (eds) Deformation of sediments and sedimentary rocks. Geol Soc Specail Publ 29:39–53

Martel SJ (1990) Formation of compound strike-slip fault zones, Mount Abbot quadrangle, California. J Struct Geol 12:869–882

McGarr A (1980) Some constraints on levels of shear stress in the crust from observations and theory. J Geophys Res 85:6231–6238

McQuillan H (1973) Small-scale fracture density in Asmari Formation of southwest Iran and its relation to bed thickness and structural setting. Am Assoc Petrol Geol Bull 57:2367–2385

Meier D, Kronberg P (1989) Klüftung in Sedimentgesteinen. Enke, Stuttgart

Mollema PN, Antonellini M (1999) Development of strike-slip faults in the dolomites of the Sella Group, Northern Italy. J Struct Geol 21:273–292

Moore DE, Summers R, Byerlee JD (1990) Deformation of granite during triaxial friction tests. In: Rossmanith HP (ed) Proceedings of the International Conference on the Mechanics of Jointed and Faulted Rock, Vienna, Austria. Balkema, Rotterdam, pp 345–352

Narr W, Suppe J (1991) Joint spacing in sedimentary rocks. J Structl Geol 13(9):1037–1048

Nur A (1982) The origin of tensile fracture lineaments. J Struct Geol 4(1):31–40

Olson JE, Pollard DD (1991) The initiation and growth of en échelon veins. J Struct Geol 13(5): 595–608

Petit J-P, Barquins M (1988) Can natural faults propagate under mode II conditions? Tectonics 7: 1243–1256

Petit J-P, Auzias V, Rawnsley KD, Rives T (2000) Development of joint sets in the vicinity of faults. In: Lehner FK, Urai JL (eds) Aspects of Tectonic Faulting. Springer-Verlag, Berlin Heidelberg, pp 167–183

Poliakov ANB, Herrmann HJ, Podladchikov YY, Roux S (1994) Fractal plastic shear bands. Fractals 2(4):567–581

Pollard DD (1973) Derivation and evaluation of a mechanical model for sheet intrusion. Tectonophysics 19:233–269

Pollard DD, Aydin A (1988) Progress in understanding jointing over the past century. Geol Soc Am Bull 100:1181–1204

Pollard DD, Müller O (1976) The effect of gradients in regional stress and magma pressure on the form of sheet intrusions in cross section. Geophys Res Letters 81(5):975–988

Pollard DD, Segall P (1987) Theoretical displacements and stresses near fractures in rocks: with applications to faults, joints, veins, dikes, and solution surfaces. In: Atkinson BK (ed) Fracture mechanics of rocks. Academic Press, London, pp 277–349

Pollard DD, Segall P, Delaney PT (1982) Formation and interpretation of dilatant echelon cracks. Geol Soc Am Bull 93:1291–1303

Price NJ (1966) Fault and joint development in brittle and semi-brittle rock. Pergamon Press, Oxford

Price NJ (1974) The development of stress systems and fracture patterns in undeformed sediments. Proceedings of the 3rd Congress of the International Society for Rock Mechanics, Denver, Colorado, pp 487–496

Price NJ, Cosgrove JW (1990) Analysis of geological structures. Cambridge University Press

Ramsay JG (1980) The crack-seal mechanism of rock deformation. Nature 284:135–139

Ramsay JG, Huber M (1987) The techniques of modern structural geology, vol. 2: Folds and fractures. Academic Press, London

Rawnsley KD, Rives T, Petit J-P, Hencher SR, Lumsden AC (1992) Joint perturbations at faults. J Struct Geol 14(8/9):939–951

Renshaw CE, Pollard DD (1994) Are large differential stresses required for straight fracture propagation paths? J Struct Geol 16:817–822

Rives T, Petit J-P (1990) Experimental study of jointing during cylindrical and non-cylindical folding. In: Rossmanith HP (ed) Proceedings of the International Conference on the Mechanics of Jointed and Faulted Rock, Vienna, Austria. Balkema, Rotterdam, pp 205–211

Rossmanith HP (1990) Mechanics of jointed and faulted rock: Proceedings of an International Conference, Vienna, Austria. Balkema, Rotterdam

Rudnicki JW, Rice JR (1975) Conditions for the localization of deformation in pressure-sensitive dilatant materials. J Mech Phys Solids 23:371–394

Secor DT Jr., Pollard DD (1975) On the stability of open hydraulic fractures in the Earth's crust. Geophys Res Letters 2(11):510–513

Segall P, Pollard DD (1983) Nucleation and growth of strike-slip faults in granite. J Geophys Res 88(B1):555–568

Sibson RH (1981) Controls on low-stress hydro-fracture dilatancy in thrust, wrench and normal fault terrains. Nature 289:665–667

Sobolev G, Spetzler H, Salov B (1978) Precursors to failure in rocks while undergoing anelastic deformations. J Geophys Res 83(B4):1775–1784

Suppe J (1985) Principles of structural geology. Prentice Hall, London

Takada A (1990) Experimental study on propagation of liquid-filled crack in gelatin: shape and velocity in hydrostatic stress condition. J Geophys Res 95:8471–8481

Tatsuoka F, Nakamura S, Huang C-C, Tani K (1990) Strength anisotropy and shear band direction in plane strain tests on sand. Soils and Foundations 30(1):35–54

Terzaghi K (1923) Die Berechnung der Durchlässigkeitsziffer des Tones aus dem Verlauf der hydrodynamischen Spannungserscheinungen. Sitzungsberichte der Akademie der Wissenschaften Wien, mathematisch-naturwissenschaftliche Klasse 132(11a):107–122

Terzaghi K (1925) Erdbaumechanik auf bodenphysikalischer Grundlage. Deutike, Wien

Thomas A L, Pollard DD (1993) The geometry of echelon fractures in rock: implications from laboratory and numerical experiments. J Struct Geology 15:323–334

Vermeer PA (1990) The orientation of shear bands in biaxial tests. Géotechnique 40(2):223–236

Vermeer PA, de Borst R (1984) Non-associated plasticity for soils, concrete, and rock. Heron 29:1–64

Vutukuri VS, Lama RD, Saluja SS (1974) Handbook on mechanical properties of rocks, vol I. Transtech Publications, Clausthal

Weertman J (1971) Theory of water-filled crevasses in glaciers applied to magma transport beneath oceanic ridges. J Geophys Res 76:1171–1183

Index

Printing: Krips bv, Meppel
Binding: Litges & Dopf, Heppenheim